Assessment of the Fate and Effects of Toxic Agents on Water Resources

NATO Security through Science Series

This Series presents the results of scientific meetings supported under the NATO Programme for Security through Science (STS)

Meetings supported by the NATO STS Programme are in security-related priority areas of Defence Against Terrorism or Countering Other Threats to Security. The types of meeting supported are generally "Advanced Study Institute" and "Advanced Research Workshops". The NATO STS Series collects together the results of these meetings. The meetings are co-organized by scientist from NATO countries and scientists from NATO's "Partner" or "Mediterranean Dialogue" countries. The observations and recommendations made at the meetings, as well as the contents of the volumes in the Series, reflect those of participants and contributors only; they should not necessarily be regarded as reflecting NATO views or policy.

Advanced Study Institutes (ASI) are high-level tutorial courses to convey the latest developments in a subject to an advanced-level audience

Advanced Research Workshops (ARW) are expert meetings where an intense but informal exchange of views at the frontiers of a subject aims at identifying directions for future actions

Following a transformation of the programme in 2004 the Series has been re-named and re-organised. Recent volumes on topics not related to security, which result from meetings supported under the programme earlier, may be found in the NATO Science Series.

The Series is published by IOS Press, Amsterdam, and Springer, Dordrecht, in conjunction with the NATO Public Diplomacy Division.

Sub-Series

A. Chemistry and Biology	Springer
B. Physics and Biophysics	Springer
C. Environmental Security	Springer
D. Information and Communication Security	IOS Press
E. Human and Societal Dynamics	IOS Press

http://www.nato.int/science
http://www.springer.com
http://www.iospress.nl

Series C: Environmental Security

Assessment of the Fate and Effects of Toxic Agents on Water Resources

edited by

I. Ethem Gonenc
IGEM Research and Consulting,
Istanbul, Turkey

Vladimir G. Koutitonsky
Institut des Sciences de la Mer de Rimouski (ISMER),
QC, Canada

Brenda Rashleigh
U.S. Environmental Protection Agency,
Athens, GA, U.S.A.

Robert B. Ambrose, Jr.
U.S. Environmental Protection Agency,
Athens, GA, U.S.A.

and

John P. Wolflin
U.S. Fish and Wildlife Service,
Chesapeake Bay Field Offices, U.S.A.

🐎 Springer

Published in cooperation with NATO Public Diplomacy Division

Proceedings of the NATO Advanced Study Institute on
Advanced Modeling Techniques for Rapid Diagnosis and
Assessment of CBRN Agents Effects on Water Resources
Istanbul, Turkey
4–16 December 2005

A C.I.P. Catalogue record for this book is available from the Library of Congress.

ISBN-10 1-4020-5527-7 (PB)
ISBN-13 978-1-4020-5527-0 (PB)
ISBN-10 1-4020-5526-9 (HB)
ISBN-13 978-1-4020-5526-3 (HB)
ISBN-10 1-4020-5528-5 (e-book)
ISBN-13 978-1-4020-5528-7 (e-book)

Published by Springer,
P.O. Box 17, 3300 AA Dordrecht, The Netherlands.

www.springer.com

Printed on acid-free paper

PREFACE

The domestic incident management landscape has changed dramatically following the terrorist attacks of September 11, 2001 in the United States of America. This has been demonstrated with attacks world-wide (e.g. Madrid, London, Tokyo). Today's threat environment includes not only the traditional spectrum of manmade and natural hazards—wild land and urban fires, floods, oil spills, hazardous materials releases, transportation accidents, earthquakes, hurricanes, tornadoes, pandemics, and disruptions to energy and information technology infrastructure—but also the deadly and devastating terrorist arsenal of chemical, biological, radiological, nuclear, and high-yield explosive weapons.

These complex and emerging 21st century threats and hazards demand a unified and coordinated approach to domestic incident management. Every country needs to establish a national strategy for homeland security and management of domestic incidents with clear objectives for a concerted national effort to prevent terrorist attacks; reduce vulnerability to terrorism, major disasters, and other emergencies; and minimize the damage and recover from attacks, major disasters, and other emergencies that occur.

Coastal lagoons are highly subject to threat from terrorist attack because they represent concentrated centers of the socio-economic system. The surrounding catchment areas are rich in natural capital, producing goods (agriculture and fisheries products) and services (industry and technology) for broad geographic areas. Lagoons are typically densely populated areas that serve as centers of domestic and international trade, tourism, and commerce. Cities around lagoons are often the center of governments (i.e. capitals) for local, regional and national legislatures. Therefore, coastal lagoon systems lend themselves well as highly subject to threat from terrorist attack.

The book Coastal Lagoons: Ecosystem Processes and Modeling for Sustainable Use and Development (CRC Press, 2005) presented models and other tools recommended for sustainable management of coastal lagoons. That book described the development of a decision support system based on modeling and a process for development of a sustainable management plan. Development of a sustainable management plan that uses the best, most current data, processes, and models is recommended for all lagoon systems in order to conserve these important ecological and socio-economic systems. A sustainable use and development plan is an important proactive and long-term tool to be implemented and updated on an annual basis for maximum effectiveness.

v

This book builds upon that previous work and describes how such a sustainable use and development plan also can be used as a foundation in response to a chemical, biological, radiological, or nuclear (CBRN) threat. In order to meet the serious challenges and growing current threats of terrorism and other anthropogenic pressures on these important systems, a sustainable management plan should include an emergency response section. Such an emergency response plan would detail the priority actions necessary for rapid response to avoid, minimize and mitigate any adverse impacts to the lagoon system. As noted in the pages to follow, a current plan also serves as preemptive protection and guidance on rapid response and recovery from such a CBRN attack.

This book is a result of a NATO sponsored Advanced Study Institute (ASI) held in Istanbul, Turkey in December 2005. The aim of the ASI was to transfer information and knowledge gained by the LEMSM group (NATO CCMS Pilot Study Group on Ecosystem Modeling of Coastal Lagoons for Sustainable Management) during their 10 years NATO-CCMS Pilot Study to young international scientists. It also encouraged the use of models as a tool for rapid response and decision making in today's climate of growing terrorism threat. The book is reflective of the instruction provided at the ASI. It specifically reframes the need for maintaining current scientific data and modeling in management of lagoon systems in the context of emergency conditions. Further, it defines the decision-making process in the context of assessment, diagnosis, and response to a terrorist threat on a lagoon system.

Background materials including tools and examples of Emergency Response and Decision Making are provided. The examples include summaries, sources of information with references to additional information. Training materials are also provided which include practice exercises and decision-making scenarios.

ACKNOWLEDGEMENTS

This book is a product of the "Ecosystem Modeling of Coastal Lagoons for Sustainable Management (LEMSM)" Pilot Study supported by NATO—Committee on the Challenges of Modern Society (CCMS), which was initiated in 1995. This book would not exist without the support of the NATO-Public Diplomacy Division in Brussels. I particularly want to thank to Dr. Deniz Beten (CCMS Program Director) for her dedication and support.

The book is based on a NATO sponsored Advanced Study Institute (ASI) held in Istanbul, Turkey, December 2005. I want to offer my deepest affection to the LEMSM family. I am indebted to my colleagues, the scientists who served as lecturers and wrote the chapters. They are truly dedicated to improve the management of coastal lagoons before, during, and after terrorist attack. In addition to me the following individuals have made-up our ASI leadership team and provided guidance and editorial review within their areas of expertise: Vladimir Koutitonsky, John Wolflin, Brenda Rashleigh, and Robert Ambrose. Particularly worth noting are my students Nusret Karakaya, Ali Erturk, Alpaslan Ekdal, and Melike Gurel who provided tremendous assistance in preparing for and implementing the ASI. I acknowledge Ms. Lynn Schoolfield, USEPA, as part of the LEMSM family and one of our most ardent supporters. I want to recognize Biimyrza Toktoraliev as co-director of the ASI and representative of the Republic of Kyrgyzstan.

This book could not have been accomplished without the dedication of many individuals and organizations. On behalf of all of the authors, we especially want to thank to Mrs. Wil Bruins and Miss Satvinder Kaur for editorial guidance. The ASI, as well as this book, would not have been accomplished without the assistance of the people from several organizations including ATS Water Technologies Group, IGEM Research, and the staff of the Hotel Marine Princess, Kumburgaz, Istanbul, Turkey.

I also want to recognize the fifty-five fine students from nine countries who participated in the ASI. We look forward to working with these young professionals in the future and their involvement in the Sustainable Ecosystem Society (*SES*). As the editor, I devote this book to young students worldwide who will be responsible for ecosystems management under the threat of terrorist attack. You are our future! I hope that valuable areas, like coastal lagoons, are conserved for future generations. This can only be achieved through the implementation of sustainable management practices that take into consideration emergency response principles such as those described herein.

 We all are particularly grateful for the summer offshore breezes of lagoons (called "meltem" in Mediterranean Countries), which provided us coolness of spirit during our comprehensive studies and the development of this manuscript.

I. Ethem Gönenç

Istanbul, 27 June 2006

CONTENTS

PART 1. INTRODUCTION

1. INTRODUCTION

I. Ethem Gonenc[1] and Biymyrza Toktoraliev[2]
[1]*IGEM Research and Consulting, Kaptan Arif sok. 12/9 Suadiye 34740, Istanbul, Turkey. e-mail: iegonenc@igemconsulting.com*
[2]*Osh Technological University, Isanova Street 81 714018 Osh, Kyrgyzstan*

The world's population increase in the last 50 years has affected water re-sources in several aspects. More water is now needed and therefore more wastewater is produced and discharged to other water bodies of the planet. These two aspects made water a limited and strategic natural resource in terms of its quality and quantity. It is then easy to conclude that water, like all the limited and vulnerable resources, has become one of the potential targets of terrorists.

Coastal lagoons are highly subject to threat from terrorist attack because they represent concentrated centers of socio-economic systems. The surround-ing catchment's areas are generally rich in natural capital, producing goods (agriculture and fisheries products), and services (industry and technology) for broad geographic areas. Coastal lagoons are typically densely populated areas that serve as centers of international and domestic trade, tourism, and com-merce. Cities around lagoons are often the center of governments (capitals) for local, regional, and national legislatures. Therefore, coastal lagoon systems lend themselves well as realistic scenarios for rapid response and decision making.

These complex and emerging 21st century threats and hazards demand a unified and coordinated approach to domestic and international incident management. Every country should establish a national strategy for homeland security and management of domestic incidents with clear objectives for a concerted national effort to reduce threats and manage incident response of any hazard, manmade or natural. Of particular concern and the subject of this book is the prevention of terrorist attacks; reduce vulnerability to terrorism, and minimize the damage and recover from attacks, in the aftermath of such.

In the last decades, Chemical, Biological and Radionuclide (CBRN) agents were used against humans, causing thousands of casualties or injured people. Once released to the environment, it is a complex task to understand the mechanisms of effect of CBRN agents on humans and the environment, and remediate the environment. Another aspect is that after the attack, human life will be in danger and a rapid diagnosis and assessment of these agents must

3

I. E. Gonenc et al. (eds.), Assessment of the Fate and Effects of Toxic Agents on Water Resources, 3–7.
© 2007 *Springer.*

be made to support the decisions made to minimize the damage on humans and the environment.

Rapid diagnosis and assessment of CBRN agents on water resources is a broad issue, which requires expertise from several disciplines like water resources management, transport of pollutants in the aquatic medium, aquatic chemistry and ecology as well as social sciences and decision making. Multidisciplinarry approaches are required to do these rapid diagnosis and assessments. In other words, groups of natural scientists, engineers, social scientists and managers have to be able to work and to exchange information among each other.

The NATO Advanced Study Institute (ASI) on Advanced Modeling Techniques for Rapid Diagnosis and Assessment of CBRN Agents Effects on Water Resources followed the multidisciplinary approach summarized in the preceding paragraph. The students were divided into three groups. The first group was trained on the transport of CBRN agents in the water environment, the second group was trained on their ecological and health effects on the water ecosystem, and the third group was trained on management and decision making. All three groups had common lectures before starting their actual training. These common lectures which aimed to put the students under the same scientific coverage and terminology, were designed to train all the students on the common and general issues of circulation and damage of substances such as CBRN agents on the environment as well as to train the groups on information exchange among each of the other groups.

The first group was intensively trained on transport processes in the aquatic environment. They had lectures on general hydrodynamics, monitoring coastal oceanography and transport, time series analyses, and numerical transport modelling as well as an introductory lecture on toxics modelling. During their training, they had two small exercises, the first on monitoring network design and the second on transport modelling using the Water Quality Analysis Simulation Program (WASP) numerical model.

The second group was first trained on aquatic ecology. The issues were plankton ecology and trophic networks (food webs). They were also lectures on long and short-term effects of chemicals and radionuclides on aquatic ecosystem, biological and chemical monitoring, and transport issues in aquatic ecosystems. Finally, lectures on ecosystem and toxics modelling were held. Modelling tools on this topic are in vast numbers, and usually more than one model is required to get the big picture in aquatic ecosystem modelling. According to this, the second group was trained on three modelling tools: WASP/TOXI, AQUATOX, STELLA and MOHID.

The third group consisted of decision makers. Decision makers have their own tools but they also need to use results from other tools (i.e. models) used and operated by the other two groups. For the decision makers, the social processes and factors are important, since they had also to prevent panic among the population. They were trained on issues such as emergency acting, social impact assessment, health assessment, environmental sensitivity indices, monitoring, risk assessment, and decision making. They were also trained on how to utilize models or model results for decision making.

In the last days of training, a case study was done. According to the scenario in the case study, a coastal lagoon was attacked by terrorists using a very toxic compound (Dieldrin). The case study was designed as a "real time" case. The first group performed a transport modelling study using WASP transport model. WASP was selected, because it is highly descriptive and free. The transport model development took one full day. Then the ecosystem group used that transport model to develop a short-term toxic chemical model using WASP/TOXI. After the development of short-term toxic chemical model, the ecosystem group also developed a long-term ecosystem model using AQUATOX. The model operations performed by the ecosystem group lasted a full day too. Then both groups presented their results to decision-making group, which also worked a full day on the outputs of them. On the next day a meeting, where all the groups, instructors and lecturers were involved, was organized. After these meetings each group prepared a report (please find www.sesonline.org), including the details of their studies, and submitted it to the lecturers.

As a result, 55 students were trained on advanced modelling techniques for rapid diagnosis and assessment of CBRN agents' effects on water resources during ten working days and 18 lecturers exchanged their knowledge for successful implementations.

It can be concluded that such courses should be repeated and more experts should be trained to be able to develop the numerical models for the strategic water resources in their countries and to construct the databases about various CBRN agents, and to establish a network among experts for being ready to mitigate the effects of any possible terrorist attack.

This book, a result of the ASI workshop, reframes the need for models and current scientific data in management of lagoon systems in the context of emergency conditions. The aim of this effort is to transfer information and knowledge gained by the LEMSM group (NATO CCMS Pilot Study Group on Ecosystem Modeling of Coastal Lagoons for Sustainable Management) during their 10 years NATO-CCMS Pilot Study, to young international scientists. Further, it is to encourage the use of models as a tool for rapid response and decision making in today's climate of growing terrorism threat.

Part 2 acts, as an overall framework for the rest of the chapters. A thorough assessment on; how ecosystem models and other tools will be employed in this process, and design, improvement, implementation phases of a response plan are provided. In other words, the entire information presented in the book after this chapter are then gathered and a discussion made on how to use all of such information for the decision-making process. This chapter builds upon that ASI Course and describes how such a plan can also be used as a foundation in response to a Chemical, Biological, Radiological or Nuclear (CBRN) threat. Consequently, the actual mission of the book could only be accomplished provided that the information and knowledge presented in this chapter are used and implemented by the reader. In order to meet the serious challenges and growing current threats of terrorism and other anthropogenic pressures on these important systems, a sustainable management plan should include an emergency response section. Such an emergency response plan would detail the priority actions necessary for rapid response to avoid, minimize, and mitigate adverse impacts to a lagoon systems and the surrounding catchment's area. A plan also would serve as preemptive protection and provide guidance for rapid response and recovery from such a CBRN attack.

In Part 3, at the first step, the physical processes that drive transport in lagoon are reviewed. Then equations defining mass, momentum, and energy transfer are stated together, with other sets of equations determining temperature, salinity, and sediment transport. This part contains valuable information that will help modelers with refined and synthesized knowledge gathered as a result of longstanding experience. The critical considerations on model implementation, stability, and accuracy problems of numerical modeling and model analysis, are major concerns of this part. Thus, even solely, this part is advised to all modelers as a valuable reference guide. Hence, this will present the reader with an excellent background on physical modeling of lagoons.

Water resources form a part of a larger system that encompasses ecological, sociological, and economic aspects at multiple scales in time and space are discussed. This larger system, the socio-ecological complex, must be understood if environmental management is to be successful and sustainable, as described in Part 4, Chapter 9. Complex interactions with this system also determine the effects of multiple stressors, including toxic agents, on aquatic ecosystems. An emerging framework for dealing with the complexity of these stressors considers Driving forces, Pressures, States, Impacts, and Responses (D-P-S-I-R). This framework is described in Chapter 10. Driving forces and pressures will determine the state of the system and overall impact, based on the ecology of the system. Chapter 11 provides a background on the ecology of algae, the base of most aquatic ecosystems. The management response usually results in changes in the driving forces and pressures, thus closing the loop.

Part 4 also describes specific modeling tools that can be used to address ecological effects of terrorist threats: the STELLA model (Chapter 12), the ECOPATH model (Chapter 13), and the AQUATOX model (Chapter 14). Each of these models has strengths and weaknesses for representing the aquatic ecosystem and the effects of toxicants on such an ecosystem. The STELLA model allows the most flexibility, where the user specifies state variables and relationships among these variables. The ECOPATH modeling system provides the opportunity to use a number of existing trophic models for the modeling of bioaccumulation of persistent pollutants in trophic food webs. AQUATOX is a complex model for simulating ecosystem dynamics and bioaccumulation. Taken together, this set of tools provides scientists with options for representing the effect of a toxic event in a range of aquatic systems. Chapter 15 surveys different types and effects of common toxic chemicals. Finally, monitoring after attack is presented in Chapter 16.

Part 5 presents a case study implemented and documented by the ASI participants. Their contribution is highly appreciated.

Naturally, readers from various disciplines being involved with different aspects of CBRN Agents Effects on water resources might not necessarily feel the need to recognize and absorb the information provided in every chapter with their full detail. Thus, with attention paid to the structure of the book and the content of the chapters, it would be advisable for readers who study hydrodynamics to focus on Part 3, whereas for those related to ecology, Part 4 would be beneficial, and finally those readers interested in management could locate relevant information in Part 2, but as necessary, they will need to refer to other chapters also. Parts 1 and 5, however, are for the common use of all readers.

PART 2. DECISION MAKING IN RAPID ASSESSMENT AND DIAGNOSIS OF CBRN EFFECTS ON COASTAL LAGOONS

2. DECISION MAKING IN RAPID ASSESSMENT AND DIAGNOSIS OF CBRN EFFECTS ON COASTAL LAGOONS

John P. Wolflin[1], Karen Terwilliger[2], and Rosemarie C. Russo[3]
[1] *U.S. Fish and Wildlife Service, Chesapeake Bay Field Offices, USA*
[2] *Terwillinger Consulting Inc., USA*
[3] *Ecosystems Res. Div. U.S.EPA, 960 College Station Road Athens, Georgia 30605-2700, USA*

2.1. Introduction

Numerous examples of response plans exist for natural emergencies, such as floods, hurricanes, fires, etc., in many areas of the world, as in the National Incident Management System in the United States which was developed so that responders from different jurisdictions and disciplines can work together to better respond to natural disasters and emergencies, and the Centre for Emergency Preparedness and Response (CEPR) in Canada (http://www. phac-aspc.gc.ca/cepr-cmiu/).[1] Plans that handle industrial chemical disasters also have been developed, such as the Seveso Directive II (named after the Seveso accident in 1976). This was developed by the European Commission in discussions with representatives from EU Member States and is an EU Directive that serves to mitigate and minimize accidents, threats, and hazards from industrial accidents involving dangerous substances (http://www.cemac.org/english/E2009002.html).[2]

A quick search of the internet on incident management and emergency response plans provides geographically diverse examples at multiple scales. Because of the infrequent and unpredictable nature of disasters, however, much improvement stands to be made in our preparedness and response. Recent examples include the tsunami in the Indian Ocean in 2004 and Hurricane Katrina in the Gulf Coast, United States of America in 2005.

This chapter reframes the decision-making process in the context of assessment, diagnosis and response to a terrorist threat on a lagoon system. It provides background materials including tools and examples of Emergency

---

[1] Canada's Centre for Emergency Preparedness and Response can be found online at http://www.phac-aspc.gc.ca/cepr-cmiu.
[2] History of the Seveso Directive II developed by the European Commission can be found online at http://www.cemac.org/english/E2009002.html.

I. E. Gonenc et al. (eds.), Assessment of the Fate and Effects of Toxic Agents on Water Resources, 11–52.

Response and Decision Making. These examples include summaries and sources of information with references to additional information. Training materials also are provided, which include practice exercises and decision-making scenarios.

2.2. Decision Making in Rapid Assessment and Response

Decision Support Systems include the following:

1. An Inventory (identify and prioritize) of the most current, accurate Information, Processes, and Infrastructures relevant to the lagoon's socioeconomic and ecological systems.
2. Tools and Resources available for Decision Making.
3. A Model for Decision Making.
4. A Response Plan.

2.2.1. INVENTORY INFORMATION, PROCESSES, AND INFRASTRUCTURES

Begin with an understanding of the System in need of protection. A robust decision support system begins with an inventory of the available data, processes, and infrastructures that exist in the area, which provide access to the most current and accurate data. It is built upon a foundation of knowledge and information about the local system. As a general rule, the more information available, the more informed and appropriate the assessments and decisions will be. Knowledge of the ecological and socioeconomic systems facilitates rapid and effective response to natural and anthropogenic emergencies. This knowledge allows for identification of priority and vulnerable resources that in turn focuses protection and response actions. If priority areas or resources are identified prior to emergencies then preemptive actions can be taken to reduce vulnerability and adverse impacts. Identification and modeling of system dynamics and interrelationships can better predict and prevent impacts to these vulnerable resource priorities.

Information on ecological and socioeconomic systems and their interrelationships is complex and sectorial. The best sources of current information from each of these disciplines need to be identified and a process established to access and input updated information into any assessment for priority setting. Rapid access to current, high quality information needs to be part of any decision-making process and long-term lagoon management plan. Monitoring mechanisms (Chapter 3) for key system parameters and priority data will need to be established so current information can be accessed efficiently.

Interdisciplinary coordination and integration are required to first inventory resources and then identify priorities across the many relevant disciplines. For example, priorities might include historical, cultural, ecological, economic, and social resources determined to be valued locally, nationally, or globally. If there is no clearinghouse for these types of data, then an effective retrieval mechanism will need to be established.

The Inventory of Information, Processes, and Infrastructures should:

- Identify critical system data, parameters, and sources.
- Identify priority/vulnerable System Resources to Protect.
- Determine the Content and Format of Information for compatibility and use.
- Identify the Flow of Information from the source to the necessary parties and determine if an improved information path and arrangements need to be established for emergency use.
- Identify and chart the relevant infrastructures (organizations/entities) managing data, personnel, resources.
- Identify the Location and Timing for Input and Output for rapid assessment.

The existing infrastructures (agencies, organizations, etc.) and processes that are presently used to collect, manage and distribute information for decision making should all be identified. A flow chart of responsible parties and authorities with contact information and access to data and personnel is essential for rapid assessment. Ideally a response plan should outline and identify all this information, as emergencies might require that a different mechanism be established. Regular meetings and coordination established with both public and private entities responsible for system resources and information flow will provide the foundation and network for the response plan.

In some cases, the existing processes and infrastructures could suffice and provide timely, quality information. However, proactively establishing a response plan with the most efficient and effective pathways for information flow will have both short- and long-term benefits for management of the system.

2.2.2. DECISION-MAKING TOOLS

Rapid assessment and response requires decision makers to have ready access to an organized assortment of tools. The more customized and prepared the toolbox is with local systems—specific information and tools—the more effective it will be. This section will focus on four major areas that are considered to be essential tools in the decision-making process: (1) modeling, (2) monitoring, (3) indicators, and (4) assessment.

2.2.2.1. *Modeling*

Modeling provides the decision maker with a prediction that in turn provides the necessary basis for assessment and analysis of possible scenarios. Therefore, different potential options can be weighed against one another before any action is taken. Models can be used as proactive, preparedness tools to predict CBNR impacts on the system. Models have the capacity to handle large numbers of variables which make them more responsive and flexible predictors. Models can be difficult and/or expensive to create, but can serve as a valuable decision-making tool. Chapter 5 provides detailed examples of models and the case study Chapters 3 and 4 demonstrates how these models were applied during the ASI decision-making training exercises.

A wide variety of models are available for decision makers. They range in scale, complexity and focus on different types of CBRN agents in the water system. Some of the most widely used models are those that were developed for petroleum spills. In the United States, NOAA and the petroleum industry have developed models for coastal areas called Spill Tools™; it is a set of three technological programs designed for oil spill planners and responders (http://www.response.restoration.noaa.gov).[3] There is also a Worldwide Oil Spill Model (WOSM) that was developed by Applied Science Associates (ASA) with a consortium of oil companies and government groups to provide the best available oil spill modeling research and development. It is a Windows application that has applications worldwide for industry and government clients. The system contains an oil spill trajectory and fates (weathering) model, stochastic and receptor models, and a search and rescue application for use in defining search areas and patterns for vessels and persons lost at sea. The package is designed to track and predict surface oil movement and weathering to assist in spill response efforts and training. This software as well as several other model applications for oil, gas, chemical, and marine industry and for freshwater and coastal environments can be downloaded on the ASA website (http://www.appsci.com/).[4]

The U.S. Environmental Protection Agency and many private utilities have developed models for inland water systems as well. The EPA Basic Oil Spill Cost Estimation Model (BOSCEM) was developed to provide their Oil Program with a methodology for estimating oil spill costs, including response costs and environmental and socioeconomic damages, for actual or hypothetical spills. EPA BOSCEM incorporates spill-specific factors that influence

---

[3] National Oceanic and Atmospheric Administration, USA, Spill Tools™; it is a set of three technological programs designed for oil spill planners and responders; it is free of charge and is available at http://www.response.restoration.noaa.gov.

[4] Several model applications for oil, gas, chemical, and marine industry and for freshwater and coastal environments can be downloaded on the ASA website http://www.appsci.com.

costs—spill amount; oil type; response methodology and effectiveness; impacted medium; location-specific socioeconomic value, freshwater vulnerability, habitat/wildlife sensitivity; and location type (http://www.epa.gov/oilspill/pdfs/etkin2_04.pdf).[5] Other countries have similar models such as the North Sea Disaster Plan presided by the Governor of Western Flanders and activated by the Belgian Navy's operational command centre (COMOPSNAV) in Zeebruges. The North Sea Disaster Plan uses the Management Unit of the North Sea Mathematical Models (MUMM). MUMM and the Scheldt estuary, is a department of the Royal Belgian Institute of Natural Sciences (RBINS), a Federal scientific establishment that comes under the Federal Science Policy (http://www.mumm.ac.be).[6] Another excellent example is the Oil Spill Trajectory Modeling (OSTM) used by the Australia Maritime Safety Authority when implementing the National Marine Oil Spill Contingency Plan (http://www.amsa.gov.au/).[7] The National Plan is a cooperative arrangement involving the States, the Northern Territory, and the petroleum, chemical and shipping industries which aims to maximise Australia's marine pollution response capability.

International Tanker Owners Pollution Federation Limited (ITOPF) is a non-profit organization, funded by the vast majority of the world's shipowners. It devotes considerable effort to a wide range of technical services, most importantly is oil spill response (http://www.itopf.com/).[8]

2.2.2.2. Graphical User Interfaces
Anything that helps the user prepare input data for a model and/or analyze and visualize the outputs of a model can be thought of as a graphical user interface (GUI). A good GUI can help the user enter the correct data in the correct format by detecting mistakes that a model cannot. A model is only as good as the data put into it. GUI's also can take model output files and make them easier to interpret. Overall, GUI's make models easier and more effective to use.

---

[5]EPA BOSCEM incorporates spill-specific factors that influence costs—spill amount; oil type; response methodology and effectiveness; impacted medium; location-specific socioeconomic value, freshwater vulnerability, habitat/wildlife sensitivity; and location type http://www.epa.gov/oilspill/pdfs/etkin2_04.pdf.

[6]Management Unit of the North Sea Mathematical Model adopts a 'triple M' strategy: Modelling, Monitoring. More information about their models can be found online at http://www.mumm.ac.be.

[7]Oil Spill Trajectory Modeling (OSTM) used in the National Marine Oil Spill Contingency Plan by the Australia Maritime Safety Authority http://www.amsa.gov.au.

[8]ITOPF is a non-profit making organisation, funded by the vast majority of the world's shipowners. We devote considerable effort to a wide range of technical services, the most important of which is responding to oil spills. Our technical advisers have attended on-site at 500 spills in 90 countries. More information can be found online at http://www.itopf.com.

There are several GUI's available as free software and can be downloaded for free on the internet. ALOHA (Arial Locations of Hazardous Atmospheres) uses information provided by its operator and physical property data from its extensive chemical library to predict rates of chemical release of leaks and evaporating puddles, and can model the dispersion of both neutrally-buoyant and heavier-than-air gases. ALOHA originated as an emergency response tool and is used for response, planning, training, and academic purposes and is distributed worldwide to thousands of users in government and industry. ALOHA has several attributes especially for emergency response, however also has several major limitations. Free download and more information can be found at http://www.epa.gov/ceppo/cameo/aloha.htm.[9] GNOME (General NOAA Oil Modeling Environment) is the oil spill trajectory model developed and used by HAZMAT responders during an oil spill. Information about GNOME and other modeling tools can be found at NOAA's Office of Response and Restoration website at (http://www.response.restoration.noaa.gov).[10]

2.2.2.3. *Monitoring*
Monitoring provides specific answers for specific questions. Designing a good monitoring system can lead to easier and better decisions by maximizing the quality of data. Choosing the best variables to measure can be complicated but will ultimately determine the quality of the monitoring system.

The following section provides a useful foundation for identifying the potential threat agents and how to monitor after a chemical attack.

2.2.2.4. *Indicators*
Indicators can be used to measure the ecological status of an ecosystem as well as the effectiveness of management efforts. Indicators include both measures of environmental quality and anthropogenic pressures resulting from social and economic activity. The existence of indicators helps to facilitate and to stimulate long-term protection of the environment and to foster sound environmental decision making. The Organization of Economic Cooperation and Development developed a systematic framework for environmental indicators commonly referred to as the "pressure-state-response." The framework is based on the theory that human activities exert pressure on the environment and change its quality and the quantity of natural resources. Society responds to these changes through environmental, general economic, and sectoral policies. More detailed information on indicators can be found in Chapter 3.

---

[9] Free download and more information about ALOHA can be found at http://www.epa.gov/ceppo/cameo/aloha.htm.
[10] GNOME and other modeling tools can be found at NOAA's Office of Response and Restoration website at http://www.response.restoration.noaa.gov.

2.2.3. ASSESSMENT TOOLS

Numerous examples exist on how to identify appropriate assessment tools. One example that has been used effectively in the U.S.A. to identify sensitive environmental resources is called the Environmental Sensitivity Index (ESI). This method has been used for more than 30 years to identify and prioritize vulnerable environmental (biological, historical, cultural) resources to minimize impacts in coastal systems for oil and chemical spills.

2.2.3.1. *Environmental Sensitivity Index (ESI)*

The Environmental Sensitivity Index (ESI), developed by RPI's senior scientists in 1976, has become an integral component of oil-spill contingency planning and response as well as coastal resource management in the USA and other countries worldwide. The first ESI maps were prepared days in advance of the arrival of the oil slicks from the IXTOC 1 well blowout in the Gulf of Mexico. Since that time, ESI atlases and databases have been prepared for most of the U.S. shoreline, including Alaska and the Great Lakes. ESI atlas and database archives include the U.S. territories and many other countries worldwide. Nearly all of the maps of the lower 48 States have been compiled at a scale of 1:24,000, using U.S. Geological Survey (USGS) 7.5-minute quadrangles as the base map. There are a few exceptions where USGS maps were available at different scales or too outdated to be of use. For work in Alaska, 15-minute USGS topographic quadrangles at a scale of 1:63,360 have been used as base maps. See atlas list.

Until 1989, ESI atlases were prepared by hand using traditional cartographic methods. Since then, Research Planning Inc. (RPI) has generated all atlases using Geographic Information System (GIS) techniques. We are consistently upgrading the resource content and mapping capabilities to keep up with rapidly changing technologies.

The ESI digital databases being developed are a subset of those needed for a wide range of natural resource management applications. ESI datasets are comprised of:

1. Shoreline Classification.
2. Biological Resources.
3. Human-Use Resources.

When a shoreline is threatened by a chemical spill, responders must quickly decide which locations to protect from the spill.

Protection Priorities. Which areas would be worst affected by the spill? Which areas can be protected?

Different kinds of shorelines are more or less sensitive (vulnerable to damage by oiling). Marshes and swamps are especially sensitive, because oil in these areas damages plants and is very difficult to clean up. Areas used by sensitive species, such as seabirds and sea otters—which are easily killed when they are coated with oil—are also especially vulnerable. Responders usually have only a limited amount of time and containment equipment, so must decide where to deploy equipment to be most effective.

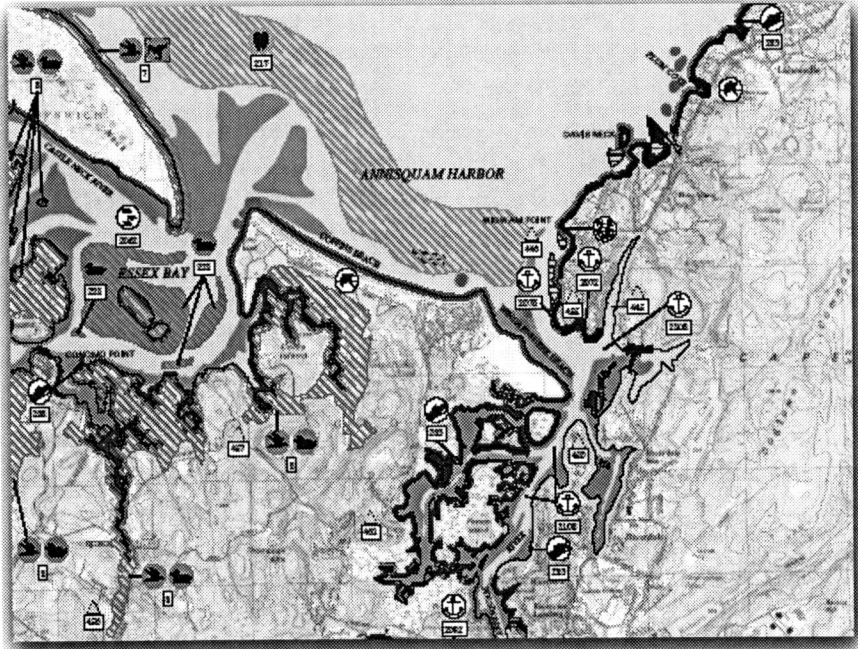

History of ESI and Environmental and Coastal Resources Management. The first ESI was developed in 1976 by Research Planning, Inc. (RPI) in preparation for oil from IXTOC in the Gulf of Mexico. Now ESI is an integral component of oil-spill contingency planning and response and coastal resource management in the USA and worldwide.

ESI Maps and Atlases

• Atlases/maps produced using Geographic Information System (GIS).
• Nearly all maps of USA have been compiled at a scale of 1:24,000, using U.S. Geological Survey (USGS) 7.5-minute quadrangles as the base map.
• ESI atlases and databases have been prepared for most of the U.S. shoreline, including Alaska and the Great Lakes.
• Constant upgrading of the resource content and mapping capabilities is needed to keep up with rapidly changing technologies.

ESI datasets are comprised of:

1. Shoreline Classification.
2. Biological Resources.
3. Human-Use Resources.

(1) Shoreline Classification. Ranked according to a scale relating to sensitivity, natural persistence of oil, and ease of cleanup.
   Sensitivity ranking is controlled by:

• Relative exposure to wave/tidal energy.
• Shoreline slope.
• Substrate type (grain size, mobility, penetration, and trafficability).
• Biological productivity and sensitivity.

   To classify shorelines, ESI map developers:

• Compile and assess information about the shorelines within each area to be mapped.
• Overfly the area and visit individual shorelines to 'ground-truth' their aerial observations.
• Use the information and observations to assign each shoreline a *ranking* between 1 and 10.

   "1" = shorelines least susceptible to damage by oiling
   "10" = locations most likely to be damaged

• "Rank 1" includes steep, exposed rocky cliffs and banks, where oil cannot penetrate into the rock and will quickly be washed off by the action of waves and tides. "Rank 10" includes protected, vegetated wetlands, such as mangrove swamps and saltwater marshes. Oil in these areas will remain for a long period of time, penetrate deeply into the substrate, and inflict damage to many kinds of plants and animals.

Rank	Type
7	Exposed tidal flats
8A	Sheltered rocky shores
8B	Sheltered artificial structures
9	Sheltered tide flats
10A	Salt to Brackish marshes
10B	Freshwater marshes
10C	Swamps
10D	Mangroves

Example: ESI = 6. Shoreline types meeting these definitions include gravel and shell beaches and riprap, such as the section of riprap shown below.

High Permeability, High Potential for Oil Penetration and Burial

- The substrate is highly permeable (gravel-sized sediments), allowing penetration to 100 centimeters.
- The slope is intermediate-to-steep (between 10 and 20 degrees).
- Rapid burial and erosion of shallow oil can occur during storms.
- There is high annual variability in degree of exposure, and thus in the frequency of mobilization by waves.
- Penetration can extend to depths below those of annual reworking.
- Sediments have the lowest trafficability of all beaches.
- Natural replenishment rate of sediments is the slowest of all beaches.
- Infauna and epifauna populations are very low, except at the lowest intertidal levels.

(2) Biological Resources—Seven Categories

- Marine Mammal.
- Terrestrial Mammal.
- Bird.
- Reptile/Amphibian.
- Fish.
- Shellfish/Insect.
- Habitat/Rare Plant.

Vulnerable Species. Many species that are vulnerable to oil are wide-ranging, and may be present over large areas at any time. These species can be especially vulnerable at particular times and places. Animals and their habitats tend to be most at risk from oil spills when:

- Many individuals are concentrated in a small area (e.g. seal haulout area, waterfowl concentration during migration).
- Early life stages are present in certain areas (e.g. seabird rookeries, spawning beds used by anadromous fish, turtle nesting beaches).
- A particular species is threatened or endangered.

Animals and their habitats tend to be most at risk from oil spills when:

- Oil affects areas important to specific life stages or important for migration (e.g. foraging or over-wintering sites.
- Specific areas are critically important for propagation of a species.

- A substantial percentage of an animal or plant population is likely to be exposed to oil.

Complete ESI Species List. ESI maps show where these most sensitive species, life stages, and locations exist, but don't necessarily show the entire area where members of a sensitive species occur. The Complete ESI Species List (download available at (http://archive.orr.noaa.gov/esi/species.pdf) lists all species included on ESI maps.

For each species, ESI maps include the following information:

- The life stage(s) present at a particular location, for each month of the year.
- The numbers of species members present at that location.
- The species' status (whether endangered or threatened, according to either State or Federal listings).
- Estimated start and end dates for specific breeding activities.

Note: Because showing the locations of certain endangered species on public maps could potentially result in increased visitation and possible disturbance or vandalism, the exact locations of such species are not shown on ESI maps.

Data collected for biological resources:

- Large of individuals concentrated in a relatively small area.
- Marine/aquatic species present during special life stages or activities (e.g. nesting, birthing, resting, molting).
- Early life stages or important reproductive activities occur in somewhat restricted areas.
- Restricted areas important to specific life stages or migration patterns.
- Specific areas known to be vital sources for seed or propagation.
- Species that are threatened, endangered, or rare.
- A significant percentage of threatened/endangered species that are likely to exist.

(3) Human-Use Resources. Specific areas that have added sensitivity and value because of their use.

Human-use resources are comprised of following *4 categories*:

1. Recreation/Access.
2. Management Area.
3. Resource Extraction Site.
4. Cultural Resource.

(1) Recreation/Access. High-use recreational use and shoreline access areas, such as boat ramps, marinas, recreational beaches, and sport-fishing and diving areas.

(2) Management Area. Officially-designated natural resource management or protected areas, such as national parks, marine sanctuaries, national wildlife refuges, preserves, and reserves.

(3) Resource Extraction Sites. Such as aquaculture sites, locations of subsistence and commercial fisheries, log storage sites, mining leases, and surface water intakes.

(4) Cultural Resource. Water-associated archaeological, historical, and cultural sites, including lands managed by Native Americans. Cultural sites located in the intertidal zone or close to the shoreline where they could be damaged by cleanup crews are at particular risk.
Notes: (1) The exact locations of some archaeological and cultural resources cannot be disclosed, because of the risk of vandalism. Either such a location is shown within a polygon enclosing a larger area, or a map symbol is placed near but not exactly at the location. (2) People using the ESI method to map human-use resources are encouraged to denote not only surface water intakes, but also groundwater recharge zones and well fields. Groundwater protection can be of particular concern when light petroleum products are spilled in rivers, and when wells are located in the floodplain and are hydraulically connected to the river.

Front of an ESI Map. ESI maps show shorelines, wildlife habitats, places important to people, and other locations that are especially sensitive to damage from a spill. People can look at an ESI map of an area threatened by a spill to quickly see the most sensitive locations.

Shoreline Color. On ESI maps, each shoreline is color-coded to indicate its sensitivity to oiling. Warm colors indicate the most sensitive shorelines, and cool colors indicate the least sensitive. Large habitat areas, such as tidal flats and wetlands, are shown as polygons filled with a pattern of the appropriate color.

Biological Resource Symbols. When a biological resource exists in a small area (an example is a bird nesting site), it is indicated on an ESI map by a *point symbol*. Most biological resources are represented on ESI maps by polygons, which encompass larger areas where particular species exist. Fish streams are usually represented as lines.

Point Symbols. The point symbols, polygons, and lines representing different kinds of animals are color-coded. Associated with each biological polygon, line or point symbol on the map is a number (usually located under the icon).

This number references the table on the reverse side of the map with a complete list of species associated with that number. Several numbers indicate several species groups. Polygons for each kind of biological resource are colored according to the following scheme:

SENSITIVE BIOLOGICAL RESOURCES

BIRD	MARINE MAMMAL	SHELLFISH AND INSECT
Alcid / Pelagic Bird	Dolphin	Bivalve
Diving Bird	Manatee	Crab
Gull / Tern	Polar Bear	Echinoderm
Passerine Bird	Sea Otter	Gastropod
Raptor	Seal / Sea Lion	Lobster/ Crayfish
Shorebird	Whale	Shrimp
Wading Bird	**REPTILE / AMPHIBIAN**	Squid/ Octopus
Waterfowl	Alligator / Crocodile	Insect
TERRESTRIAL MAMMAL	Turtle	**HABITAT**
	Other Reptiles / Amphibians	Coral/ Hardbottom Reef
Bear	**FISH**	Floating Aquatic Vegetation
Deer	Fish	Rare Plant
Small Mammal	Nursery Area	Submerged Aquatic Vegetation

Human Resources Symbols. Important human-use resources, such as water intakes, marinas, and swimming beaches are also depicted with *point symbols.*

Much of the information on an ESI map is shown in a table on the back of the printed copy of the map (some older maps do not include this table of information).

The back of the map is also where you'll find contact information for important human-use locations shown on the map.

The table shows when each species is present in the area, and what that species is doing during different seasons (for example, a bird species may be nesting, laying eggs, hatching, or fledging young birds). The "back-of-the-map" table lists sensitive plant and animal species that live in the area shown on the map, their concentrations, and the seasons of the year in which they are breeding.

Back-of-the-map table also give species designations:

S/F = State or Federally listed,
T/E = Threatened or Endangered.
Concentrations are measured either by: (depending on the taxonomic group)
High/Medium/Low, or
Rare/Occasional/Common.
Example: Lewes, Delaware, USA

Exercise: area around Lewes, Delaware, USA
Oil Spill Scenario
Exercise Tasks

Tasks:

Identify resources at risk as follows:

1. Identify the biological resources (animals, plants, or their habitats) that
   may be at risk from the spilled oil (at sea as well as on shore, if oil reaches
   shore). List these resources.
2. Prioritize the list of resources at risk according to (a) sensitivity to the
   oil spilled and (b) the likelihood of being contacted by the spilled oil.
   In general, species in early life stages (i.e., nesting, laying, hatching, or
   fledging) or present in large numbers are at the greatest risk. Justify and
   explain your priorities.
   (i) Identify Biological Resources
   (ii) Prioritize list of resources at risk

(a) Establish Protection Priorities. Looking at the shoreline habitats affected
by the spill, identify areas that would receive high priority for protection
(assuming that protection is possible). In general, the habitats shown in red
and orange on ESI maps (and given ESI rankings of 9 or 10) are the most
sensitive to spills. Discuss your rationale for choosing these areas over others
and the tradeoffs your choices involve. Discuss any potential difficulties you
anticipate.

(b) Identify Resources at Risk. Identify the biological resources (animals,
plants, or their habitats) that may be at risk from the spilled oil (at sea as well
as on shore, if oil reaches shore). List these resources.
    Prioritize the list of resources at risk according to (a) sensitivity to the oil
spilled and (b) the likelihood of being contacted by the spilled oil. In general,
species in early life stages (i.e., nesting, laying, hatching, or fledging) or
present in large numbers are at the greatest risk. Justify and explain your
priorities.

Establish Protection Priorities. Looking at the shoreline habitats affected
by the spill, identify areas that would receive high priority for protection
(assuming that protection is possible). In general, the habitats shown in red
and orange on ESI maps (and given ESI rankings of 9 or 10) are the most
sensitive to spills. Discuss your rationale for choosing these areas over others

and the tradeoffs your choices involve. Discuss any potential difficulties you anticipate.

NOAA and EPA have additional resources including Coastal Protection and Restoration Division (CPRD) Tools, which is a collection of Database and GIS tools created by NOAA to support the protection and restoration of coastal species and habitats. NOAA CPRD and project partners can take advantage of these tools to support analysis, mapping, and communication of complex spatial contaminant and restoration issues in a catchment/watershed context.

2.2.3.2. *Social Impact Assessment (SIA)*

Social Impact Assessment (SIA) includes impacts to health and welfare, recreational and aesthetic values, land and housing values, job opportunities, community cohesion, life styles, governmental activities, physiological wellbeing, and behavioral response by individuals, groups, and communities. SIAs should be used to identify, quantify, and interpret the significance of the anticipated social changes and decide whether or not they are acceptable. Social impact assessment includes the processes of analyzing, monitoring, and managing the intended and unintended social consequences, both positive and negative, of planned interventions (policies, programs, plans, projects) and any social change processes invoked by those interventions. The primary purpose is to bring about a more sustainable and equitable biophysical and human environment.

SIA is much more than the prediction step within an environmental assessment framework. Social impacts are much broader than the limited issues often considered in EISs (such as demographic changes, job issues, financial security, and impacts on family life). A convenient way of conceptualizing social impacts is as changes to one or more of the following:

- *way of life*—how people live, work, play, and interact with one another on a day-to-day basis;
- *culture*—people's shared beliefs, customs, values, and language or dialect;
- *community*—human and cohesion, social stability, character, services, and facilities;
- *political systems*—the extent to which people are able to participate in decisions that affect their lives, the level of democratization that is taking place, and the resources provided for this purpose;
- *environment*—the quality of the air and water; the availability and quality of the food; the level of hazard or risk, dust and noise exposure; adequacy of sanitation, physical safety, and access to and control over resources (natural, capital);

- *health and wellbeing*—health is a state of complete physical, mental, social, and spiritual wellbeing and not merely the absence of disease or infirmity;
- *personal and property rights*—particularly whether people are economically affected, or experience personal disadvantage which may include a violation of civil liberties;
- *fears and aspirations*—perceptions about safety, fears about the future of community, and aspirations for the future (self, family, friends).

The International Association for Impact Assessment is the leading global authority on the best practice in the use of impact assessment for informed decision making regarding policies, programs, plans, and projects; they have published Guidelines and Principles for Social Impact Assessment.[11]

The U.S. Agency for International Development has a procedure and guidelines called Social Soundness Analysis, as does the World Bank.[12] Social Soundness Analysis is a form of qualitative stakeholder analysis, specifying who wins and who loses as a result of proposed developmental projects. It has three distinct but related aspects: (1) the compatibility of the activity with the sociocultural environment in which it is to be introduced (its sociocultural-feasibility); (2) the likelihood that the new practices or institutions introduced among the initial activity target population will be diffused among other groups (i.e., the spread effect); and (3) the social impact or distribution of benefits and burdens among different groups, both within the initial activity population and beyond.

2.2.3.3. *Environmental Assessment (EA)*
In the U.S.A. there is a National Environmental Policy Act (NEPA) that each Federal agency must be in compliance with as they institute new plans, projects, or programs in order to ensure that any Federal action is evaluated for potential environmental impacts. NEPA is strictly enforced and each Federal agency has responded to these requirements by developing and publishing thorough and extensive procedural guidelines. In the U.S.A. there is a technical difference between an Environmental Assessment (EA) and an Environmental Impact Assessment (EIA). Projects whose impacts are not known or not expected to have any significant impacts go through an Environmental Assessment (EA). However, if the impacts are known then, an EIA needs to be furnished, which would include identifying various alternatives to the proposed

---

[11] The Interorganizational Committee on Guidelines and Principles for Social Impact Assessment. 1994. Guidelines and Principles for Social Impact Assessment. U.S. Department of Commerce, National Oceanic and Atmospheric Administration, National Marine Fisheries Service.
[12] U.S. Agency for International Development. 1978. Social Soundness Analysis, Appendix 4A, dated 2/15. Washington, DC.

action and the identification of a preferred alternative based on impacts to the environment. The environment is defined as the natural capital of an area and includes potable water, subsistence and commercial fisheries, endangered and threatened species, and other attributes of natural capital that may be affected.

Like NEPA in the U.S.A., the European Union has the Strategic Environment Assessment Directive (SEAD), which assesses the effects of certain plans and programs on the environment. SEAD is a European Union Directive 2001/42/EC, which all State members must be in compliance with. For example, UK's Environment Agency is the leading public body for protecting and improving the environment in England and Wales and is one of the designated environmental authorities that must be consulted during the process of Strategic Environmental Assessment (SEA). The Environment Agency has published Good Practice Guidelines for Strategic Environmental Assessment, which provides guidance on the SEA Directive (http://www.environment-agency.gov.uk).[13] This document provides a SEA Toolkit (see below),which outlines guidelines and tools and techniques to improve efficiency and promote good practice in Strategic Environmental Assessment.

SEA is a process designed to ensure that significant environmental effects arising from proposed plans and programs are identified, assessed, subjected to public participation, taken into account by decision makers, and monitored. SEA sets the framework for future assessment of development projects some of which require Environmental Impact Assessment (EIA). More information can be found on (http://www.sea-info.net).[14]

There are numerous helpful agencies and organizations that aid in fulfilling the requirements of SEA, including the Institute of Environmental Management and Assessment (IEMA), which is a not-for-profit organization that was established to promote best practice standards in environmental management, auditing and assessment. http://www.iema.net/).[15] With over 9,000 individual and corporate members, the IEMA is now also the leading international membership-based organization dedicated to the promotion of

---

[13] U.K. Environmental Agency. 2005. Good Practice Guidelines for Strategic Environmental Assessment. http://www.environment-agency.gov.uk/aboutus/512398/830672/?version=1& lang=e.

[14] SEA sets the framework for future assessment of development projects some of which require Environmental Impact Assessment (EIA); more information can be found on http://www.sea-info.net.

[15] Institute of Environmental Management and Assessment (IEMA) is a not-for-profit organization that was established to promote best practice standards in environmental management, auditing and assessment. More information can be found online at http://www.iema.net.

sustainable development, and to the professional development of individuals involved in the environmental profession, whether they be in the public, private, or non-governmental sectors.

2.2.3.4. *SEA Toolkit*

Assessing cumulative effects on the environment. This tool is used to identify and address cumulative effects of proposed policies, plans, or programs. While individual effects may be insignificant, interactions between them can be significant and they can affect the same location and/or different locations and times. Examples include loss of wetland habitats, increased air pollution, climate change, increased risk of flooding, and other issues that may affect quality of life. There are various tools and techniques that may be used and examples are given in the Office of Deputy Prime Minister Guidance for Planning Authorities SEA: A Practical Guide (2004) and SA of RSS and LDF (2004).

Best Available Technique (BAT) and Best Practicable Environmental Option (BPEO). Best Available Techniques (BAT) are required to be considered (under EC Directive 96/61) in order to avoid or reduce emissions resulting from certain installations and to reduce the impact on the environment. BAT takes into account the balance between the costs and environmental benefits. While BAT refers primarily to installations, it may be more meaningful to consider the BPEO procedure, which establishes, for a given set of objectives, the option that provides the most benefits or the least damage to the environment as a whole, at acceptable cost, in the long term as well as in the short term (http://www.rcep.org.uk/).[16] An example to use the BPEO procedure would be how to strategically establish municipal waste management system.

Constraints and opportunities mapping. Approach is traditionally used by land use planners and ensures that development takes place in the least sensitive areas and is useful for both regional and local applications. Geographic Information System (GIS) is a valuable tool to understand spatial constraints and opportunities to particular activities and can include environmental, social, economic and technical considerations. For example, typical constraints on the development of an industrial facility would include the exclusion of urban areas, wooded areas, areas of outstanding natural beauty, ancient woodlands, proximity to settlements and airports, sites of recreational use, etc.

---

[16]The Royal Commission on Environmental Pollution is an independent standing body established in 1970 to advise the Queen, the Government, Parliament and the public on environmental issues. They use this technique. http://www.rcep.org.uk.

An example of opportunities mapping would be the identification of priority areas for river restoration, habitat connectivity to ameliorate fragmentation problems, increased access to waterside amenity. The usefulness of this technique depends on the scale and quality of data used, and how constraints are selected.

Consultation and participation. Consultation and participation affords public and stakeholders an opportunity to interact with the process and influence the outcome. Numerous publications are written about this process, for example: Perspectives: Guidelines on participation in environmental decision making by the Institute of Environmental Management and Assessment (IEMA).[17]

The following are examples of tools that can be used to involve stakeholders and the public:

- *Information feedback*—surveys, staffed exhibits and displays, staffed telephone lines.
- *Involvement and consultation*—workshops, focus groups, open house.
- *Extended involvement*—stakeholder groups, forums.

Cost–benefit analysis techniques. CBA aims to translate all environmental and social impacts into monetary values, in order to assist decision makers consider all impacts in measuring units they already use to make decisions. While it may not always seem appropriate to assign monetary values to environmental and social costs, CBA promotes consideration of environmental and social impact in decision making where such values have not traditionally been considered. The following are example of techniques used to translate values:

- *Dose–response approach*—determines the links between pollution (dose) and its impacts (response), and values the final impact at a market or shadow price (e.g. the cost of crop/forest damage from air pollution).
- *Replacement cost approach*—ascertains the environmental damage done and then estimates the cost of restoring the environment to its original state.
- *Avertive expenditures*—measures expenditures undertaken by households which are designed to offset some environmental risk (e.g. noise abatement).
- *Travel cost method*—detailed sample survey of tourists to a site determines how they value the mainly recreational characteristics of the site and the time spent traveling to the site (e.g. visiting a nature area).

---

[17] Institute of Environmental Management and Assessment (IEMA). 2002. Perspectives: Guidelines on participation in environmental decision-making. Available online at http://www.iema.net/shop/product_info.php?cPath=27_26&products_id=56.

- *Hedonic price methods (house prices approach)*—applies to environmental attributes which are likely to capitalize into the price of housing and/or land. Involves assembling cross-sectional data on house prices, together with factors likely to influence these prices, and analyzing these prices using multiple regression techniques.
- *Contingent valuation*—involves asking people for their willingness to pay and/or accept compensation for changes in environmental resources.
- *Contingent ranking*—individuals are asked to rank several alternatives rather than express a willingness to pay.

Ecological footprinting. The ecological footprint (EF) provides an aggregated indicator of natural resource consumption (energy and materials) in much the same way that economic indicators (such as Gross Domestic Product or the Retail Prices Index) have been adopted as a way of representing dimensions of the financial economy.[18] This technique, which excludes the social and economic dimensions of sustainability, estimates the bioproductive area that would be required to sustainably maintain current levels of consumption, using current technology. Non-renewable resources are included via their impact on renewable biological activity. The European Commission's Common Indicators Programme has adopted the EF as an indicator of regional environmental sustainability and the methodology has support from many in the public, private and civil sectors worldwide.[19]

---

[18] Barrett, John and Craig Simmons. 2003. Ecological footprint of the UK: providing a tool to measure the sustainability of local authorities. Stockholm Environment Institute and Best Foot Forward Ltd, Stockholm, Sweden.

[19] World-Wide Fund for Nature International, United Nations Environment Programme, World Conservation Monitoring Centre, Redefining Progress & Center for Sustainability Studies. 2002. Living planet report 2002. World-Wide Fund for Nature, Gland, Switzerland. Other references for ecological footprinting include:

Chambers, N., C. Simmons, and M. Wackernagel. 2000. Sharing Nature's Interest: Ecological footprints as an indicator of sustainability. Earthscan, London.

Lewan, L. and C. Simmons. 2001. The use of ecological footprint and biocapacity analyses as sustainability indicators for sub-national geographical areas: a recommended way forward. http://www.prosus.uio.no/ english/sus_dev/tools/oslows/2.htm.

Aall, C. and I. Norland. 2002. Ecological footprint for the municipality of Oslo—results and suggestions for use of ecological footprint as a sustainability indicator. (Det økologiske fotavtrykk for Oslo kommune—resultater og forslag til anvendelse av økologisk fotavtrykk som styringsindikator). Report. Western Norway Research Insitute/ProSus. (English summary can be downloaded from http://www.vestforsk.no/publikasjonar.asp?gruppe=Miljøgruppa.

Simmons, C., K. Lewis and J. Barrett. 2000. Two feet—two approaches: a component-based model of ecological footprinting. Ecological Economics, 32: 375–380.

Barrett, J., H. Vallack, A. Jones, and G. Haq. 2002. A material flow analysis and ecological footprint of York: technical report. Stockholm Environment Institute, Sweden.

Horizon scanning. The logic of horizon scanning is that, by "scanning" the future for emerging issues (like foot and mouth disease, flooding, fuel crisis), agencies will be better armed to deliver both adaptive and preventative policies, to identify opportunities, and ensure that plans are underlain by a more long-term perspective.

Sustainability Appraisal and Integrated Appraisal. Sustainability Appraisal can be defined as a single appraisal tool which provides for the systematic identification and evaluation of the economic, social, and environmental impacts of a proposal. Integrated Appraisal can be defined as the process of assessing the performance of options of a proposal in terms of their economic, social, and environmental implications.

Modeling. Modeling is the process by which conceptual models are developed into formal, consistent descriptions of the relationship between important aspects of a system (variables, their dependencies and consequences). The aim of modeling is to identify and predict the environmental impacts associated with actions anticipated under a plan or strategy. Models do not guarantee an outcome; they are always an abstraction of reality. A variety of models of oil, chemical, and gas spills were presented in the previous section called Modeling.

Multi Criteria Analysis (MCA). MCA involves a variety of decision-making techniques that incorporate different criteria on which to base a decision, rather than techniques based solely on, for example, financial analysis. Its main role is to deal with large amounts of complex information in a consistent way, which can otherwise create difficulties for decision makers. MCA techniques generally include the use of weighted and scored matrices, and hence require the establishment of measurable criteria, whether qualitative or quantitative, to assess the extent to which objectives may be fulfilled. MCA emphasizes the judgment of the decision-making team, in the selection of objectives, criteria for selection, estimating weightings, and in assessing the contribution of options to each performance criterion. The associated subjectivity can be a matter of concern, but consultation may be used to debate, and where possible, agree to these somewhat subjective scores and weights. However, the method is open in that the criteria, weights, and scores used are not hidden.[20]

---

[20] The MCA manual is available online at http://www.communities.gov.uk/index.asp?id=1142254.

Network (Casual Chain) analysis. This approach can be used to ascertain the probable impacts and benefits on the sustainability of actions by identifying their outcomes via the development of a chain of causation. This technique also may be used in the opposite way to identify actions that may be able to achieve desired objectives. It can be a useful technique to identify appropriate parameters for monitoring and is an established technique in transport policy assessment and monitoring

Quality of life capital. This technique is based on deciding what matters and why. It aims to identify priorities for guiding land-use planning and management decisions. This technique is particularly useful for engaging public participation as it seeks to maximize social, economic, and environmental benefits. An example would be considering the value of a small woodland to a suburban community—benefits of the woodland, in the form of recreation, wildlife habitat, soil stability, water retention, absorption of carbon dioxide and improving air quality, and the economic benefit of timber and charcoal.

Risk assessment. Risk assessment informs management decisions taken at a number of levels including strategic decisions on priorities and resources allocation; licensing of activities such as waste management and the clean-up of contaminated land. Risk management also is being increasingly used in sustainability studies, particularly where there is uncertainty in the long term. Risk assessment can provide a forum for stakeholders to make collective decisions about uncertain future outcomes. It can assist by providing a framework to evaluate economic, social, and physical outcomes (including impacts on human health and the environment) of proposed policies, plans, and programs. A particular strength of risk assessment is its ability to explicitly recognize uncertainty surrounding future predictions.

Scenario testing. Scenario testing is a method of forecasting possible states of the environment under a range of plausible future conditions. Participation of a range of stakeholders is required. Global climate change and impact are examples when scenario testing is widely used.

Sustainability Threshold Assessment (STA). Sustainability Threshold Assessment is a pragmatic approach that builds on existing tools used by planning authorities with an emphasis on policy, management, and options for mitigation. STA was developed by the Environment Agency, the planning authority, and developers.

2.2.4. DECISION-MAKING APPROACH (MODEL)

There are numerous approaches and models for decision-making and it has been the source of many books, seminars, workshops, dissertations, and theories that discuss the pros and cons of each. There are three generally accepted models for decision-making:

1. Classical Rational Decision-making model.
2. Organizational Process Model.
3. Bureaucratic Politic Model.

2.2.4.1. *Classical Rational Decision-Making Model*

A rational decision-making process is one that is logical and follows an orderly path from problem identification through to a solution. Rationality refers to consistent, value-maximizing choice within specified constraints. The core concepts of this model of rational action are[21]:

- *Goals and objectives*—understanding the desired goal, and translating interests and values into payoff, utility, or preference function, which represents the desirability or utility of alternative sets of consequences.
- *Alternatives*—all alternative actions must be considered.
- *Consequences*—each action must be paired with consequences.
- *Choice*—rational choice consists simply of selecting the alternative whose consequences rank highest in the decision-maker's payoff function.

All possible options or approaches to solving the problem are identified and the costs and benefits of each option are assessed and compared with each other. The option that promises to yield the greatest net benefit is selected.

2.2.4.2. *Organizational Process Model*

Where formal organizations are the setting in which decisions are made, the particular decisions or policies chosen by decision makers can often be explained through reference to the organization's particular structure and procedural rules. Such explanations typically involve looking at the distribution of responsibilities among organizational sub-units, the activities of committees and ad hoc coordinating groups, meeting schedules, rules of order, etc. The notion of fixed-in-advance standard operating procedures

---

[21] Allison, G. and P. Zelikow. 1999. Essence of decision—explaining the Cuban Missile Crisis, second edition. Addison Wesley Educational Publishers Inc.

typically plays an important role in such explanations of individual decisions made.[22]

2.2.4.3. *Bureaucratic Politic Model*

Politics is about bargaining along regular circuits among players positioned hierarchically within the government. Government behavior can be understood accordingly not as organizational outputs but as a result of bargaining games. Outcomes are formed by the interaction of competing preferences. Politics models sees no unitary actor but rather many actors as players: players who focus not on a single strategic issue but on many diverse international issues as well; players who act in terms of no consistent set of strategic objectives but rather according to various conceptions of national, organizational, and personal goals; players who make government decisions not by a single rational choice but by the pulling and hauling that is politics (see footnote 25).

2.2.4.4. *Decision-Making Process for Rapid Assessment*

In addition to the conceptual model types above, many applications exist for rapid assessment and response, like New Zealand Environmental Risk Management Authority's Decision Making A Technical Guide to Identifying, Assessing and Evaluating Risks, Costs and Benefits.[23] The approach selected for the ASI training exercise was simple, straightforward and used for emergency response in the U.S. Federal Emergency Management Agency (FEMA) Emergency Response training modules. The model of decision making used in this case study was the five step model shown in Figure 1.[24] It has proven to be effective in emergency situations. Every step of this model was followed and used to identify short, medium, and long term decisions and recommendations.

1. *Identify the problem*
   It is the most important but the most difficult step in the process, because all the other steps will be based on it. This includes delineating the problem parameters such as who is involved and what is at stake.

---

[22] Johnson, P. M. 2005. A glossary of political economy terms. Department of Political Science, Auburn University, Auburn, Alabama. Available online at http://www.auburn.edu/~johnspm/gloss.

[23] Environmental Risk Management Authority. 2004. Decision-making: a technical guide to identifying, assessing and evaluating risks, costs and benefits. ER-TG-05-1 03/04. ERMA, New Zealand.

[24] Federal Emergency Management Agency's (FEMA's) Independent Study Program. 2002. Decision-making and problem solving. Emergency Management Institute, Emmitsburg, Maryland.

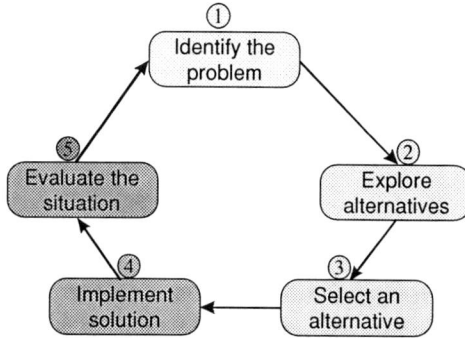

Figure 1. Decision-making model

2. *Explore alternatives*

   It consists of two parts: generating alternatives (through brainstorming, surveys and/or discussion groups) and evaluating alternatives (identify constraints, determine appropriateness, verify adequacy, evaluate effectiveness, evaluate efficiency, and determine side effects).

3. *Select an alternative*

   After all the alternatives are evaluated, one alternative should stand out as coming closest to solving the problem with the most advantages and fewest disadvantages.

4. *Implement the solution*

   This step consists of five parts: develop an action plan, determine objectives, identify needed resources, build a plan, and implement the plan.

5. *Evaluate the situation*

   This involves two parts: monitoring the process and evaluating the results.

The stage at which an emergency is identified can define the decision-making ability and opportunities. There are three basic decision-making stages:

- *Future*—planning, preparedness, prevention. (This is the ideal option to proactively prepare for CBRN incidents. Every effort should be made to prepare at this stage; the investment has a large potential benefit for avoidance or minimization of impacts.)
- *Present*—immediate response (implementing a response plan).
- *Past*—recovery and mitigation.

Investing time, energy, and money in establishing a proactive response plan for potential future emergencies may preclude investing more costly investments in the reactive stages. Being adequately prepared can eliminate the necessity of maintaining a clear mind in a crisis situation and having to respond to unfamiliar situations. The unique aspects of decision-making in rapid assessment and crisis management are:

- Timeliness of decision making.
- Inflated consequences of poor decisions.
- Changes in authorities and chains of command.
- Changing role of public in decision-making process.
- Reduction in the number of plausible decision-making tools.

Once the decision support system has been used for rapid assessment, it is time to make the decisions and implement a rapid response.

2.2.5. DEVELOPMENT AND IMPLEMENTATION OF A RESPONSE PLAN (RP)

Emergency response plans have proven their effectiveness and have been used in a wide variety of ways from routine fire drills to chemical or oil spill response. The value of being adequately prepared is common knowledge. However, there is a big difference between a fire in an office building and a chemical release into a city municipal drinking water system. Large-scale emergencies require large-scale response plans for which a structured response team network is required, oftentimes on a national scale. The purpose of a response plan is to establish a comprehensive, national, all-hazards approach to domestic incident management across a spectrum of activities including prevention, preparedness, response, and recovery. The following example illustrates the interconnectedness and step-down approach from a national to local level. This example was used in the ASI training.

The National Response Plan (NRP) needs to integrate the capabilities and resources of various governmental jurisdictions, incident management and emergency response disciplines, nongovernmental organizations (NGOs), and the private sector into a cohesive, coordinated, and seamless framework for incident management. At the same time it also recognizes and incorporates the various jurisdictional and functional authorities of each in domestic incident management. Planning tools that are used in the development of the response plan are social impact analysis and environmental impacts assessment procedures.

NRP establishes interagency and multijurisdictional mechanisms for the Federal Government's involvement in, and coordination of, incident management operations. This includes coordinating structures and processes for incidents requiring:

- Federal support to State, local, and tribal governments;
- Federal-to-Federal support;

- The exercise of direct Federal authorities and responsibilities, as appropriate under the law; and
- Public and private-sector domestic incident management integration.

NRP is applicable to all U.S. Federal departments and agencies that may be requested to provide assistance or conduct operations in the context of actual or potential incidents of national significance. And, it also includes the International Red Cross, which functions as an Emergency Support Function (ESF) primary organization in coordinating the use of mass care resources in a declared disaster or emergency.

The key concepts reflected throughout the NRP include systematic and coordinated incident management, including protocols for:

- Incident reporting.
- Coordinated action.
- Alert and notification.
- Mobilization of Federal resources to augment existing Federal, State, and local capabilities.
- Integration of crisis and consequence management functions.
- Proactive notification and deployment of Federal resources in anticipation of or in response to catastrophic events in coordination and collaboration with state and local governments and private entities when possible.
- Organizing interagency efforts to minimize damage, restore impacted areas to pre-incident conditions if feasible, and/or implement programs to mitigate vulnerability to future events.
- Coordinating incident communication, worker safety and health, private-sector involvement, and other activities that are common to the majority of incidents (see Support Annexes).

The NRP focuses on those activities that are directly related to an evolving incident or potential incident rather than steady-state preparedness or readiness activities conducted in the absence of a specific threat or hazard.

Examples of incident management actions include:

- Increasing public awareness;
- Assessing trends that point to potential terrorist activity;
- Coordinating Federal support to State and local authorities in the aftermath of an incident;
- Providing strategies for coordination of Federal resources required to handle subsequent events;
- Restoring public confidence after a terrorist attack; and
- Enabling immediate recovery activities, as well as addressing long-term consequences in the impacted area.

2.2.5.1. *Steps in the Response Planning Process*
Incident management actions range from initial threat notification to early coordination efforts to assess and disrupt the threat, to preparatory activation of the response team structure, to deployment of Federal resources in support of incident response and recovery operations. These actions do not necessarily occur in sequential order; many may be undertaken concurrently in response to single or multiple threats or incidents.

Notification and Assessment. Federal, State, local, private-sector, and non-governmental organizations report threats, incidents, and potential incidents using established communications and reporting channels. Environmental Impact Assessments are executed at this point to assess the appropriate actions necessary to minimize damage.

Activation. Early-stage efforts to activate and deploy response plan organizational elements and Federal resources, including special teams, emergency facilities, and other response resources.

Response Actions. Once an incident occurs, the priorities shift from prevention, preparedness, and incident mitigation to immediate and short-term response activities to preserve life, property, the environment, and the social, economic, and political structure of the community. In the context of a terrorist threat, simultaneous activities are initiated to assess regional and national-level impacts, as well as to assess and take appropriate action to prevent and protect against other potential threats.

Reinforcing the initial response to an incident, some Federal agencies may operate in the Incident Command Post as Federal first responders and participate in the Unified Command structure. Once the joint field operation (JFO) is established, its Coordination Group sets Federal operational priorities. The joint field operation provides resources in support of a unified command and incident management team conducting on-scene operations through the State and local emergency operation centers. Depending upon the scope and magnitude of the incident, the National Response Coordination Center (NRCC) and/or the Regional Response Coordination Centers (RRCC) activate the appropriate Emergency Support Functions, as needed, to mobilize assets and the deployment of resources to support the incident. The NRCC and/or the RRCCs facilitate the deployment and transportation of the Emergency Response Teams and specialized capabilities such as, but not limited to, teams under the National Disaster Medical System (NDMS), the Health and Human Services (HHS) Secretary's Emergency Response Team, the Epidemic Intelligence Service, HHS behavioral health response teams, the U.S. Public Health Service Commissioned Corps, and Urban Search and Rescue teams.

Other response actions include the establishment of the JFO and other field facilities and providing a wide range of support for incident management, public health, and other community needs.

Response actions also include immediate law enforcement, fire, ambulance, and emergency medical service actions; emergency flood fighting; evacuations; transportation system detours; emergency public information; actions taken to minimize additional damage; urban search and rescue; the establishment of facilities for mass care; the provision of public health and medical services, food, ice, water, and other emergency essentials; debris clearance; the emergency restoration of critical infrastructure; control, containment, and removal of environmental contamination; and protection of responder health and safety.

During the response to a terrorist event, law enforcement actions to collect and preserve evidence and to apprehend perpetrators are critical. These actions take place simultaneously with response operations necessary to save lives and protect property, and are closely coordinated with the law enforcement effort to facilitate the collection of evidence without impacting ongoing life-saving operations.

In the context of a single incident, once immediate response missions and life-saving activities conclude, the emphasis shifts from response to recovery operations and, if applicable, hazard mitigation. The JFO Planning Section develops a demobilization plan for the release of appropriate components.

Recovery Actions. Recovery involves actions needed to help individuals and communities return to normal when feasible. The JFO is the central coordination point among Federal, State, and local agencies and voluntary organizations for delivering recovery assistance programs.

Mitigation Actions. Hazard mitigation involves reducing or eliminating long-term risk to people and property from hazards and their side effects. The JFO is the central coordination point among Federal, State, local, and tribal agencies and NGOs for beginning the process that leads to the delivery of mitigation assistance programs.

Planning Assumptions and Considerations. The response plan is based on the following planning assumptions and considerations. Incidents of national significance may:

- occur at any time with little or no warning;
- require significant information-sharing at the unclassified and classified levels across multiple jurisdictions and between the public and private sectors;
- involve single or multiple geographic areas;

- have significant international impact and/or require significant international information sharing, resource coordination, and/or assistance;
- require all stages of incident management to include prevention, preparedness, response, and recovery;
- involve multiple, highly varied hazards or threats on a local, regional, or national scale;
- result in numerous casualties; fatalities; displaced people; property loss; disruption of normal life support systems, essential public services, and basic infrastructure; and significant damage to the environment;
- impact critical infrastructures across sectors;
- overwhelm capabilities of State, local, and tribal governments, and private-sector infrastructure owners and operators;
- attract a sizeable influx of independent, spontaneous volunteers and supplies;
- require extremely short-notice Federal asset coordination and response timelines; and
- require prolonged, sustained incident management operations and support activities.

Thus the top priorities for incident management are to:

- Save lives and protect the health and safety of the public, responders, and recovery workers;
- Ensure security of the homeland;
- Prevent an imminent incident, including acts of terrorism, from occurring;
- Protect and restore critical infrastructure and key resources;
- Conduct law enforcement investigations to resolve the incident, apprehend the perpetrators, and collect and preserve evidence for prosecution and/or attribution;
- Protect property and mitigate damages and impacts to individuals, communities, and the environment; and
- Facilitate recovery of individuals, families, businesses, governments, and the environment.
- Deployment of resources and incident management actions during an actual or potential terrorist incident are conducted in coordination with the Department of Justice (DOJ).

2.2.5.2. *Roles and Responsibilities*
The roles and responsibilities of Federal, State, local, private-sector, and non-governmental organizations and citizens involved in support of domestic incident management are critical in the execution of the U.S.A. National Response Plan. Information regarding other countries' strategies of collaborative response plan management can be found on ITOPF webpage (www.itopf.com)

(See footnote 8). Although the information on that site focuses on petroleum spills, it offers a profile of the resources and scope of each country's readiness.

State and Local Governments. State and local governments provide support to emergency response through police, fire, public health and medical, emergency management, public works, environmental response, and other personnel, and often these are the people first to arrive and the last to leave an incident site. When State resources and capabilities are overwhelmed, Governors may request Federal assistance under a Presidential disaster or emergency declaration. Summarized below are the responsibilities of the Governor and local chief executive officer.

Governor. As a State's chief executive, the Governor is responsible for the public safety and welfare of the people of that State or territory. The Governor is responsible for coordinating State resources to address the full spectrum of actions to prevent, prepare for, respond to, and recover from incidents in an all-hazards context to include terrorism, natural disasters, accidents, and other contingencies; and under certain emergency conditions, typically has police powers to make, amend, and rescind orders and regulations. The Governor provides leadership and plays a key role in communicating to the public and in helping people, businesses, and organizations cope with the consequences of any type of declared emergency within State jurisdiction.

Local Chief Executive Officer. A mayor or city/county manager, as a jurisdiction's chief executive, is responsible for the public safety and welfare of the people of that jurisdiction. The Local Chief Executive Officer is responsible for coordinating local resources to address the full spectrum of actions to prevent, prepare for, respond to, and recover from incidents involving all hazards including terrorism, natural disasters, accidents, and other contingencies; and dependent upon State and local law, has extraordinary powers to suspend local laws and ordinances, such as to establish a curfew, direct evacuations, and, in coordination with the local health authority, to order a quarantine. The Local Chief Executive provides leadership and plays a key role in communicating to the public, and in helping people, businesses, and organizations cope with the consequences of any type of domestic incident within the jurisdiction.

Federal Government in the USA

Department of Homeland Security. The Homeland Security Act of 2002 established DHS to prevent terrorist attacks within the United States; reduce the vulnerability of U.S.A. to terrorism, natural disasters, and other emergencies; and minimize the damage and assist in the recovery from terrorist attacks, natural disasters, and other emergencies. The Secretary of Homeland Security is responsible for coordinating Federal operations within U.S.A. to prepare for, respond to, and recover from terrorist attacks, major disasters, and other emergencies. The Secretary of Homeland Security is the "principal Federal official" for domestic incident management. In this role, the Secretary is also responsible for coordinating Federal resources utilized in response to or recovery from terrorist attacks, major disasters, or other emergencies

Department of Defense. DOD has significant resources that may be available to support the Federal response to an incident of national significance, including military and medical trained personnel.

Other Federal Agencies. During an incident of national significance, other Federal departments or agencies may play primary, coordinating, and/or support roles based on their authorities and resources and the nature of the incident. Some of these agencies include the Federal Emergency Management Agency, the National Oceanic and Atmospheric Administration, and the Environmental Protection Agency.

Tribal Chief Executive Officer. The Tribal Chief Executive Officer is responsible for the public safety and welfare of the people of that tribe. The Tribal Chief Executive Officer, as authorized by tribal government:

Is responsible for coordinating tribal resources to address the full spectrum of actions to prevent, prepare for, respond to, and recover from incidents involving all hazards including terrorism, natural disasters, accidents, and other contingencies. Has extraordinary powers to suspend tribal laws and ordinances, such as to establish a curfew, direct evacuations, and order a quarantine. Provides leadership and plays a key role in communicating to the tribal nation, and in helping people, businesses, and organizations cope with the consequences of any type of, domestic incident within the jurisdiction. Negotiates and enters into mutual aid agreements with other tribes/jurisdictions to facilitate resource sharing. Can request State and Federal assistance through the Governor of the State when the tribe's capabilities have been exceeded or exhausted. Can elect to deal directly with the Federal Government. (Although a State Governor must request a Presidential disaster declaration on behalf of a tribe under the Stafford Act, Federal agencies can work directly with the tribe within existing authorities and resources.)

Nongovernmental and Volunteer Operations. NGOs collaborate with first responders, governments at all levels, and other agencies and organizations providing relief services to sustain life, reduce physical and emotional distress, and promote recovery of disaster victims when assistance is not available from other sources. For example, the International Red Cross is an NGO that provides relief at the local level. Community-based organizations (CBOs) receive government funding to provide essential public health services. In the U.S., The National Voluntary Organizations Active in Disaster (NVOAD) is a consortium of more than 30 recognized national organizations of volunteers active in disaster relief. Such entities provide significant capabilities to incident management and response efforts at all levels. For example, the wildlife rescue and rehabilitation activities conducted during a pollution emergency are often carried out by private, nonprofit organizations working with natural resource trustee agencies.

Citizen Involvement. Citizen groups and organizations provide crucial and hands-on support for incident management prevention, preparedness, response, recovery, and mitigation. A basic premise of the National Response Plan is that incidents are generally handled at the lowest jurisdictional level possible, including police, fire, public health and medical, emergency management, and other personnel. This is where strong partnerships with citizens groups and organizations can provide both practical and moral support for affected areas.

Overall Coordination of Federal Incident Management Activities. In the U.S.A., during actual or potential incidents, the overall coordination of Federal incident management activities is executed through the Secretary of Homeland Security. Other Federal departments and agencies carry out their incident management and emergency response authorities and responsibilities within this overarching coordinating framework.

2.2.5.3. Case Studies
National Response Plan—Example: U.S.A.

National Response Plan (NRP) Training (IS-800).[25]
    The following topics are covered in the NRP training:

• National Response Plan Overview (and correlation to NIMS).
• Roles and Responsibilities of Federal, State, local, and tribal governments, and private sector organizations.
• Coordinating structures in the field and at the regional and national levels.

---

[25] More information about the training courses, activities, and programs offered by U.S. Department of Homeland Security, Federal Emergency Management Agency, Environmental Management Institute can be found on their training site http://www.training.fema.gov/EMIWEB/IS/is800.asp.

- Field-level organizations and teams that support an incident.
- Incident management actions, including notification and assessment, activation, deployment, and demobilization.

All Federal, State, territorial, tribal, and local emergency managers or personnel whose primary responsibility is emergency management must complete must this training. Individuals who must take part in the training are:

Federal Level—Officials in Federal government departments and agencies with emergency management responsibilities under the NRP.

State/Territorial Level—Officials in State and territorial governments with emergency management responsibilities to include personnel from State and territorial emergency management agencies and from agencies who support and interact with the 15 Emergency Support Functions (ESF) in the NRP.

Tribal/Local Level—Officials in tribal and local jurisdictions with overall emergency management responsibilities as dictated by law or ordinance; those officials with overall emergency management responsibilities through delegation; and those officials primarily involved in emergency planning.

National Incident Management System (NIMS) Training (IS 700). [26] NIMS is a comprehensive, national approach to incident management that is applicable at all jurisdictional levels and across functional disciplines. The intent of NIMS is to be applicable across a full spectrum of potential incidents and hazard scenarios, regardless of size or complexity; and improve coordination and cooperation between public and private entities in a variety of domestic incident management activities.

NIMS is comprised of several components that work together as a system to provide a national framework for preparing for, preventing, responding to, and recovering from domestic incidents. These components include:

- Command and management.
- Preparedness.
- Resource management.
- Communications and information management.
- Supporting technologies.
- Ongoing management and maintenance.

---

[26]More information about the training courses, activities, and programs offered by U.S. Department of Homeland Security, Federal Emergency Management Agency, Environmental Management Institute can be found on their training site http://www.training.fema.gov/emiweb/is/is700.asp.

Individuals who must take part in the training are:

Executive Level—Political and government leaders, agency and organization administrators and department heads; personnel that fill ICS roles as Unified Commanders, Incident Commanders, Command Staff, General Staff in either Area Command or single incidents; senior level Multi-Agency Coordination System personnel; senior emergency managers; and Emergency Operations Center Command or General Staff.

Managerial Level—Agency and organization management between the executive level and first level supervision; personnel who fill ICS roles as Branch Directors, Division/Group Supervisors, Unit Leaders, technical specialists, strike team and task force leaders, single resource leaders and field supervisors; midlevel Multi-Agency Coordination System personnel; EOC Section Chiefs, Branch Directors, Unit Leaders; and other emergency management/response personnel who require a higher level of ICS/NIMS Training.

Responder Level—Emergency response providers and disaster workers, entry level to managerial level including Emergency Medical Service personnel; firefighters; medical personnel; police officers; public health personnel; public works/utility personnel; and other emergency management response personnel.

Note: Multi-agency Coordination System personnel include those persons who are charged with coordinating and supporting incident management activities. These emergency management personnel typically may function from an emergency operations center or similar facility.

Organizational Structure. The national structure for incident management establishes a clear progression of coordination and communication from the local level to regional to national headquarters level. The local incident command structures are responsible for directing on-scene emergency management and maintaining command and control of on-scene incident operations. The support and coordination components consist of multiagency coordination centers/emergency operations centers (EOCs) and multiagency coordination entities and they provide central locations for operational information-sharing and resource coordination in support of on-scene efforts. Multiagency coordination entities aid in establishing priorities among the incidents and associated resource allocations, resolving agency policy conflicts, and providing strategic guidance to support incident management activities.

Emergency Response and Support Teams (Field Level). Various teams on the field level are available to deploy during incidents or potential incidents to assist in incident management, set up emergency response facilities, or provide specialized expertise and capabilities.

External Affairs. Mechanisms for ensuring accurate, consistent, and timely communications with all of the critical external audiences—the general public, media, congressional and governmental leaders, and the international community are handled through external affairs. Informing the public of emergencies is a critical component of emergency preparedness and response, as the negligence of this can result in large amounts of fatalities and/or injuries.

There are two training sessions through FEMA called IS-700 and IS-800 that must be completed by persons who have a direct role in emergency preparedness, incident management or response. More information on these training courses can be found online at www.training.fema.gov. Summaries of who must complete each of the training courses has been stated above in the appropriate section.

State/Regional Plans—Example: Rhode Island. All 50 States in the U.S.A. are required to have a State emergency response plan. A great example is the Emergency Response Plan for Rhode Island's Department of Environmental Management (RI DEM).[27] This comprehensive plan serves to guide RI DEM in a safe, timely, and effective response to incidents that threaten the State's environment and public health, safety, or welfare. It is also intended to promote coordination among Federal, Sate and local, public and private responders. RI DEM needs to take the lead when the incident involves

- release of hazardous materials,
- oil spill,
- wildfire,
- epidemic of livestock or zoonotic disease,
- pathogenic insect infestation, or
- failure of a dam.

RI DEM manages an incident with the following general principles:

- Preserve life and minimize risks to the health and safety of emergency responders and the public;
- Identify and protect sensitive habitats and wildlife;
- Contain and/or control the release of pollutants;
- Identify the cause of the incident and the source of environmental hazards;
- Remediate environmental impacts;

---

[27] Rhode Island Department of Environmental Management. 2004. Emergency response plan for Rhode Island. Coastal Institute, Emergency Response Administrator.

- Collect and preserve evidence;
- Assist with the apprehension of perpetrators.

The ERP also serves to complement and support the National Oil and Hazardous Substances Contingency Plan and the Region One Oil and Hazardous Substances Regional Contingency Plan. One final caveat is that in the event of a disaster, as defined by R.I. General Laws Section 30-15-1 et seq., the Department's emergency powers will be supplemented, and in some cases superseded, by the emergency powers of the Rhode Island Emergency Management Agency. Additionally, in the event of a disaster, the Governor has the authority to issue executive orders, proclamations, and regulations pursuant to R.I. Gen. Laws Section 30-15-7. More information regarding this comprehensive plan can be found online at (http://www.dem.ri.gov/topics/erp/titlepage.pdf)

Local Response Plans. There are also numerous local emergency response plans for incidents regarding contamination of local drinking water, pathogenic outbreaks, fire, etc. These plans are created by local businesses, infrastructures, universities and other academic establishments, and even by national organizations for localities, like EPA's

Large Water System Emergency Response Plan Outline: Guidance to Assist Community Water Systems in Complying with the Public Health Security and Bioterrorism Preparedness and Response Act of 2002.[28] The EPA has a Local Emergency Planning Committee Database which currently has over 3000 listings, and it looks to each of these local agencies to help them with this water emergency response plan. The database can be found online at (http://yosemite.epa.gov/oswer/lepcdb.nsf/HomePage?openForm)

2.3. Preparation of a Continuity of Operations Plan

In any event, a plan will be needed for continuity of operations at all these levels. The next section describes one approach to establishing such a plan. The following is provided as a summary statement of implementation of rapid incidence response.[29]

The workplace may be vulnerable to natural disasters such as floods, tornados, earthquakes, severe thunderstorms, or other adverse weather conditions,

---

[28] Environmental Protection Agency. 2003. Large water system emergency response plan outline: guidance to assist community water systems in complying with the public health security and bioterrorism preparedness and response act of 2002. EPA 810-F-03-007. Office of Water, Office of Ground Water and Drinking Water.

[29] Information for this section was extracted from the following two websites: http://www.ready.gov/business/st1-planning.html and http://www.fema.gov/onsc/dhscoop.shtm.

and to malicious attack from vandals and subversive organizations such as chemical, biological, or radiological attacks that would make it impossible to conduct business at its current location. In a worst case scenario, transportation routes may be blocked, services such as electrical power, natural gas, water, and communications may be unavailable, inoperable, or severely restricted.

In order to deal effectively with such circumstances, it is important that the workplace develop and implement a "Continuity of Operations Plan" (COOP) so that employees are safeguarded, critical needs are met, and essential functions continue to be performed. The COOP should also include plans for the restoration of all operations in a timely manner.

The objectives of a COOP are to:

- Ensure the safety of employees.
- Establish the capability to communicate.
- Ensure that essential functions are performed.
- Protect essential equipment and records and minimize damage.
- Provide organizational and operational stability.
- Facilitate decision making during the emergency.
- Have an orderly recovery from the emergency.
- Identify emergency needs before an emergency occurs.

In order to prepare a COOP, one must first identify site security concerns through vulnerability assessments, then prioritize potential risks and take measures to enhance site security and reduce potential hazards. Then one develops a COOP plan to deal effectively with an emergency situation. Staff should work with local emergency units (police, medical personnel, etc.) as well as nearby companies or institutions that it would be valuable to cooperate with.

In the event of a disaster, the first thing that must be done is to communicate to all employees as soon as possible, first to determine if they and their families are okay or whether they require assistance, and second to provide them with an assessment of damages at the facility and give them various assignments as needed and appropriate. Such communications with employees should be made daily, by telephone or Email if possible or by public news media otherwise. Other organizational units should also similarly be kept informed. Responders outside of the organization, such as police, fire, and other professional emergency personnel should be contacted, and efforts should be coordinated with them as appropriate. Ensure that clear structures are in place to exchange relevant information with security, law enforcement, and health service entities. To this end, it is important to ensure in advance that facilities personnel and first responders have the information necessary to respond to emergencies safely and efficiently.

In advance of any emergency, it is important to develop procedures for protection of sensitive records and data, and to establish a capability for continuous and reliable communications (including voice and data) with employees, responders, and decision makers.

It is important to clearly state the roles and responsibilities of specific individuals for particular critical operations. An example of typical critical operations would comprise human resources (employees and their safety), infrastructure and operations (such as safety, health, security, facilities, mail, and transportation), information resources (Email and telephone capability), and acquisitions and payments (purchasing needed items and paying vendors). Generally the highest-ranking individual at a given workplace (senior official) would be designated as the person having overall responsibility for the COOP, and other individuals would be designated for specific critical activities. For example, individuals in charge of the critical operations listed above would include the chief administrative person, the chief information technology person, the facilities manager, the health and safety manager, the purchasing agent, the funds control agent, and a personnel manager. Different workplaces may have additional or different specific critical operations that should be taken into consideration. All of these individuals should be specified by name, they should all be well aware of the COOP provisions and their respective roles, and their contact information should be readily available. The usual chain of command should be adhered to, to eliminate confusion and miscommunication.

Because the goal is to be prepared to continue essential functions during an emergency, the critical personnel who are key to operations must be able to be brought quickly to effective action. Consequently, during an emergency such personnel need to be at alternative location(s) so they can ensure the effective resuming of normal operations. The capability to communicate and coordinate operations must be established from the alternate facility, as well as being able to obtain or to access needed information, records, and data. The ability to obtain needed materials and services for continuing operations must also be provided for.

In order to ensure continuity of operations, alternative site(s) should be identified and readied for possible use by critical staff in the event of an emergency. It is useful to identify both a "hot site" and a "cold site". A hot site is a relocation site that would be available for immediate occupancy as it is equipped with the necessary capabilities and equipment to permit rapid resumption of essential functions. A cold site is a site reserved for emergency use, but one that requires provision of additional capability and equipment before it is totally operational as a functioning site.

In addition to critical personnel, essential equipment should also be identified. This is equipment that is essential to the ability of workers to perform the identified essential functions in case of emergency. In addition to the equipment itself, critical support items for the equipment should also be identified, such as supplies, repair parts, operating and maintenance manuals, tools, etc. Once identified, it is important that this equipment is provided the highest level of protection in terms of physical security, maintenance, and location. For example, the designated equipment should be kept in continuous good repair and any relocation to a different part of the building for any reason should be carefully considered and immediately made known to the facilities manager or other appropriate COOP critical position.

In implementing the COOP, it is useful to consider it in five steps. Step 1 is a "pre-emergency" phase, when plans are reviewed regularly, updated as needed, staff is trained, and exercises are held. Step 2 is "activation", when the workplace is evacuated and critical staff move to the alternate site. Step 3 is "operations at relocation site", when critical operations are performed at the alternate site. Step 4 is "sustaining and expanding operations at relocation site". Step 5 is "ending relocation site operations", when operations are transitioned from the relocation site to the original workplace or to a new home base.

Step 1. (Prior to plan activation.). This is carried out prior to any emergency. It involves locating alternative work site(s), and ensuring that essential operations can be carried out at that site. Special attention should be given to ensure that the alternative site provides for the health and safety of those who will work there. Step 1 also includes identifying COOP points of contact and essential equipment, and the periodic review and updating of the COOP plan. It includes training all workers to carry out the plan.

Step 2. (Implemented 1–24 hours from plan activation.). Depending upon whether or not there is a warning prior to the emergency and whether the emergency occurs during working hours or non-working hours, required actions for Step 2 will be somewhat different. If there is a warning prior to the emergency, staff should assemble documents required for performing essential functions at the relocation site. It may also be possible, depending on time available, to prepare essential equipment for relocation as well as to take appropriate protective measures for equipment not designated for relocation. The facility itself should be secured insofar as possible. Critical individuals move to the designated relocation site, taking critical records and equipment insofar as possible. All other personnel will be instructed whether to move to the relocation site, go home, or report to some other safe location. If there is no warning prior to the emergency and it occurs during non-working hours, the senior official will initiate the planned contact mechanism for notifying all employees and directing them to remain at home, to report to the home site to assist in transporting essential records and equipment, or to report to the relocation site to assist in operations. If there is no warning prior to the emergency and it occurs during working hours, relocation actions are initiated. This includes backing up data and preparing to relocate people, records, and equipment insofar as time permits. Depending on the nature of the emergency, the senior official may direct employees to shelter in place or for essential employees to move to the relocation site and other employees to go home. In all of these situations, the senior official of the organization will make the appropriate notifications to local emergency officials and other appropriate individuals.

When essential individuals have arrived at the designated relocation site, the senior official gathers everyone together to assess the situation and provide any needed instructions with regard to operations at the alternate site. Workers perform their essential and other assigned functions.

Step 3. (24 hours through day five.). Essential workers continue to perform their functions. In addition to performing their designated essential functions, this group also monitors the emergency situation, reports to staff and outside emergency officials on status and, plans schedules relocation site operations, and provides support for all personnel at other sites, as appropriate. The essential personnel also must be cognizant of the possible need for evacuation of the alternate site itself, depending on the nature of the emergency and the site's location.

Step 4. (Day six until termination.). During this time critical personnel sustain the performance of those duties necessary to maintain essential operations. Additional workers may be relocated to resume their functions, as required and feasible. Operational capabilities from the relocation site may be expanded.

Step 5. (Return to normal operations/site.). When the senior official determines that the emergency has ended and is unlikely to recur, and that the home site can be reoccupied or a new site occupied, then operations at the relocation site can be terminated. Before terminating operations at the relocation site, a basic level of operations must be re-established at the home (or new) site. Then all employees can return to the home (or new) site, and the alternate site operations can be stopped. If necessary, this relocation can be accomplished gradually, although it should be accomplished as quickly as feasible, and it is important to maintain at the alternate site and then at the home (or new) site all critical operations throughout the move.

During the emergency period, the senior official serves as the designated spokesperson for all communication and coordination between the facility and its employees, and for communication and coordination with all external entities, such as governmental emergency personnel and news media.

All managers and employees must be educated and trained regarding the COOP. It is essential to have a communications process for reaching employees at work and at home. Develop and maintain personnel emergency plans, train employees in implementing the plans (including holding training exercises and drills), and update the plans periodically to reflect new and emerging issues or any appropriate changes needed. It is useful for individuals to develop emergency plans for their homes as well, and it is suggested that they do so.

2.4. Conclusion

Rapid decision making is a crucial aspect to effective emergency response. There are several different decision-making models available; however, the most fundamental requirement of rapid decision making for incident management is that a framework and methodology be established well in advance of any potential emergencies. The importance of being prepared far exceeds the usefulness of any number of last-minute reactive actions that can be taken. Having a solid network on national, regional, state, and local levels crossing all forms of institutions and organizations is ideal to be able to minimize the effects on people, infrastructure, and the environment.

PART 3. PHYSICAL PROCESSES AND MODELING

3. PHYSICAL PROCESSES IN LAGOONS

Irina Chubarenko

Atlantic Branch, P.P. Shirshov Institute of Oceanology, Russian Federation

3.1. Introduction

The laws of nature are simple and objective: they work inevitably and do not depend on an observer. They will work exactly the same way, even if this observer never existed: they are just absolute truth. At present, we can say already, that the main lows driving water motion in large basins are well known. However, it is still a very long and challenging way from a basic understanding to effective use of these laws in applications! Equations, written in order to express these laws for geophysical environment, are objective—but not at all absolute: they depend on observer (like equations of Eulerian/Lagrangian representation), on choosing an inertial/non-inertial system of coordinates, on our knowledge of the processes under description etc. Moreover, solutions of these equations for real basins are always difficult to obtain: first, because of the equations are complicated, second, because they are simply always incomplete! Indeed, even though the Laws are simple, they play in a very complicated environmental systems, so that we, objectively, can never take into account *all* the processes involved; we principally cannot provide for a large geophysical system (like sea, lake or lagoon) *full, exact* and *complete* initial and boundary conditions.

To obtain a solution for complicated lagoon/sea/lake conditions, it is convenient to use numerical modeling. No other method in environmental hydrodynamics is as informative at present as numerical simulation, especially for prognostic purposes. Numerics, however, in its turn, adds to the list of approximations its own specific distortions: continuous in space and time media and processes are presented now only discretely, with certain time and space step, and experience additional modifications due to specific features of numerical scheme. To illustrate this complexity, a general sketch of a lagoon as a part of a complicated environmental system is presented in Figure 1.

Thus, we have a puzzling and exiting situation: knowing very well the laws of physical motion, having collected lots of detailed field data on particular basin (its bathymetry, inflowing rivers, winds, heat exchange, etc), we still have to examine the final model solution in order to reveal its limitations and deviation from real behavior. In fact, using any equation or model, we have actually to *foresee*, far in advance, what kind of a result we should obtain under

I. E. Gonenc et al. (eds.), Assessment of the Fate and Effects of Toxic Agents on Water Resources, 55–81.
© 2007 *Springer.*

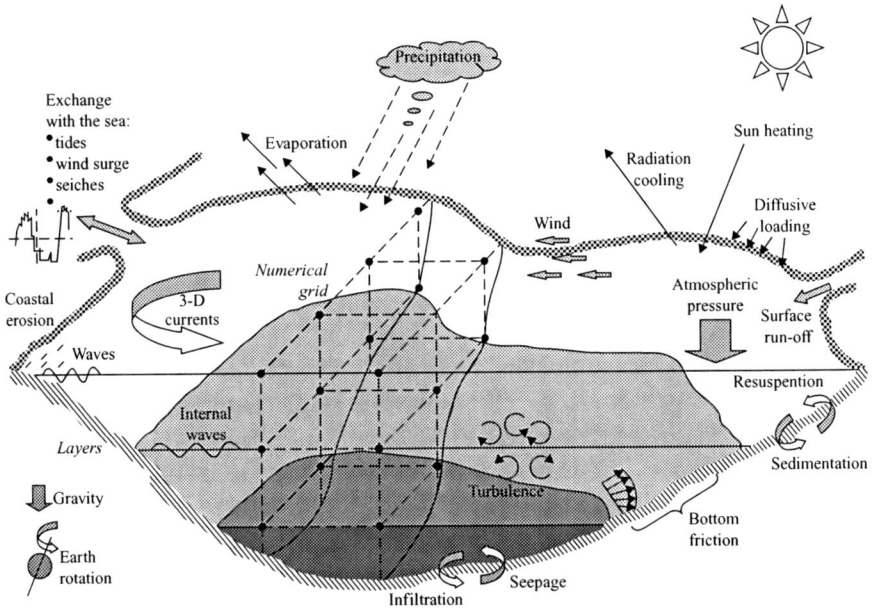

Figure 1. Sketch of a lagoon as a part of complicated hydrophysical environment. Various external factors and internal processes are shown schematically, as well as a discrete representation of a water body in numerical modeling

given conditions. This means: we should have certain experience (the wider the better) on real motions, typically observed situations, and their reasons. To provide some of this experience—is the aim of this lecture.

The lecture has two parts. Part I describes external conditions for water motions in natural basins: the Earth rotation, gravity, wind and solar radiation, and some properties of natural water. Part II presents the information about typical internal structure of natural water bodies (vertical and horizontal water stratification) and shows various examples of internal motions, important on lagoon/lake scale like surface and internal waves, wind effects, Langmuir circulations, tidal motions and exchange with the sea, vertical and horizontal convection, gravity currents. The necessity of coupling of biological and physical processes is briefly discussed.

3.2. Part I: External Conditions

3.2.1. SYSTEM OF COORDINATES: THE ROTATING EARTH

Water in lakes, seas and oceans is never at rest. Even if any external forcing were stopped, the mass of water continues to move due to its inertia. If the

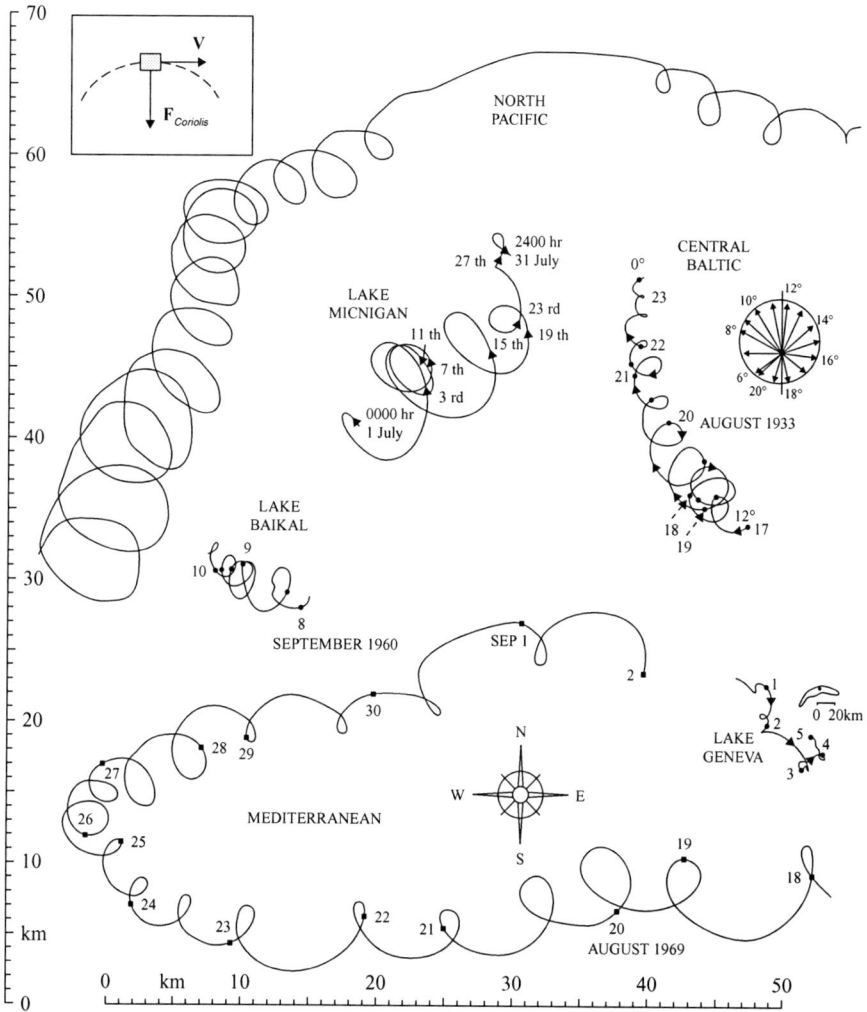

Figure 2. Inertial currents observed in basins of Northern hemisphere (compilation of examples from listed below text books, presented in unified scale). The scheme in left upper corner shows the Coriolis force $\vec{F}_C$ acting on a body moving with the speed $\vec{V}$

water were in space, it would move in a straight line, according to Newton's second law. We, however, observe everything while rotating together with the Earth, and so this motion *looks* much different, see Figure 2: while moving *by inertia,* water parcel now describes circles and loops. The phenomenon can be described in equations by introducing a deflecting force, which is always perpendicular to the speed of motion. This additional deflecting force, appearing due to the Earth rotation, is called *the Coriolis force* (see inset in Figure 2). In Northern hemisphere, it tends to turn the moving body to

TABLE 1. Inertial oscillations, calculated for the
speed $V = 20$ cm s^{-1}

Latitude, φ	T_{inertial} (h)	D (km)
90°	11.96	2.7
30°	23.92	5.5
10°	68.90	15.8

the right, and its value depends on the speed of motion and latitude. The resulting rotation is *anti-cyclonic*, i.e., clockwise in the Northern hemisphere and counter-clockwise in the Southern one.

It is easy to show, that the motion trajectory is a circle with the diameter $D_{\text{inertial}} = 2V/f$ and period $T_{\text{inertial}} = 2\pi/f$, where V is the speed of motion, $f = 2\Omega \sin \varphi$ is the Coriolis parameter, φ is latitude, and $\Omega = 7.292 \times 10^{-5}$ radians s^{-1} is the frequency of the Earth rotation. Table 1 shows the values of the inertial period and the diameter of the circle for some particular latitudes, calculated for the speed of motion of 20 cm s^{-1}.

This gives us a very simple time and space scales, showing when the Earth rotation must be taken into account. Indeed, if the process under consideration is much shorter than the inertial period, the Earth rotation has "not enough time" to deflect the motion significantly. On the other hand, if the size of a basin L is smaller than $R = D/2$, the Coriolis force has *"not enough space"* to valuably turn the velocity vector. In these cases, the motion could be described by simpler equations—without taking into account the Earth rotation. Note, however, that the inertial radius depends on the speed of the very motion as well, so, for smaller velocities, the inertial radius is smaller. For example, for the speed of 7 mm s^{-1} (which is a typical scale for coastally induced gradient currents in nice summer day), at some latitude, say, 30°, it makes up only

$$R = \frac{D_{\text{inertial}}}{2} = \frac{V}{f} = \frac{V}{2\Omega \sin \varphi} = \frac{7 \times 10^{-3}\,\text{m s}^{-1}}{2 \times 7.3 \times 10^{-5} \sin 30\,\text{s}^{-1}} \sim 100\,\text{m}\,!$$

Since various motions exist in natural basins at any time, the Coriolis force in fact can never be ignored. Thus, even though the Coriolis force is very small, it changes the very type of the flow trajectory, and is important for almost all environmental applications.

For further reading, we recommend here some text books on general physical oceanography: Cushman-Roisin (1994), Davis (1987), Defant (1961), Neumann and Pierson (1966), Open University 1989(a), Pedlosky (1987), Pickard and Emery (1990), Pinet (2000), Stewart (2003), Stommel and Moore (1989), Thurman (1985), Tomczak and Godfrey (1994).

3.2.2. MAIN EXTERNAL FORCING

3.2.2.1. *Gravity*

For a large water body on the Earth, the gravity force is important in two aspects: it drives *tides* and causes *buoyancy forces* inside fluid.

Tides Tide-generating forces result from the gravitational attraction between the Earth, the Sun and the Moon (and also other planets, to much lesser extent). After works of Galileo Galilei and Sir Isaac Newton (1642–1727), who discovered the law of gravity, the effect of the Sun and the Moon on the tides was fully understood. As well as all the bodies at the Earth, water of the Earth's oceans is pulled toward the Moon and Sun. As the Moon rotates around the Earth, it pulls water on the nearest side of the Earth outward into a bulge (Figure 3). The water being thrown outward by the planet's spin causes a similar bulge on the opposite side of the Earth. These two bulges travel around the globe, producing two high tides each day. During time of the new moon and full moon, when the Sun and the Moon are in line with the Earth, their gravitational attraction combine and produce *spring tides* (Figure 3(a)); at this time the high tides are very high and the low tides are very low. When the Sun and the Moon are at right angles from the Earth (during the quarter phases of the Moon) the gravitational pull on the oceans is less, producing a smaller difference between high and low tide known as a *neap tide* (Figure 3(b)).

An additional non-astronomical factors such as configuration of the coastline, local depth of the basin, floor topography, and other hydrographic and meteorological influences may play an important role in altering the range, interval between high and low water, times of tides arrival. Some locations have much bigger tides than others. Tidal ranges are usually small in the middle of the ocean but can be very large where tidal waters are funneled into a bay or river estuary. For example, Hawaii has hardly any tidal range at all

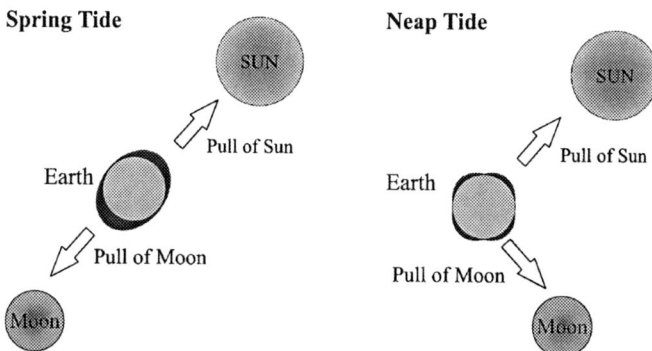

Figure 3. Spring and neap tide generation

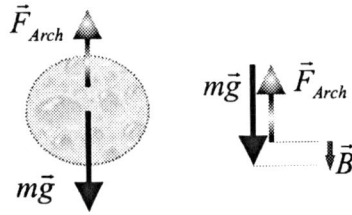

Figure 4. The buoyancy force is the vector sum of the gravity force and the Archimedean force; it may be directed both upward and downward

while in the Bay of Fundy, in Canada, it is of the order of 10 meters and can reach 16 meters when various factors affecting the tides are in phase.

For coastal lagoons, not the very tides inside them are important, but the associated water level rise at the lagoon entrance. This periodic level variation causes specific currents in the lagoon and establishes certain water exchange with open sea/ocean area, often principally important for lagoon water dynamics. Thus, knowledge of the times, heights, and extent of inflow and outflow of tidal waters becomes of great importance for a wide range of practical applications.

Buoyancy Gravitational attraction of a water parcel of a mass m to *the Earth* is the cause of the *gravity force*, $m\vec{g}$ (see Figure 4). This force acts on every water parcel in a basin, and each of them tends to move down—to be "closer to the Earth". Most heavy (dense) parcels win this competition: subject to bigger gravitational force, they move down, *pushing up* lighter (less dense) water parcels. This upward-directed force, exerted on the water parcel *by surrounding water*, is called the *Archimedean buoyancy force*, $\vec{F}_{Arch}$ (Figure 4). Now, every water parcel experiences two forces: the gravity force, which depends on the parcel mass (density), and the Archimedean force, which depends on *difference* in density between the parcel and its neighbors. The sum of the two gives what we call *the buoyancy*, $\vec{B} = m\vec{g} + \vec{F}_{Arch}$; it is finally the *upward or downward* directed force acting on a parcel of water that is more or less dense than other parcel at its level. For example, if cold air blowing over the sea cools surface waters and causes them to be more dense than the water beneath, one says that surface waters get *negative buoyancy*. In this example, the difference in density results in a force that causes surface water to sink. On the contrary, warm spring welling out of the bottom into a colder environment, has a *positive buoyancy*, which causes its water to rise up.

Even though both gravity force and Archimedean force are vertical, they generate in hydro-physical environment both vertical and horizontal motions. Some common examples are gradient currents and coastal heating/cooling processes.

In addition to the books on general oceanography mentioned above, the reader may also use here Batchelor (1967), Cartwright (1999), Gill (1982), Kundu (1990), Neumann and Pierson (1966), Open University (1989c), Pugh (1987), Richardson (1961).

3.2.2.2. *Winds*

Wind blowing over water surface is often the main cause of currents. To know its distribution is therefore vital for an adequate understanding of the circulation pattern in a basin. The global atmospheric circulation on the Earth has a clear structure of a convective transport from one pressure zone to another, which is driven by solar heating and the Earth rotation. On the length scales of the ocean, knowledge of this general air circulation is quite enough for the description of the main features of water circulation beneath; dynamics of atmospheric cyclones and anticyclones provides its smaller details. At smaller scale—for lagoons, lakes, and even some seas, the situation is much different because many meso-scale processes become of importance for wind generation. Finally, wind over such water bodies is the result of interference of (i) the general atmospheric circulation, (ii) the passing-by cyclonic/anty-cyclonic structures and (iii) local processes, induced by surrounding topography. Figure 5 illustrates this idea. Since we deal in this book with meso-scale

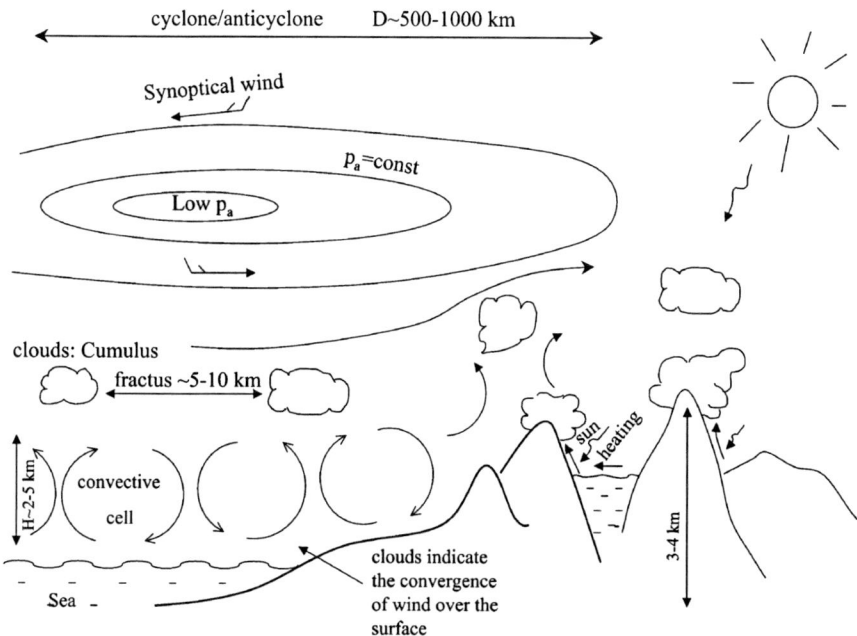

Figure 5. Wind field structure above large water surfaces as it is commonly formed by a general atmospheric transport and local topography

processes, we concentrate on the last two items, and send the reader for further information on the global atmospheric circulation to text books on meteorology and oceanography: Houghton (1977), Davis (1987), Defant (1961), Stewart (2003), Open University (1989a), Dritschel et al. (1999).

Atmospheric gyres It is well known that about 90% of mass of the atmosphere is concentrated in the lowest 20 km. Thus, in comparison with the radius of the Earth, $R_{Earth} = 6400$ km, the atmosphere is a very thin layer of relative thickness of $\frac{H_{atmosphere}}{R_{Earth}} \sim \frac{20}{6400} \sim 0.003$, i.e., just a thin film near the solid boundary! The same applies to the ocean: its mean depth is about 6 km, so that $\frac{D_{ocean}}{R_{Earth}} \sim \frac{6}{6400} \sim 0.001$. Typical bottom slope inclination in natural basins is also very small. It is usually characterized by the *aspect ratio A* — ratio of characteristic vertical and horizontal scales. The slopes with $A \sim 0.001$–0.0001 are the typical ones, whilst the slopes with $A \sim 0.01$ (or about $1°$ only!) are considered as steep ones and 3–$5°$-slopes are exceptional. As we will see further in Chapter 4.5, this fact is of utmost importance for mathematical description of water motion in natural basins: vertical velocities are much smaller than the horizontal once, so that *shallow water* equations work very well in geo-physical hydrodynamics.

Due to different heating at equator and at poles, atmospheric air cannot be motionless—we must observe *some* winds. Here we find another effect of smallness of geo-physical aspect ratio, which is quite surprising: both in atmosphere and in ocean, all (more or less) large-scale motions are *structured in gyres*. To understand this deeper, let us take an example. Imagine that our Earth is a small haired ball in your hands, and every hairline models a vector of wind speed at the surface. We should comb it now in a way, that all the hair is in a very thin layer, smoothed closely to the very ball' surface. As you will shortly discover, it is simply topographically impossible to comb hair in a perfectly smoothed way: the hair-layer MUST have vortices, at least one pair, right and left swirled. (Note: perfectly smoothed "coiffure" could be possible, for example, if the Earth were tore-shaped!). These gyres are models of cyclones and anticyclones in atmosphere: they occur—inevitably— on our planet, simply because it is round. In real atmosphere, many pairs of cyclones/anticyclones occur, with a typical diameter of a few hundreds of kilometers. Thus, even without local topographic effects, the opposite ends of large and middle-sized lagoons/lakes (of a length of ca. 100 km and more) may experience winds of the opposite directions.

The Earth rotation (the Coriolis force) plays a very important role in the formation of atmospheric gyres. Even their names reflect this: *"cyclone"* is, by definition, a gyre, rotating *in the same direction* as the Earth does, while *"anti-cyclone"* rotates in the opposite (to the Earth) direction. Note here: it also means, that cyclone of the Northern hemisphere rotates in the opposite direction to the cyclone of the Southern hemisphere. Indeed, if you look at the Earth from its Northern pole, it rotates counter-clockwise, and so does any cyclone in Northern hemisphere. Now look once again from the Northern pole—through the Earth—to the surface of the Southern hemisphere, and draw another counter-clockwise gyre (=cyclone) *there*. It is rotating in the same direction with the Earth, but *for the observer in the Southern hemisphere*—it has *clockwise* rotation. In other words, for the observer in Southern hemisphere (or if you look at the Earth from the Southern pole), the *clockwise rotation* corresponds to the rotation of the Earth, and cyclones of the Southern hemisphere are of *clockwise* rotation. In particular, this is the reason why any geophysical gyres cannot cross the Earth's equator.

This terminology, however, says not much on the physics of air motion in atmospheric gyres. This motion is driven by the pressure gradient between the center of a gyre and its periphery. Common feature of all the cyclones (in both hemispheres) is lower air pressure in their middle, whilst anti-cyclones have a higher pressure there. So, general air dynamics is as follows: if air pressure in the center is low (cyclone), surrounding air masses tend to flow towards it. Being deflected to the right (in Northern hemisphere) by the Earth rotation, they form general counter-clockwise circulation (i.e. the *cyclone*). The same logic applies for anti-cyclones or the Southern hemisphere. Cyclones are more intense gyres (larger pressure gradients, higher wind speed), perhaps, exactly because the very Earth favors their rotation.

Local winds In coastal zone, where water meets land, various local winds are formed. They are always the result of difference in pressure (temperature) between neighbouring air masses. These differences are highly dependent on local land topography, relative areas of water and land, and their properties. Typical for coastal zone are various *breezes*—land breeze, sea breeze (Figure 6). Near mountainside, the foehn, valley and mountain breezes and other local winds are often observed. Wind forcing is extremely important for the development of water currents. However, it is rather difficult to acquire sufficient information on wind distribution above a certain basin, with its particular surrounding topography. This point is often crucial for numerical modelling and forecasts.

3.2.2.3. Radiation
Solar radiation The ellipticity of the Earth's orbit around the Sun and the tilt of Earth's axis of rotation to the plane of the Earth's orbit lead to an unequal distribution of heating and finally—to the existence of seasons. For a water basin, the incoming solar radiation is primarily determined by *latitude, season, time of a day, and cloudiness*. This way, polar regions are heated less than the tropics; areas in winter time are heated less than in summer, in early

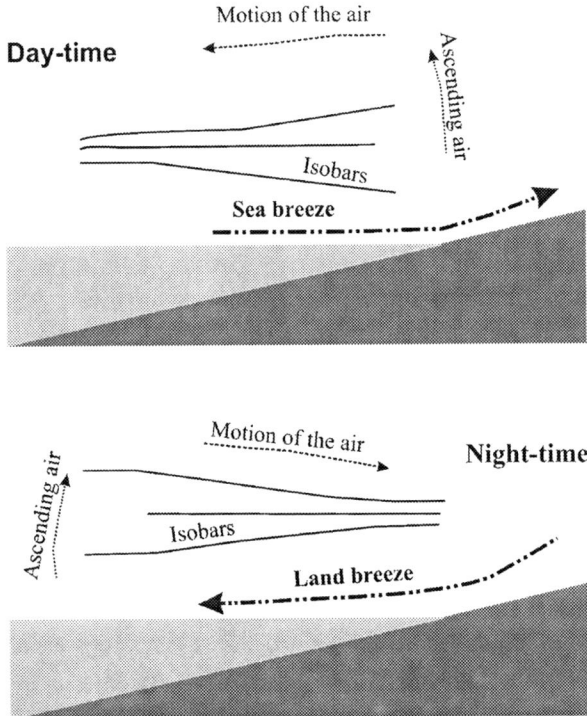

Figure 6. Example of local winds: sea-land and land-sea breezes

mornings—less than at noon; cloudy days have obviously less sun than sunny days. These variations drive many and many processes in water bodies. From physical point of view these variations are the part of a *heat budget* of a basin. The major terms of the heat budget at water surface are:

- Solar radiation, or the flux of sunlight into the water;
- Net infrared radiation, or net flux of infrared radiation from the water;
- Sensible heat flux, or the flux of heat between air and water due to difference in their temperatures;
- Latent heat flux, or the flux of heat carried out of water by evaporation;
- Advection, or heat carried in/away a certain water volume by currents.

Units for heat fluxes are Watts m^{-2}. The exact calculation of heat fluxes is really complicated problem, since every term depends on many features, conditions and ongoing processes. For environmental numerical modeling, however, bulk estimations are often quite acceptable. Indeed, in the list above: (i) the insolation can be (in general) estimated from the area' latitude and the date; typical range of its natural annual variation is 30 W m^{-2} < Q < 260 W m^{-2};

(ii) net infrared radiation is almost constant, related to the mean water temperature; (iii) cloudiness, sensible/latent heat fluxes and many other features cannot influence much water currents; (iv) advection is taken into account by the numerical model automatically. So, typically, the description of this part of external conditions is not very crucial. For further reading we recommend the mentioned above books on meteorology and oceanography.

3.2.3. WATER PROPERTIES

Natural water is so familiar to everybody, that we are apt to overlook the fact that it is an altogether peculiar substance; its properties and behavior are quite unlike those of any other liquid. To take just one example, water has the rare property of being denser as a liquid than as a solid, and its greatest density is a few degrees above the freezing point. The consequence of this is of greatest importance for life on our planet. When ice forms on water surface, it keeps floating, and acts as an insulating cover to retard further cooling of water beneath. As a result, ice-covered basins do not freeze solid to the very bottom but leave a space for the winter survival of aquatic life. Here, we will only briefly describe some of geo-physically important properties and anomalies of water (specific structure of its molecule, density anomaly, anomaly of volume change when freezing) and just list some other. We encourage the reader to learn more and more on this topic in books Open University (1989b), Millero (1996), Stewart (2003).

3.2.3.1. *Structure of Water Molecule*

Molecule of water consists of one oxygen atom and two hydrogen atoms (see Figure 7), what we usually denote as H_2O, only rarely mentioning that it can be 33 different combinations of H and O isotopes. Having comparatively large positive charge ($+8$), the oxygen atom nucleus attracts all electrons, leaving distant the hydrogen nuclei (positively charged protons). These two protons, carrying one positive charge each, experience mutual repulsion. Surprisingly, they do not take the most distant positions: the angle between them is $105°$ instead of $180°$. In such water-molecule arrangement, electrical charges are not fully balanced in space: oxygen atom carries finally more negative charges (electrons) than its nucleus has, whilst hydrogen atoms have lack of electrons, and thus have uncompensated positive charge. As a result of this, pure water

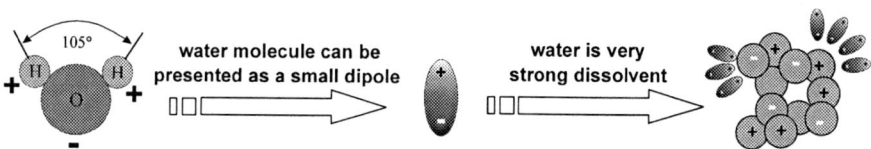

Figure 7. Structure of water molecules and their interaction with other substances in solution

can exist only in laboratory, and for a few milliseconds (!) only, whilst natural water is always a solution of "surroundings": gases, organic/inorganic
substances, etc.

There are other important consequences of the molecular structure of pure
water. The major of them are:

- Water dipoles form aggregations of molecules (clusters, polymers), of on
  average 6 molecules at 20 °C. Therefore, water reacts slower to changes
  than individual molecules.
- Since water has an unusually strong disassociative power, i.e. it splits dissolved material into electrically charged ions, the dissolved materials greatly
  increase the electrical conductivity of water.
- The angle 105° is close to the angle of a tetrahedron, i.e. a structure with
  four arms emanating from a centre at equal angles (109° 28′). As a result,
  oxygen atoms in water try to have four hydrogen atoms attached to them in
  a tetrahedral arrangement.
- Tetrahedrons are of a more wide-meshed nature than the molecular closest
  packing arrangement. They form aggregates of single, two, four and eight
  molecules. At high temperatures the one- and two-molecule aggregates
  dominate; as the temperature falls the larger clusters begin to dominate.
  The larger clusters occupy less space than the same number of molecules
  in smaller clusters. As a result, the density of water shows a maximum at
  4 °C.

3.2.3.2. *Water Density*

Oceanic water, with its typical salinity of 35 psu, is thus a mixture of 96.5%
pure water and 3.5% other material, such as salts, dissolved gases, organic
substances, and undissolved particles. So, its physical properties are mainly
determined by the 96.5% of pure water. As it was described above, the density
of pure fresh water varies with temperature in a very unusual way: there is
a range of temperature—from 0 °C to 4 °C—where the density of (fresh)
water decreases together with rising temperature, so that fresh-water density
is maximum at 4 °C. At the same time, freezing point for fresh water is 0 °C,
i.e., freezing water is not the densest one. This feature exists for water salinity
up to 24.7 psu. Thus, water with higher salinity become just "normal liquid"—
with their density increasing till the very freezing point. This feature affects the
thermal convection (see examples in Part II), and is so important for a basin,
that have deserved even special terminology: in contrast to *salty* waters, ones
with the salinity less than 24.7 psu are called *brackish* waters. When cooling,
brackish waters mix throughout the entire basin' depth until they reach the
maximum density; then, the cooling is restricted to the wind-mixed layer,

Figure 8. Floating ice is 90% under water

which eventually freezes over. Deep fresh or brackish basins are typically filled in winter with water of maximum density.

Density is one of the most important parameters in the study of the basin' water dynamics. Small horizontal density differences (caused for example by differences in surface heating) can produce very strong currents. The determination of density field is therefore one of the most important tasks. The density of sea-water, ρ, depends on its temperature T, salinity S, and pressure p. This dependence is known as the Equation of State of Sea Water $\rho = \rho$ (T, S, p); it has been obtained as the result of many careful laboratory determinations (see for example, Millero and Poisson, 1981; JPOTS Editorial panel; 1991; JPOTS, 1981). Density increases with increasing salinity and pressure (depth), and decreasing temperature (except for the temperatures below that of density maximum in brackish waters).

Liquid water is denser than the ice. This property is also unusual: typically, the density of substances in solid state is higher than that in liquid one; for example, solid metals sink in their own melt. The reason is again in water molecule structure. When freezing, water molecules form tetrahedrons. This leads to a sudden expansion in volume, i.e., decrease in density. Solid phase of water is therefore lighter than the liquid phase, which is a rare property. One of the most important consequences is that *ice floats*. This is extremely important for life in water basins, since the ice acts as an insulator against further heat loss, preventing water from freezing from the surface to the bottom. The density of ice is ∼10% less than the density of water. In particular, this means, that floating ice is 90% under water (Figure 8).

3.3. Part II: Internal Structure and Water Dynamics

Now, we take *some* real basin, say, lagoon or lake, and discuss briefly the most common *internal structure* and *motions*, generated by the described above external forcing: wind, water exchange with the sea, inflowing rivers, day/night or seasonal heating/cooling processes, surface and internal waves and some other processes. Again, following the general idea of this course, we pay attention mainly to the processes, which generate water currents and thus contribute to the matter transport.

3.3.1. STRATIFICATION

Stratification of water means that it is "structured in layers"; these layers need not to be strictly horizontal. In oceanography, this term is used to denote *any vertical or horizontal variations* in fields of temperature, salinity, density, etc. You can imagine, that under real wind, weather, fresh river inflows, tidal exchange with the sea and other external forcing, natural water bodies cannot be just homogeneous: they are inherently stratified. For different particular processes, stratification in different fields is important: temperature and oxygen stratification is important for biological processes, whilst temperature, salinity and suspended matter content jointly contribute to the field of density. Water density variations are the cause of many kinds of currents, so, we pay attention to the processes, contributing to variations of temperature, salinity and suspended matter fields. Water density stratification strongly influences the internal motions, so that water currents in a homogeneous basin may be completely different from that with stratification. Consider first a principal seasonal behavior of vertical and horizontal density stratification in deep basin (lake), taking place due to seasonal variations in solar heating.

3.3.1.1. *Seasonal Cycle of Water Stratification*
In summer time, an intense solar heating causes in deep water body so-called direct thermal summer stratification, i.e., the very top surface layer is the warmest, and water temperature decreases with depth (see Figure 9). The more intense and long solar heating is the lighter is surface water, what makes the stratification stronger. In early autumn, even though the solar heating goes on, it is not strong enough to overshadow the cooling due to heat exchange with the cool atmosphere. The top layer looses its heat, becomes colder and thus heavier, than the water below, what causes vertical turbulent mixing. With time of cooling (usually accompanied by autumnal wind strengthening), the thickness of this mixed layer increases, see Figure 9. Its thickness in the ocean makes up to several tens to hundreds of meters. In lakes without vertical salinity stratification, autumnal mixing often reaches the very bottom. This time period is called autumnal overturn, and it is very important for lukastrine life: mixing throughout the depth ventilate the entire lake, enriches water body with both oxygen from the surface and nutrients from the bottom.

It is typical for winter cooling at mid-latitudes, that surface waters in brackish basins reach the temperature of maximum density (4 °C for fresh

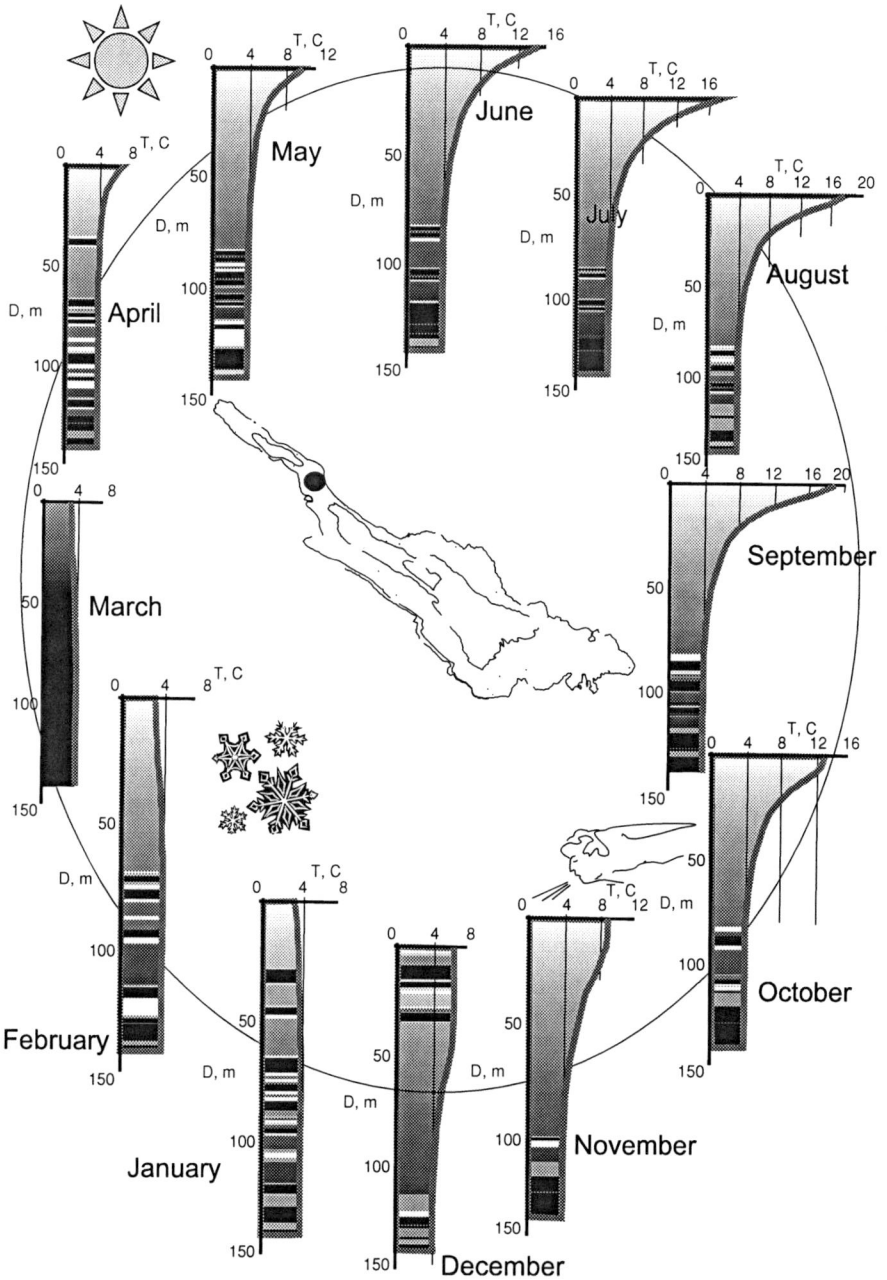

Figure 9. Seasonal variations of vertical temperature profile in deep basin. Lake Constance, monthly mean data of year 1987

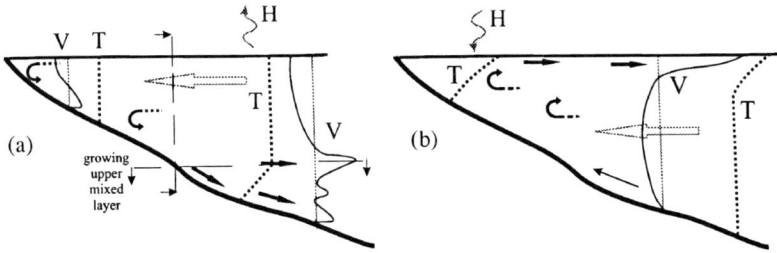

Figure 10. Seasonal coastal cooling (a) and coastal heating (b) processes generate specific water circulation in coastal zone: cold down-slope gravity cascades or warm surface jet transport littoral waters off-shore, being accompanied by distributed over the depth return flow

water). With further cooling, surface water becomes *lighter* again, so that an *inverse winter stratification* is now established. When spring comes then, the mixed upper layer develops once again, then *spring overturn* mixes water body down to the bottom, and solar heating establishes, step-by-step, the direct thermal summer stratification. Important to note here, that all the variations in temperature profiles are in fact *driven by combined effect of solar heating and gravity*: both the direct and inverse thermal stratifications are gravitationally stable, so that lighter water overlies the denser one, while the upper mixed layer is uniform in density—and lighter, than underlying deeper layers.

Near coasts, the described above seasonal variations generate specific *horizontal* temperature gradients. The reason for them is the time lag in response of deep and shallow areas to seasonal variations in heat exchange: shallow areas respond faster, whilst deeper parts are more inertial. This way, seasonal horizontal temperature gradients occur, which generate specific system of currents in the basin (Figure 10): during periods of cooling, slow (scale of current speed is millimeters to units of centimeters per second) down-slope cascades transport cooled littoral waters off-shore, and during periods of heating—spring and summer—a warm surface jet drives the circulation, carrying more light littoral waters sea-ward. Both down-slope cold currents and warm surface jets are accompanied by even slower return transport, which is distributed over the rest of water depth.

Even though this circulation is rather weak in comparison to wind or wave induced motions, they persist throughout the whole year, forming the basic "flow climate" for strong but short-lived wind-induced or other episodes. The adjustment to the Earth rotation turns these thermally-induced flows to the right (left) in Northern (Southern) hemisphere, so that so-called *thermal wind* is established along the coast. This thermal wind seems to be a significant part of general basin-wide cyclonic circulation, reported to be typical in many large lakes and seas.

Many other physical processes like wind mixing, internal waves, horizontal intrusions, etc., can cause changes in vertical temperature profile. Rather

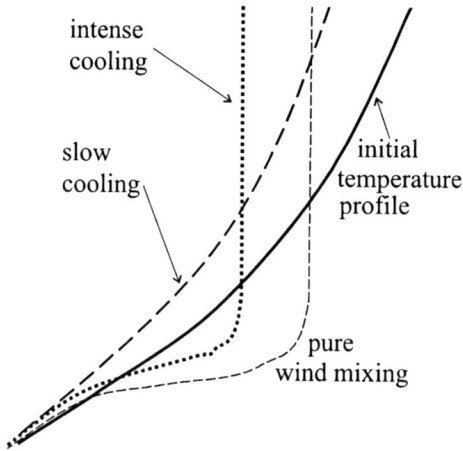

Figure 11. Vertical temperature profile how it is changed by different physical processes

often, these changes can shed some light onto the background processes. As an example, Figure 11 shows some initial vertical temperature profile (solid line)—and how it changes under particular processes. So, slow cooling from the surface causes temperature decrease diffusing throughout the depth, with maximum cooling at the surface (long-dashed line). Wind mixing without heat exchange (short-dashed line) makes the upper layer colder—but lower layer warmer than initial ones, so that the heat lost by the upper layer is equal to that gained by the lower layer, and total heat loss (proportional to the area between temperature profiles) is zero. Strong (rapid) cooling causes so-called *penetrative convection*: below the upper iso-thermal layer, the final (dotted) line crosses the initial (solid) temperature profile, i.e. cold thermals from the surface penetrates so deep, that entrains into turbulent mixing some "additional" water, which was even *colder* before; therefore, unexpectedly, the process of strong cooling causes the *heating* of a certain layer. Of course, the relation between the ongoing processes and temperature profile variations in real situations can be much more complicated and often ambiguous. For example, the last profile—characteristic to *vertical* penetrative convection— can also be produced by *horizontal* intrusion, propagating below the upper mixed layer, what is characteristic to coastal cooling process. The more additional information do we have on a particular situation, the better result we can obtain from such analysis.

3.3.2. WIND-INDUCED MOTIONS

Wind is often the main source of currents in water basins: blowing over large water surfaces, air masses transmit their momentum to upper water layer. Near boundaries, wind-induced currents are even stronger, because water mass from the entire basin' surface is to be deflected in a rather thin along-the shore zone. We consider first the structure of wind-induced current in an open deep area, and then show examples of its modifications due to presence of boundaries.

The most surprising feature of wind-induced currents is that they are not at all directed down-wind. It was Fridtjof Nansen (1861–1930), who noticed first that wind tended to blow ice at an angle of 20–40° to the right of the wind in the Arctic, by which he meant that the track of the iceberg was to the right of the wind, looking downwind. Walfrid Ekman, young PhD student, was asked to describe this motion theoretically. Ekman assumed a *steady, homogeneous, horizontal* flow *with friction* on a *rotating* Earth, first without boundaries. Later, Ekman expanded the study to include the influence of coasts and differences of water density (Ekman, 1905). So, he discovered, that wind-induced current at the surface is, in general, 45° *to the right of the wind* when looking downwind in the northern hemisphere (45° to the left of the wind in the southern hemisphere). Below the surface, the velocity decays exponentially with depth, what can be described by formulae

$$u = V_0 \exp(az) \cos(\pi/4 + az),$$
$$v = V_0 \exp(az) \sin(\pi/4 + az).$$

Here, u and v are the horizontal velocity components, V_0 is the water velocity at the surface, and z is vertical coordinate, positive downward. Graphically, the current is presented in Figure 12.

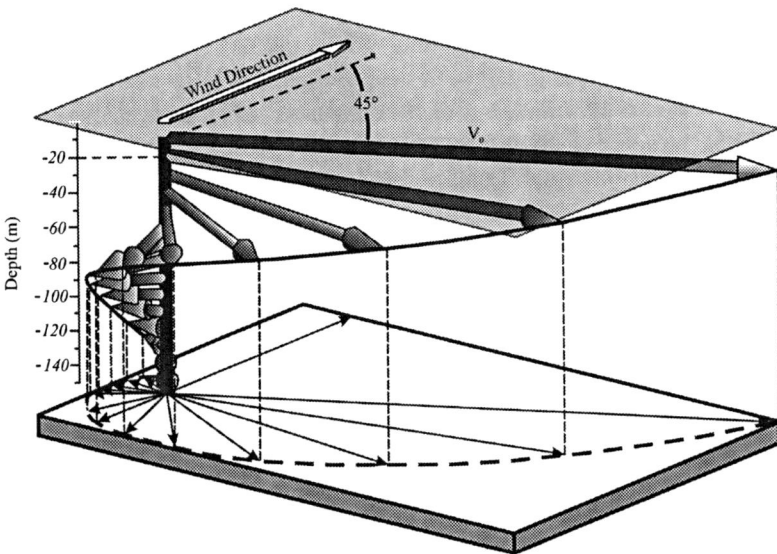

Figure 12. Distribution of currents in the Ekman spiral

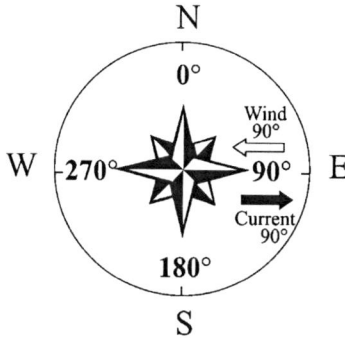

Figure 13. Wind direction and current direction are expressed in different ways: "wind direction" is where wind comes *from*, while "current direction" is where the current goes *to*

So, the structure is clear, but how to obtain the V_0, current velocity at the surface? Physically, it depends on wave height, wind fetch, air density and many other conditions! This question is still a very challenging problem for every particular situation, and the common thumb rule here is just to take as the scale for surface current the value of *about 3% of wind speed* at the height of 10 m above the water surface. Important to note that in field practice there are different ways of expression of wind direction and current direction: "eastern wind" means wind *from* the east, while "eastern current" means current *towards* the east (see Figure 13).

Wind-induced currents are much influenced by presence of boundaries. In closed basins, even in shallow ones, these currents are significantly 3-dimentional, because the down-wind transport in the surface layer, which is directly pushed by the wind, must be compensated by upwind gradient flow in deeper layers. So, one always has in real basins an up-wind return flow along bottom depressions and deep channels, and often also at lee sides of coasts and islands. More information can be found in Horikawa (1988), Gregg (1991), Munk (1950) and many other books.

Figure 14 illustrates the result of wind action on a closed deep basin with vertical density stratification. Before the all, wind pushes the upper water layer, so that finally water level in the basin becomes inclined. For the constant wind speed, this inclination is also constant. If the basin is homogeneous or has a weak vertical stratification, the generated circulation has a single one cell in vertical plane: upper water layer moves (approximately) down-wind, while the lower layer goes in the opposite direction (Figure 14(a)). If, however, the basin has well-pronounced layers of different density, then separated circulation cell is formed in every such layer. Common for lakes, lagoons, estuaries is a two-layered vertical density structure: warm epilimnion/cold hypolimnion in lakes or freshened upper layer/more salty bottom layer in lagoons and estuaries.

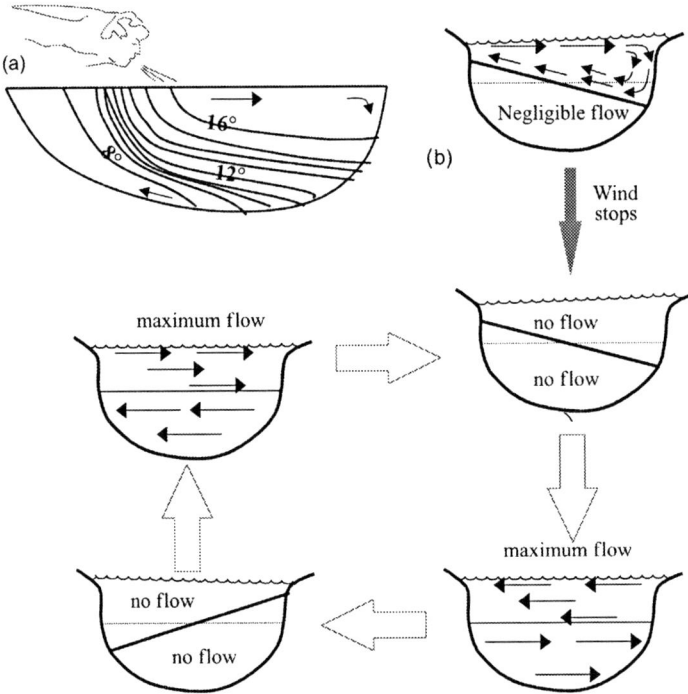

Figure 14. One-cell and two-cells wind-induced water circulation in closed basin. After wind stops blowing, internal oscillations can persist for many days.

Principal water circulation during wind forcing and after wind stops blowing is presented in Figure 14(b).

The described above example of so-called *wind surges* brought us close to the discussion of internal waves and their role in the basin-wide water dynamics. Exactly like surface waves appear at the air-water interface, any interface inside water body is subject to wavy motions. Since the density differences between water layers are much smaller, than between air and water, the amplitudes and periods of internal waves are much bigger than that of the surface waves. Both surface and internal wave breaking causes mixing in water body, especially in surf zone, however, on a basin-wide scale, wave motions do not produce an effective transport mechanism. So, we omit more detailed description of waves here, sending the reader for further details to another books, for example, to Hasselmann (1970), Horikawa (1988), Open University (1989c).

The only mechanism connected to wave motions will be described: the one known as the *Langmuir circulation* (after Irving Langmuir, see Langmuir, 1938). On windy days, one can often observe on water surface of large basins a long parallel streaks of foam, flotsam, algae, called windrows (see Figure 15). These windrows make visible a pattern of parallel pairs of large alternate left-handed and right-handed horizontal rolls, or circulatory cells, called *Langmuir circulation*. It is now considered as one of important processes

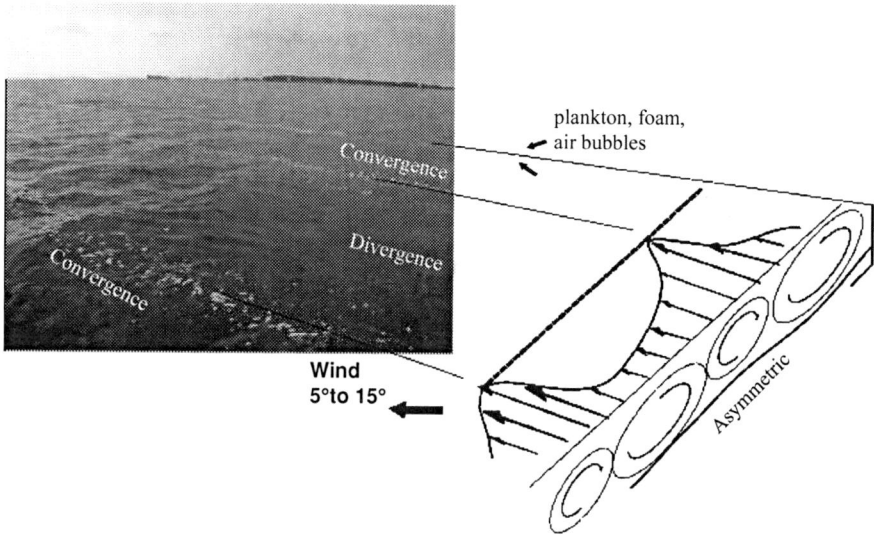

Figure 15. (a) Langmuir circulation in wide and flat area of Darss-Zingst Bodden Chain (the Baltic sea). Local depth is 1.7–2 m, while the distance between the streaks is 3.5–5 m. (b) Sketch of Langmuir circulation

in upper layers of large water bodies, influential in producing and maintaining the uniform surface mixed layer, in driving dispersion of buoyant particles, oil droplets, or bubbles, in transporting organisms between high and low light levels, promoting patchiness of swimming or buoyant algae, or affecting their exposure to light and pollution. These rolls are the result of highly nonlinear interaction between wind-induced current and wave-induced drift, i.e. the presence of both current and waves are important.

System of currents, convergence and divergence zones, produced by such coherent structures, actually limits rather than favours the spreading of floating material, especially in transverse direction. It can easily be observed for the material, floating on the surface: with time, it is accumulated in convergence zones and kept there for a long time, until this zone is destroyed. After that, the material is trapped by another convergence zone, and again kept there until the next breakdown or re-configuration of streaks. This way, the problem of spreading of a material is closely connected with a kinematics of convergence and divergence zones of Langmuir circulation. Observations show, that in deep basins the diameters of the rolls (and thus, the distance between lines of foam) grow with time, simultaneously with the thickness of the upper mixed layer, favouring roll amalgamation. In shallow areas, bottom depth limits the growth, so that picture becomes more or less regular soon after the beginning of the wind action, with the distance between lines roughly equal to doubled local depth. Now, in this fully-developed picture, the transport of material depends on the rolls amalgamation/destruction process. On average, the lifetime of the roll is reported to be in the ocean of about 15–20 min. The process of re-organization forms at the water surface so-called *Y-junctions* of four different kinds; they persist for a few minutes only—and regular picture is re-established again. One more feature typical for the Langmuir circulation in closed basins or coastal areas, in contrast to open waters, is *a drift* of windrows. This happens because coasts deflect currents while wave propagation is much influenced by bottom bathymetry. As the result, the lines of windrows, formed due to interference of current and waves, are (roughly) parallel to the current direction, and fetched down by cross-current wave propagation. For the dispersion of the floating material this means again an increase of dispersion as compared to that in the open ocean.

This example emphasises the complexity of mixing and transport processes in large closed/semi-closed basins: different physical mechanisms play at the same scene, and their interaction produces significantly three-dimensional, variable in time flow pattern. In particular, the formation of circulation, structured in large rolls, is rather common in fluid/air motion. This way, vertical convection plus horizontal wind produce cloud streets in atmosphere; vertical wind mixing plus down-slope seasonal currents generate large sediment furrows at the lake bottom; currents plus waves in sea surf zone generate parallel to the coast underwater bars, etc.

3.3.3. WATER EXCHANGE WITH THE SEA, TIDES, RIVER INFLOWS

For lagoons, estuaries, semi-closed seas and bays, water-exchange with an open marine area is a very important factor. With decreasing level of seawater influence, such basins can be subdivided into *leaky, restricted* and *choked* ones. Leaky basin responds very quickly to any variations in sea coastal zone, while in choked basin the response of distant areas to sea level variations or current changes comes with significant time lag, and has much reduced amplitude. In non-tidal areas, water level variations at the entrance are mainly related with the wind action, thus, rather random, while in tidal areas they are regular, and often become the main driving factor for water circulation, extremely influential in water renewal and the entire basin ventilation. Since water masses in open sea and in lagoon/estuary are much different, region where they interact often has a character of a barrier zone, with active hydrological fronts and intense mixing. For phyto- and zooplankton, this area is both attractive because of strong mixing and distractive, because fresh-water species, being carried away into the sea, cannot survive there, and marine creatures often are lost in freshened waters.

As a whole, conditions of water exchange at the lagoon/estuary entrance are of utmost importance for all the range of physical, chemical, biological processes there. At the same time, this exchange is always complicated, because it depends on many different factors at once: time-variable difference in water levels, currents in near shore zone, particular bathymetry of the entrance, density difference etc. This is the reason why often in modeling it is convenient to move the lagoon/estuary open boundary out into the open sea area, far from the very entrance. In estuaries and lagoons with large inflowing rivers, the difference in densities between marine and riverine waters causes the formation of a specific quasi-stationary front—the *salt wedge*: more light freshened riverine waters move seaward in upper layer, while, at the same time, heavier salty sea waters move upstream along the bottom. Due to strong density jump at the interface, water circulation of any origin becomes complicated in such basins; both currents and matter transport in upper and lower layers are completely different.

Various interesting processes may occur where river comes to some fresh lake. Water dynamics depends here only on difference in densities between river' and lake' waters; they, however, in their turn, depend on temperature, dissolved and suspended matter, so that finally various different situations may occur. If just fresh and warm rifer comes to colder lake—its waters flow above lake water, forming a thin ($A \sim 10^{-3}$) lens, often mushroom-shaped in horizontal plane and tending to turn to the right (in Northern hemisphere) due to the Earth rotation. If the river is colder (denser) than lake, like rivers of glacial origin inflowing Alpine lakes, its water sinks down along the lake slope—to the very bottom or to the level of corresponding density, separates from this bottom and moves off-shore as isopycnic intrusion. One more interesting example: when river water carries lots of sediments to the lake, the inflowing water mass can be split into many thin layers, with warm lens on the lake surface, isopycnic intrusions into lake body and rushing down-slope turbidity current.

3.3.4. GRAVITY CURRENTS

Generally speaking, the gravity current is any flow of denser water mass along the slope into the lighter water body, which is driven by gravity force. There could be different reasons why the fluid at the surface or at some depth becomes more heavy than surrounding water. Along with mentioned above river inflow, common example is wind-wave re-suspension of sediment near the coast and formation of a *turbidity current*. Another typical examples when inflowing water mass is heavier than that in the receiving basin are: out-flowing lagoon water is saltier than the sea, wastewater plume is heavier than the environment. The speed of propagation of such currents depends mainly on their (relative) density, thickness and the bottom slope.

Even though such currents are typically pulsatory (not permanent), they can be a very effective transport mechanism under certain conditions, especially in stratified basin. Their behavior in stratified ambient is driven by inertia and local density gradients. Initially, gravity current moves along the slope down to the level of its density (and by inertia—just slightly deeper). There, it becomes of neutral buoyancy, no longer experiences the downward force, but has significant speed of propagation. So, by inertia, it continues its motion: separates from the slope, roughly around the level, where its density equals the density of ambient fluid, and propagates further *horizontally* into the main water body. Field experience shows, that such intrusions do not mix much with ambient fluid, and thus can propagate very far from the initial position. As for example, Mediterranean waters flowing out of strait of Gibraltar are heavier (saltier) than water at the surface of the Atlantic ocean; so, they plunge along the continental slope down to the level of their density, continue to move horizontally—and can be identified even hundreds and thousands kilometers apart from Gibraltar. Thus, in stratified basin, this transport mechanism links the coastal zone with the very distant basin interior.

We feel guilty having not discussed complicated phenomenon of *turbulence*. We send the reader here to books on it: Bowden (1962), Hinze (1975), Thorpe (2005), Holloway (1986) and other.

3.3.5. COUPLING PHYSICS AND OTHER PROCESSES

Physical processes or properties of water can often be influenced by chemical or biological conditions in a basin. For example, a root vegetation near coasts significantly decreases currents and water exchange in the area, and makes it different in different seasons. Figure 16 presents another example—an intense

8/8/02 15:29

Figure 16. Summer blooming in Curonian lagoon (the Baltic sea)

summer blooming in Curonian lagoon (the Baltic sea). Obviously, many processes are much different under such "cover"! From physical point of view, at least the "water" viscosity is changed. In this context, one should also mention various films at water surface, which significantly reduce the wind action, and suspended or dissolved materials, which contribute to water density, and jelly-like/oozy bottom sediments, changing bottom friction, and many others.

3.4. Conclusions

3.4.1. IMPORTANT CONCLUSIONS OF "EXTERNAL CONDITIONS"

- Gravity, wind and radiation are the main driving factors for natural water bodies.
- On the rotating Earth, water moves by inertia (without external forces) not straight ahead, but in circles and loops.
- Gravity causes tides and buoyancy forces.
- At a scale of lagoon, lake, sea coastal zone, local winds are important for water dynamics.
- Water density (depending mainly on salinity, temperature, pressure and suspended matter) is the most important parameter for water motion.

3.4.2. IMPORTANT CONCLUSIONS OF "INTERNAL STRUCTURE AND WATER DYNAMICS"

- Natural basins are vertically and horizontally stratified, what influences water motions in them.
- Wind-induced currents and motions due to sea level variations at the entrance are the main driving factors in lagoons.
- River inflows, internal and surface waves, gravity currents, as well as the processes, generated by their interference, are of big importance for water dynamics in (semi-)closed basins.
- Chemical and biological processes are able to influence physical conditions.

3.4.3. CONCLUDING REMARKS

Motion of fluid in natural basins is complicated, influenced by many external conditions like winds, solar radiation, the Earth rotation and gravity, tides and sea level variations, inflowing rivers, etc. In order to

- foresee possible propagation and influence of CBRN on water resources,
- select the most important physical forcing factors for effective numerical modelling,
- apply correctly the results of numerical modelling for decision making,

one needs to use extensively a wide field experience of physical phenomena, observed in different real basins under various field conditions. Thorough analysis of current processes and possible development of the situation in the very beginning not only allows *more exact prediction*: it *saves time*, what is always crucial in the case of serious danger. The wider is our experience on natural processes, the more *effective decision* can be made in order to minimize the CBRN effects on water resources.

References

Batchelor, G.K., 1967. An Introduction to Fluid Dynamics, The University Press, Cambridge.

Bowden, K.F., 1962. Turbulence, in The Sea Volume 1, edited by M.N. Hill, Interscience Publishers/John Wiley and Sons, New York, pp. 802–825.

Cartwright, D.E., 1999. Tides: A Scientific History, The University Press, Cambridge.

Gonenc, I.E., and J. Wolflin (eds), 2004. Costal Lagoons: Ecosystem Processes and Modelling for Sustainable Use and Developments, CRC Press.

Cushman-Roisin, B., 1994. Introduction to Geophysical Fluid Dynamics, Prentice-Hall, Englewood Cliffs.

Davis, R.A., 1987. Oceanography: An Introduction to the Marine Environment, Wm. C. Brown Publishers, Dubuque.

Defant, A., 1961. Physical Oceanography, Macmillan Company, New York.

Dritschel, D.G., M. de la T. Juarez, and M.H.P. Ambaum, 1999. The three-dimensional vortical nature of atmospheric and oceanic turbulent flows, Phys. Fluids, 11 (6), 1512–1520.

Ekman, V.W., 1905. On the influence of the Earth's rotation on ocean currents, Arkiv for Matematik, Astronomi, och Fysik, 2 (11).

Gill, A.E., 1982. Atmosphere-Ocean Dynamics, Academic Press, New York.

Gregg, M.C., 1991. The study of mixing in the ocean: a brief history, Oceanography, 4 (1), 39–45.

Hasselmann, K., 1970. Wind-driven inertial oscillations, Geophys. Fluid Dyn., 1 (1), 463–502.

Hinze, J.O., 1975. Turbulence, 2nd edn, McGraw-Hill, New York.

Holloway, G., 1986. Eddies, waves, circulation, and mixing: statistical geofluid mechanics. Ann. Rev. Fluid Mech., 18, 91–147.

Horikawa, K., 1988. Nearshore Dynamics and Coastal Processes, University of Tokyo Press, Tokyo.

Houghton, J.T., 1977. The Physics of Atmospheres, Cambridge University Press, Cambridge.

JPOTS, 1981. The Practical Salinity Scale 1978 and the International Equation of State of Seawater 1980, UNESCO, Paris, 25.

JPOTS Editorial Panel, 1991. Processing of Oceanographic Station Data, UNESCO, Paris.

Kundu, P.K., 1990. Fluid Mechanics, Academic Press, San Diego.

Langmuir, I., 1938. Surface motion of water induced by wind, Science, 87, 119–123.

Millero, F.J., and A. Poisson, 1981. International one-atmosphere equation of state of seawater, Deep-Sea Res., 28A (6), 625–629.

Millero, F.J., 1996. Chemical Oceanography, 2nd edn, CRC Press, New York.

Munk, W.H., 1950. On the wind-driven ocean circulation, J. Meteorol., 7 (2), 79–93.

Neumann, G., and W.J. Pierson, 1966. Principles of Physical Oceanography, Prentice-Hall, New Jersey.

Open University, 1989a. Ocean Circulation, Pergamon Press, Oxford.

Open University, 1989b. Seawater: Its Composition, Properties and Behaviour, Pergamon Press, Oxford.

Open University, 1989c. Waves, Tides and Shallow Water-Processes, Pergamon Press, Oxford.

Pedlosky, J., 1987. Geophysical Fluid Dynamics, 2nd edn, Springer Verlag, Berlin.

Pickard, G.L., and W.J. Emery, 1990. Descriptive Physical Oceanography: An Introduction, 5th edn (enlarged), Pergamon Press, Oxford.

Pinet, P.R., 2000. Invitation to oceanography, 2nd edn, Jones and Bartlett Publishers, Sudbury, Massachusetts.

Pugh, D.T., 1987. Tides, Surges, and Mean Sea-Level, John Wiley & Sons, Chichester.

Richardson, E.G., 1961. Dynamics of Real Fluids, 2nd edn, Edward Arnolds, London.

Stewart, R.H., 2003. Introduction to Physical Oceanography. Department of Oceanography, A & M University, Texas, 344 pp.

Stommel, H.M., and D.W. Moore, 1989. An Introduction to the Coriolis Force, Cambridge University Press, Cambridge.

Thorpe, S., 2005. The Turbulent Ocean, Cambridge University Press, Cambridge.

Thurman, H.V., 1985. Introductory Oceanography, 4th edn, Charles E. Merrill Publishing Company, Columbus.

Tomczak, M., and J.S. Godfrey, 1994. Regional Oceanography: An Introduction, Pergamon, London.

4. HYDRODYNAMIC EQUATIONS

Georg Umgiesser
ISMAR-CNR, Venezia, Italy

4.1. Introduction

The hydrodynamic equations are a set of equations that describe the movement of a water body through a set of variables, the so-called state variables. The basic equations that govern the development of these variables can be derived in its full form through the application of conservation laws. In sea water 7 variables completely define the state of the fluid. These variables are the density ρ of the water, the three velocity components u, v, w in the direction of x, y, z, the pressure p, the temperature T and the salinity S. If only fresh water systems are concerned, there is no salinity as a variable, reducing the number of state variables to 6.

In meteorology, similar variables can be used also for the description of the atmosphere. In this case humidity (water vapor content) is used instead of salinity as a state variable. The equations governing atmospheric motion differ only slightly from the ones used in the oceans.

It is believed that the hydrodynamic equations are exact, and that they describe all processes going on in the oceans and the atmosphere. Only the fact that no analytic solutions exist for these equations prevent a complete description of the dynamics. Moreover, in order to solve these equation, initial and boundary conditions are needed, and in no case do we have the complete knowledge of these values. The only way out of this problem is the application of numerical methods to solve the equations and the (incomplete) application of boundary conditions to these equations.

In the following the hydrodynamic equations are derived from first principles and their applicability is discussed. We will then introduce some basic simplifications to the equations and also discuss boundary processes.

4.2. The Eulerian and the Lagrangian View

In classical mechanics all equations are normally derived by looking at a particle that is subject to external forces and is dynamically adjusting and moving due to these forces. Therefore, the particle is followed while it moves through the three-dimensional space.

83

I. E. Gonenc et al. (eds.), Assessment of the Fate and Effects of Toxic Agents on Water Resources, 83–107.
© 2007 *Springer.*

In continuum mechanics this viewpoint changes a little. Now it is more convenient to fix a point in space and see what is happening in this one place. This means that now many particles pass by the fixed point. We do not follow anymore just one particle that moves through the fluid but concentrate on one point and try to deduce what is happening to all the particle passing by.

The first point of view that follows the particle during its journey is called the Lagrangian view. In this case the local coordinate system is changing in time and remains attached to the particle and the changes occurring in the fluid is computed for each individual fluid particle. In the second case, the Eulerian view, the coordinate system remains fixed in space and the change of the fluid is computed in every fixed location. Usually the second approach is used in fluid dynamics.

Since it is our aim to describe the change of the fluid while it is dynamically adjusting to the applied forces we must be able to switch between the two coordinate systems. In the Lagrangian view the change of a quantity is simply written as dC/dt where d/dt is denoting the change of C following the fluid particle, here also called the total derivative.

In the Eulerian coordinate system the quantity C is not only depending on time but also on the spatial coordinates, e.g., the place where the particle is located. Therefore we can write

$$\frac{dC(t, x, y, z)}{dt} = \frac{\partial C(t, x, y, z)}{\partial t} + \frac{\partial C(t, x, y, z)}{\partial x}\frac{dx}{dt}$$
$$+ \frac{\partial C(t, x, y, z)}{\partial y}\frac{dy}{dt} + \frac{\partial C(t, x, y, z)}{\partial z}\frac{dz}{dt}$$

since the coordinates of the particle also depend on time. This is the same as

$$\frac{dC(t, x, y, z)}{dt} = \frac{\partial C(t, x, y, z)}{\partial t} + u\frac{\partial C(t, x, y, z)}{\partial x}$$
$$+ v\frac{\partial C(t, x, y, z)}{\partial y} + w\frac{\partial C(t, x, y, z)}{\partial z}$$

where $\partial/\partial t$ is the local or partial derivative where only the explicit dependence on time is taken into account, x, y, z are the spatial coordinates and u, v, w are the velocity components in the three coordinate directions. As can be seen now the change of the quantity C is not only depending on time but depends both on the local change $\partial/\partial t$ and the advective change and therefore reads

$$\frac{dC}{dt} = \frac{\partial C}{\partial t} + u\frac{\partial C}{\partial x} + v\frac{\partial C}{\partial y} + w\frac{\partial C}{\partial z} = \frac{\partial C}{\partial t} + \vec{u} \cdot \nabla C. \qquad (1)$$

The term $\vec{u} \cdot \nabla C$ is called advective contribution and its effect is a change due to a gradient of C and its advection to the point where the total derivative is evaluated.

$$\frac{du}{dt} = \frac{\partial u}{\partial t} + u\frac{\partial u}{\partial x}$$

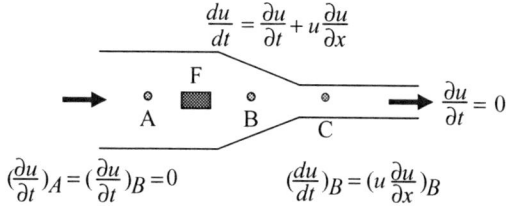

$$\frac{\partial u}{\partial t} = 0$$

$$(\frac{\partial u}{\partial t})_A = (\frac{\partial u}{\partial t})_B = 0 \qquad (\frac{du}{dt})_B = (u\frac{\partial u}{\partial x})_B$$

Figure 1. Total and partial derivative. Fixed and floating observers measure different accelerations

The above equation is true also for C being substituted by one of the velocity components, e.g., u. The equation then reads

$$\frac{du}{dt} = \frac{\partial u}{\partial t} + u\frac{\partial u}{\partial x} + v\frac{\partial u}{\partial y} + w\frac{\partial u}{\partial z}.$$

If this is done for all three components the following vector equation results

$$\frac{d\vec{u}}{dt} = \frac{\partial \vec{u}}{\partial t} + \vec{u} \cdot \nabla\vec{u}.$$

The difference of the temporal and the advective change can be explained better if we consider a practical case. If we look at a channel flow, where only the component of velocity along the channel direction is different from zero the total derivative can be written as

$$\frac{du}{dt} = \frac{\partial u}{\partial t} + u\frac{\partial u}{\partial x}$$

where the x direction has been chosen as the direction along the channel. Here it can be seen that the total acceleration is made out of the local acceleration and the advection of change in velocity in the x direction.

Suppose a channel as in Figure 1 where the flow field is in steady state. The velocity field of this channel flow will increase along the x axis. This is due to the fact that the section of the channel is getting narrower and the same quantity of fluid must go through the different sections. This can be only achieved by increasing the velocity.

The steady state assumption translates mathematically into the assumption that the local time derivative $\partial u/\partial t$ is zero. In this case an observer **B** (as well as **A** or **C**) that is fixed in one place will always see and measure the same velocity value, and he would conclude that there is no local acceleration of the fluid.

However, an observer **F** that is drifting with a boat along the channel would come to a different conclusion. Since the velocity is increasing along the x axis this observer would experience a steady acceleration of her boat while the boat is floating along the channel. Therefore this (total) acceleration is entirely due to the advective acceleration that is caused by the velocity gradient of the flow field in the x direction: the equation can be written as

$$\frac{du}{dt} = u\frac{\partial u}{\partial x}.$$

Another situation arises if we have a situation of a uniform channel with constant section, but at one end a sluice gate is opened slowly. So the flow is accelerating steadily and the acceleration should be homogeneous all over the length of the channel. In this case a fixed observer **A** will observe a certain acceleration $\partial u/\partial t$. On the other hand a observer on a boat will notice the acceleration du/dt, but since the flow in the channel is homogeneous, the advective acceleration du/dx is zero and therefore

$$\frac{du}{dt} = \frac{\partial u}{\partial t}.$$

This means that in this case both observers will see the same acceleration.

4.3. Conservation Laws

The basic hydrodynamic equations can be relatively easily deduced from conservation equations of the single state variables. These conservation laws can be deduced from the conservation of

- *mass:* In this case the equation is called continuity equation.
- *momentum:* these are the Euler or Navier–Stokes equations.
- *energy:* temperature equation.
- *salt:* salinity equation.

The same set of equations is valid also for meteorological purposes. In this case the equation of salinity has to be substituted with the conservation of water vapor. All other equations retain their validity.

Only one equation remains, that cannot be reduced to a statement of conservation. This is the equation of state that is a thermodynamic property of the fluid or gas under consideration.

We will not always show a rigorous deduction of these equations but will present the equations in the original form and then comment on their meanings and implications.

4.3.1. CONSERVATION OF MASS

The first equation can be deduced through the application of the well known law of mass conservation. Consider a infinitesimal control volume (a small

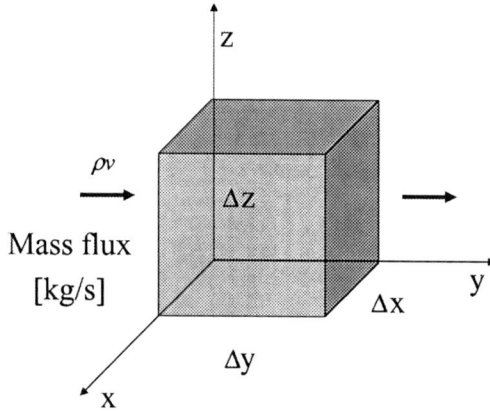

Figure 2. Derivation of the continuity equation

cube with $dV = dx\,dy\,dz$) moving inside the fluid (Figure 2). The mass flux per unit volume in the center of the cube in the x-direction can be expressed as the product of the velocity in x times the density times the section passing through $\rho u\,dy\,dz$.

The mass flux on the left side of the cube with the help of a Taylor series development can be written as

$$(\rho u)\left(x - \frac{dx}{2}\right) dy\,dz = (\rho u)(x) - \frac{\partial \rho u}{\partial x}\frac{dx}{2}dy\,dz.$$

Similarly the mass flux on the right side of the cube is

$$(\rho u)\left(x + \frac{dx}{2}\right) dy\,dz = (\rho u)(x) + \frac{\partial \rho u}{\partial x}\frac{dx}{2}dy\,dz$$

and the total mass flux into the volume can therefore be written as the difference of both fluxes as

$$-\frac{\partial \rho u}{\partial x}\frac{dx}{2}dy\,dz - \frac{\partial \rho u}{\partial x}\frac{dx}{2}dy\,dz = -\frac{\partial \rho u}{\partial x}dx\,dy\,dz.$$

The other sides of the cube give similar contributions so that we can finally write for the mass flow per unit volume into the infinitesimal cube

$$-\frac{\partial \rho u}{\partial x} - \frac{\partial \rho v}{\partial y} - \frac{\partial \rho w}{\partial z} = -\nabla(\rho \vec{u})$$

The mass flux into this volume is therefore given by the negative divergence of the mass flow per unit volume $-\nabla(\rho \vec{u})$. The conservation of mass tells us that the mass flux into the volume must equal the accumulation of mass inside the test volume. This rate of change of mass per unit volume can be expressed by $\partial\rho/\partial t$. Therefore, the conservation equation for mass reads

$$\frac{\partial\rho}{\partial t} + \nabla(\rho\vec{u}) = 0. \tag{2}$$

The conservation equation of mass is also called *continuity equation*.

This equation can be transformed with the help of (1) easily. In our case (1) reads

$$\frac{d\rho}{dt} = \frac{\partial\rho}{\partial t} + \vec{u}\cdot\nabla\rho$$

and inserting this into (2) gives

$$\frac{1}{\rho}\frac{d\rho}{dt} + \nabla\vec{u} = 0. \tag{3}$$

4.3.2. CONSERVATION OF MOMENTUM

Conservation of momentum is in its simplest form described by the law of Newton:

$$\vec{F} = \vec{a}m$$

where $\vec{F}$ is the force acting on a fluid volume, $\vec{a}$ the acceleration and m the mass of the fluid particle.

Using the density as usual in fluid dynamics instead of the mass of a particle we can write

$$\rho\frac{d\vec{u}}{dt} = \vec{f} \tag{4}$$

where $\vec{f}$ is the force per unit volume acting on the fluid volume.

4.3.2.1. *The Euler Equations*

The forces acting on a fluid body may be divided conveniently into two classes: the first are the volume forces, and the others are the interface forces. An example of the first one is the gravitational force, whereas the second one is the pressure gradient force or the wind stress.

The pressure gradient forces can be derived like this (see Figure 3 for reference): given an infinitesimal volume $dV = dx\,dy\,dz$, the force exerted by the pressure on the left side of the cube in x direction is given by $p(x)\,dy\,dz$ (pressure is force per area). Similar the force exerted on the right side of the cube is

$$-p(x + dx)\,dy\,dz = \left(-p(x) - \frac{\partial p}{\partial x}\,dx\right)\,dy\,dz.$$

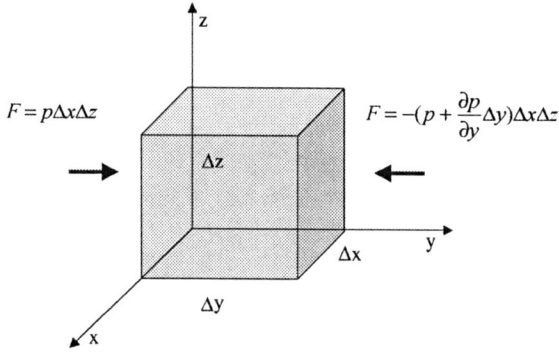

Figure 3. Derivation of the pressure gradient term

Here a Taylor series expansion has been applied to $p(x + dx)$. So the net force on the volume in x direction, after adding the two contributions, is $-(\partial p/\partial x)dV$ and the force per unit volume is just $-(\partial p/\partial x)$. The same analysis can be carried out for the other two directions, giving in the end the total pressure gradient force per unit volume of $-\nabla p$.

This result can be substituted into (4). Including also the gravitational force mg in their form per unit volume ρg we finally have, after dividing by the density

$$\frac{d\vec{u}}{dt} = -\frac{1}{\rho}\nabla p + \vec{g}.$$

In this form the equations are called the *Euler equations* or more precisely the Euler equations in a non-rotating frame of reference. Note that the gravitational acceleration is a vector with only the vertical component different from zero

$$\vec{g} = (0, 0, -g)$$

and g is 9.81 m s^{-2}. The gravitational force is therefore pointing downwards.

4.3.2.2. *The Euler Equations in a Rotating Frame of Reference*
The above derived equations are not suitable for their application to mesoscale or basin wide scale. This is because the earth is not an inertial frame of reference, but is rotating around itself. Whereas this has no impact for water bodies that are small in size, for larger applications it is very important to take account of this effect.

The influence of the earth's rotation can be described through the introduction of a new apparent volume force, called the Coriolis force. If this is

$$\left(\frac{d\vec{u}}{dt}\right)_{abs} = \left(\frac{d\vec{u}}{dt}\right)_{rel} + \underbrace{2\vec{\Omega}\times\vec{u}}_{\dot{C}_o} + \underbrace{\vec{\Omega}\times\vec{\Omega}\times\vec{r}}_{\dot{C}_e}$$

Coriolis	Centrifugal
acceleration	acceleration

$$\vec{u}=\begin{pmatrix}u\\v\\w\end{pmatrix}\quad \vec{\Omega}=\begin{pmatrix}0\\ \dot{U}\cos\varphi\\ \dot{U}\sin\varphi\end{pmatrix}\quad 2\vec{\Omega}\times\vec{u}=2\dot{U}\begin{vmatrix}0 & u & \vec{i}\\ \cos\varphi & v & \vec{j}\\ \sin\varphi & w & \vec{k}\end{vmatrix}$$

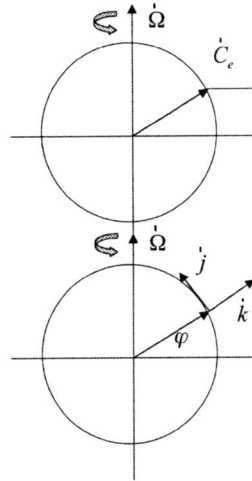

Figure 4. The Coriolis force

done the new equations (not deduced) can be written as

$$\frac{d\vec{u}}{dt} + 2\Omega\vec{k}\times\vec{u} = -\frac{1}{\rho}\nabla p + \vec{g} \tag{5}$$

where $\Omega = 2\pi/24h$ is the angular frequency of rotation of the earth, $\vec{k}$ the vertical unity vector and $\times$ denotes the vector product. In this form the equations are called the *Euler equations in a rotating frame of reference*. Please see Figure 4 for reference.

Besides the gravitational force the Coriolis force is the only other important volume force that is acting in a fluid body. Because of the vector product the Coriolis force is always acting perpendicular to the current velocity, in the northern hemisphere to the right and in the southern hemisphere to the left of the fluid flow. It is the Coriolis force that is responsible for all the meso-scale structure we can see on the weather charts that contain cyclones and anti-cyclones. However, the Coriolis force is important only for large-scale circulations.

4.3.2.3. *The Navier–Stokes Equations*

The above-derived Euler equations describe the flow of a fluid without friction. This may be sometimes a good approximation, especially if no material boundaries (lateral and vertical) are close to the area of investigation. Internal friction in a fluid is normally very small and the Euler equations provide a satisfactory simplification.

However, once the fluid is close to a boundary, friction becomes more important and another area force has to be introduced: the stress tensor. A moving fluid layer exerts a force on the neighbouring fluid layers. The strength

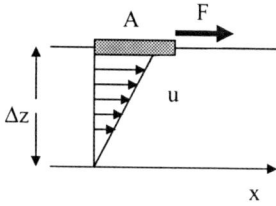

$$F \approx \frac{Au}{\Delta z} \rightarrow F = \mu \frac{Au}{\Delta z}$$

μ is dynamic viscosity

$$\tau = \frac{F}{A} = \mu \frac{u}{\Delta z}$$

$$\tau_{xx} = \mu \frac{\partial u}{\partial x} \quad \tau_{xy} = \mu \frac{\partial u}{\partial y} \quad \tau_{xz} = \mu \frac{\partial u}{\partial z}$$

ν is kinematic viscosity

$$\frac{1}{\rho}\left(\frac{\partial \tau_{xx}}{\partial x} + \frac{\partial \tau_{xy}}{\partial y} + \frac{\partial \tau_{xz}}{\partial z}\right) = \frac{\mu}{\rho}\left(\frac{\partial^2 u}{\partial x^2} + \frac{\partial^2 u}{\partial y^2} + \frac{\partial^2 u}{\partial z^2}\right) = \nu\left(\frac{\partial^2 u}{\partial x^2} + \frac{\partial^2 u}{\partial y^2} + \frac{\partial^2 u}{\partial z^2}\right)$$

Figure 5. Introduction of frictional forces

of this force is directly proportional to the area of the fluid layer and the velocity difference between these layers and inversely proportional to the distance of the layers (Figure 5):

$$F = \mu A \, du/dz$$

where A is the area, du the velocity difference from one layer to the other and dz the distance of the fluid layers. The parameter μ is the constant of proportionality and is called *dynamic viscosity coefficient*. It is a parameter that only depends on the type of fluid and its temperature and salinity contents. However, it does not depend on the fluid dynamics.

In the limit the stress (force per area) can be written as

$$\tau_{zx} = \mu \frac{\partial u}{\partial z}$$

Here the subscript zx means the force exerted by the moving fluid layer in direction x due to the velocity gradient in direction z. Therefore the stress tensor has 9 components, three terms for every spatial direction.

The effect of this stress tensor on the equation of motion can be easily derived. If we look at a fluid element with volume $dV = dx \, dy \, dz$, the stress that is exerted in the x direction by the fluid layer above can be expressed approximately as

$$\tau_{zx}\left(z + \frac{dz}{2}\right) = \tau_{zx}(z) + \frac{\partial \tau_{zx}}{\partial z}\frac{dz}{2}$$

where $\tau_{zx}(z)$ is the value of the stress tensor at the center of the fluid element. The force due to the bottom of the fluid element can be written

$$\tau_{zx}\left(z - \frac{dz}{2}\right) = \tau_{zx}(z) - \frac{\partial \tau_{zx}}{\partial z}\frac{dz}{2}.$$

The net viscous force on the fluid element in x direction due to shear in z direction can then be written as

$$\left(\tau_{zx}(z) + \frac{\partial \tau_{zx}}{\partial z} \frac{dz}{2} \right) dx\, dy - \left(\tau_{zx}(z) - \frac{\partial \tau_{zx}}{\partial z} \frac{dz}{2} \right) dx\, dy.$$

After dividing by the mass $\rho\, dx\, dy\, dz$ of the volume element we end up with the viscous force F_{zx} per unit mass

$$F_{zx} = \frac{1}{\rho} \frac{\partial \tau_{zx}}{\partial z} = \frac{1}{\rho} \frac{\partial}{\partial z} \left(\mu \frac{\partial u}{\partial z} \right) = \nu \frac{\partial^2 u}{\partial z^2}$$

where we have used the *kinematic viscosity coefficient* $\nu = \dfrac{\mu}{\rho}$ to write the equation more compactly.

In the same way the viscous forces in x direction due to shear in the other directions can be derived giving

$$F_x = \nu \left(\frac{\partial^2 u}{\partial x^2} + \frac{\partial^2 u}{\partial y^2} + \frac{\partial^2 u}{\partial z^2} \right)$$

and the viscous forces in the other directions read

$$F_y = \nu \left(\frac{\partial^2 v}{\partial x^2} + \frac{\partial^2 v}{\partial y^2} + \frac{\partial^2 v}{\partial z^2} \right)$$

and

$$F_z = \nu \left(\frac{\partial^2 w}{\partial x^2} + \frac{\partial^2 w}{\partial y^2} + \frac{\partial^2 w}{\partial z^2} \right).$$

Including the friction term into the Euler equations we end up with the so-called *Navier–Stokes equations*:

$$\frac{d\vec{u}}{dt} + 2\Omega \vec{k} \times \vec{u} = -\frac{1}{\rho}\nabla p + \vec{g} + \nu \nabla^2 \vec{u} \qquad (6)$$

where the Laplace operator $\Delta = \nabla^2$ is defined as

$$\frac{\partial^2}{\partial x^2} + \frac{\partial^2}{\partial y^2} + \frac{\partial^2}{\partial z^2}.$$

It may be interesting to note that the physical process responsible for molecular friction is the molecular diffusion of the fluid particles. If a faster fluid layer moves above a slower fluid layer some of the particles from the slower fluid layer will diffuse into the faster layer slowing the upper layer down. On the other side, faster fluid particles diffusing into the fluid layer below will accelerate the slower fluid layer. From both sides of the layer this results in an effective friction, either slowing down or accelerating the other layer. Therefore, the operator ∇^2 is also called the diffusion operator.

4.3.3. CONSERVATION OF ENERGY

Conservation of energy can be formulated as a conservation equation for temperature. The total change of temperature of a fluid parcel is given by two processes: the rate of heating of the volume and diffusion. The change of temperature with the heating rate can be expressed as

$$\frac{1}{\rho c_p} \frac{\partial Q_s}{\partial z}$$

where Q_s is the solar radiation (W/m^2), ρ the density of water (kg/m^3) and c_p is the specific heat of water (J kg^{-1} K^{-1}).

The other process that is changing the heat content of a fluid volume is molecular diffusion. In this case fluid particles that are warmer and that diffuse into colder fluid because of molecular motion will contribute to the warming of the colder fluid. As explained before the diffusion can be described through the Laplace operator ∇^2, and the whole energy conservation of temperature can be written as

$$\frac{dT}{dt} = \frac{\partial T}{\partial t} + \vec{u} \cdot \nabla T = v_T \nabla^2 T + \frac{1}{\rho c_p} \frac{\partial Q_s}{\partial z} \tag{7}$$

where T is the temperature of the fluid and v_T is the molecular diffusivity for temperature, a parameter that is similar to the viscosity in the Navier–Stokes equations and that depends only on the properties of the fluid. The total derivative has been expanded into the local and advective contributions.

4.3.4. CONSERVATION OF SALT

In the oceans and coastal lagoons the water has a certain salt content that varies from nearly zero high up in the estuaries and rivers to values of around 10 psu in brackish water and to over 30 psu for ocean waters. In lagoons where evaporation is important these values can be even higher. In all cases the constituents of salt are nearly constant in all places of the world. It is therefore possible to express the salt content through a concentration value called salinity. The salinity is measured in grams per kilogram, parts per thousand or psu (practical salinity unit).

It is therefore important to consider the salt content and its variation through a conservation equation for salt. The salt balance can be described by a similar equation as in the case of temperature. The salinity can be changed only through advective and diffusive processes and mass fluxes like evaporation and river input. The equation for salinity is therefore

$$\frac{dS}{dt} = \frac{\partial S}{\partial t} + \vec{u} \cdot \nabla S = v_S \nabla^2 S + (E - P) \tag{8}$$

where S is the salinity of the fluid and v_S is the molecular diffusivity for salt, a parameter that again depends only on the properties of the fluid. The term $E - P$ indicates the source and sinks for the equation given by evaporation and precipitation.

4.3.5. EQUATION OF STATE

The equation of state is a thermodynamic equation that relates different quantities in a fluid in equilibrium to each other. An example is the well-known equation of state for an ideal gas

$$p = \rho RT$$

that relates the pressure p, the volume V (through the density ρ) and the temperature T to each other. Here R is the gas constant per unit mass that still depends on the specific property of the gas. Its value for dry air is 287 J kg^{-1} K^{-1}.

In the case of a fluid the situation is complicated by the fact that salinity is also influencing the equation of state. In this case the equation of state can be formally written as

$$\rho = f(T, S, p)$$

so that the density can be computed given a set of values of temperature, salinity and pressure. The equation itself is a complicated polynomial function of these quantities that has been empirically deduced by different organizations (e.g., UNESCO). The formula is not given here in its general formulation. However, for some applications the equation of state can be simplified to

$$\rho = \rho_0 - \alpha_T (T - T_0) + \alpha_S (S - S_0)$$

where the dependency on pressure has been neglected and a simple linear dependence on T and S has been used. Here ρ_0 is the reference density at the reference temperature T_0 and the reference salinity S_0. The parameter α_T is called the thermal expansion coefficient and α_S the expansion coefficient due to salinity changes. Both coefficients are positive. As can be seen denser (heavier) water may result either by lowering the temperature or by rising the salinity.

4.4. Simplifications and Scale Analysis

The equations given above, especially the equation for momentum equations, are still too complicated to be used for modeling. This is mainly true because of the variability of spatial and temporal scales that these equations are applicable to. Since we did not make yet any assumptions these equations describe the flow for the global ocean circulation as they describe the flow down to scales where molecular effects become important. Therefore, these equations must describe scales from 10^6 meters on the ocean scale down to about 10^{-6} meters (micrometers) where dissipation becomes important.

The same is true if we talk about the time scales to which the equations apply. These time scales go from years for the general circulation down to microseconds if we deal with dissipation effects due to friction. Since we are interested to describe motions that take place in coastal lagoons or estuaries, not all processes included in the equations above are equally important and simplifications have to be made.

4.4.1. INCOMPRESSIBILITY

Water, compared to air, turns out to be a relatively incompressible media. That there is a certain compressibility of the water is clearly seen by the fact that acoustic waves can travel in water. These waves depend completely on the compressibility of the media. But it can be shown that, excluding acoustic waves, the effect of compressibility is negligible on the dynamics of the oceans.

If we therefore assume water as an incompressible media, this can be written mathematically as $d\rho/dt = 0$. If we substitute this into the continuity equation (3) we have the simplified continuity equation

$$\nabla \vec{u} = 0.$$

In this version the continuity equation states that the divergence of the mass flow is zero. This is the form of the mass conservation normally used in oceanography. The main effect in neglecting the compressibility effect is that acoustic waves cannot be described in the water. Since acoustic waves are of minor importance in oceanographic applications, neglecting these terms is justified.

4.4.2. THE HYDROSTATIC APPROXIMATION

If we consider a basin of water in rest ($\vec{u} = 0$), the Navier–Stokes equations reduce to

$$0 = -\frac{1}{\rho}\nabla p + \vec{g}.$$

The z component of this equation may me written as

$$\frac{\partial p}{\partial z} = -\rho g.$$

This equation is called the *hydrostatic equation*, because it is exactly valid in a static fluid with no motion. If we integrate this equation from a depth $-h$ up to the surface that is supposed to be at $z = 0$, we have

$$\int_{-h}^{0} \frac{\partial p}{\partial z} \, dz = p_0 - p(-h) = -g \int_{-h}^{0} \rho \, dz$$

or

$$p(-h) = p_0 + g \int_{-h}^{0} \rho \, dz$$

where p_0 is the surface or atmospheric pressure. In this form the hydrostatic equation states that the pressure at depth $-h$ is just due to the weight of the water column above it.

It turns out that this equation is also a very good approximation of the vertical component of the momentum equation in the case of a situation where the velocity vector $\vec{u}$ is not zero. Only in regions of very strong vertical convection due to cooling of surface water the vertical acceleration may have appreciable effects on the total momentum equation.

4.4.3. THE CORIOLIS FORCE

As specified above the Coriolis force is an apparent force that is due to the rotation of the earth. The complete expression $2\vec{\Omega} \vec{k} \times \vec{u}$ can be further simplified in remembering that only the horizontal part of the Coriolis force is important. In fact, the vertical momentum equation has been substituted by the hydrostatic assumption.

If the horizontal components are evaluated there is one term that is multiplied with the vertical velocity w. This term is nearly always much smaller than the other terms due to the smallness of the vertical velocity. If this term is neglected the remaining Coriolis vector can be written as

$$\vec{F}^{C} = \left(F_x^C, F_y^C, F_z^C \right) = (-fv, +fu, 0)$$

where f is called the Coriolis parameter that can be written as

$$f = 2\Omega \sin(\varphi)$$

with φ the latitude of the point where the equation is evaluated. For applications in coastal lagoons the value of f does not vary very much and therefore is very often kept constant. In this case φ is the average latitude of the basin to be investigated.

As can be seen from the equation the Coriolis force is always perpendicular to the flow, e.g., for a flow that has only a component in the x direction ($v = 0$), the Coriolis force is only acting in the y direction. For the northern hemisphere f is positive and therefore the force is always to the right of the direction of flow. For this reason the Coriolis force does not contribute to the energy budget and does not do work on the fluid.

Please note also that close to the equator the Coriolis force is smaller and close to the poles it is maximum. Therefore, equatorial lagoons are hardly influenced by the Coriolis force.

4.4.4. THE REYNOLDS EQUATIONS

As mentioned before the Navier–Stokes equations describe all possible water motions from scales of the order of the oceans down to scales where molecular friction is active and dissipation is important. However, for the modeling purpose this is not acceptable. For example, the fact that the very small scales are not modeled directly (because the computational grid is just too coarse) means that molecular friction will be never important in the Navier–Stokes equations.

However, the fact that molecular friction is not important on the scales we are considering is somehow misleading. Since there will be an input of energy in our basin through solar heating, tides and wind, there must be also some way to convert this energy and eventually dissipate it, otherwise the total energy will continue to increase. But the only way energy can be dissipated is through the molecular friction term. All other terms only take part in re-distributing the energy inside the basin. Therefore, there must be some place in the basin where friction becomes important and the molecular forces may do their work.

This obvious paradox has been resolved by the British physicist O. Reynolds that has applied an averaging technique to the full Navier–Stokes equations. He described the flow as one that can be divided into a slowly varying part and a fluctuation around this. So, e.g., the velocity u is represented as $u = \bar{u} + u'$ where $\bar{u}$ is the slowly varying part and u' the fluctuations (Figure 6). If all variables are represented like this and are introduced into the

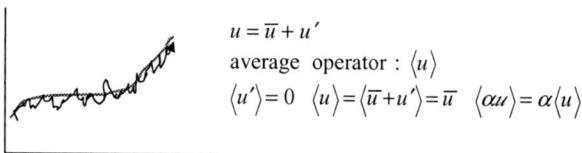

$u = \bar{u} + u'$

average operator : $\langle u \rangle$

$\langle u' \rangle = 0 \quad \langle u \rangle = \langle \bar{u} + u' \rangle = \bar{u} \quad \langle \alpha u \rangle = \alpha \langle u \rangle$

Figure 6. Turbulence: average values and fluctuations

hydrodynamic equations and these equations are averaged over a suitable time
interval then new equations result for the slowly varying part of all variables.

The structure of these averaged equations is very similar to the original
equations. However, due to the non-linear nature of the advective terms that are
contained implicitly in the total derivative, there appear some new terms in the
conservation equation for momentum, temperature and salt. These new terms
are called Reynolds fluxes and they are averages of products of fluctuating
variables. Without going into too much detail, these terms can be written in
the same way as the friction terms (diffusion terms) in the Navier–Stokes
equations. For example, the new terms now read

$$\vec{F}^t = v_t^H \left(\frac{\partial^2 \vec{u}}{\partial x^2} + \frac{\partial^2 \vec{u}}{\partial y^2} \right) + v_t^V \frac{\partial^2 \vec{u}}{\partial z^2} \tag{9}$$

where instead of the molecular viscosity v two new parameters v_t^H, v_t^V have
been substituted.

It is important to note the different physical processes that lead to these
equations. In the case of the Navier–Stokes equations the friction term was
due to the molecular (Brownian) motion of the fluid particles. This motion
was of statistical nature and the diffusivity was only a function of the fluids
static properties.

In the case of the Reynolds equations, the additional term is now due
to the fluctuating part of the variable that is under consideration (here the
velocity). This fluctuating part is also called the turbulent part. It actually
is not statistical in nature since the Navier–Stokes equations could at least
in principle be used to compute all the small-scale motions. However, as in
thermodynamics it would in principle be possible to describe the motion of
10^{23} molecules that are in one mole of air, practically it is not possible and
only a statistical representation of the motion is given.

Therefore the diffusion of the fluid particles is now due to the turbulence
that is acting on scales smaller than the ones that have been retained after the
averaging procedure. This turbulent motion is similar to the molecular motion
of the fluid particles. However, there are two main differences. Since the turbulent motions (fluctuation) depend on the averaging scale, also the turbulent
viscosities v_t^H, v_t^V will depend on the scale of averaging. Even worse, since
the turbulence depends on the dynamic state of the fluid, also the turbulent
viscosities depend on the dynamics of the fluid. Various factors such as the
local buoyancy or the velocity shear will influence these parameters.

It is clear now that there is a fundamental difference between the molecular
and the turbulent diffusion term. The first one depends only on the static
properties of the fluid (type of fluid, temperature, salinity), whereas the second
one depends also on the dynamic flow field itself. The molecular viscosity

can be given a very accurate value that can be measured or computed in statistical mechanics. However, the turbulent parameter varies over several orders of magnitude depending crucially on the flow field itself. Because of the dynamic nature of the turbulent motion it is expected that the values for the viscosities will be different in the horizontal and the vertical direction. This fact has been accounted for letting the viscosity take different values v_t^H, v_t^V for the horizontal and vertical dimension.

Comparing the range of values for the two types of parameters it can be seen that the turbulent diffusion term is some orders of magnitude bigger than its molecular counterpart. Therefore the molecular friction is almost always neglected and only the turbulent terms are retained. The value for the turbulent viscosity parameter is often set to a constant average, one that best represents the physical processes to be described. In the more general case a turbulence closure model must be used that will actually compute the parameters v_t^H, v_t^V in every point of the water body. However, the description of these turbulence closure models (e.g. $k - \varepsilon$, Mellor-Yamada) is beyond the scope of this book.

4.4.5. THE PRIMITIVE EQUATIONS

In this section the simplified three-dimensional equations are given one more time as reference for the next part. These equations are also called primitive equations, not because they are primitively easy, but because only basic simplifications have been applied to them and they have been left in their primitive structure.

The conservation of mass leads to the continuity equation

$$\frac{\partial u}{\partial x} + \frac{\partial v}{\partial y} + \frac{\partial w}{\partial z} = 0.$$

The two horizontal components of the Reynolds equations read

$$\frac{\partial u}{\partial t} + u\frac{\partial u}{\partial x} + v\frac{\partial u}{\partial y} + w\frac{\partial u}{\partial z} - fv = -\frac{1}{\rho}\frac{\partial p}{\partial x} + v_t^H\left(\frac{\partial^2 u}{\partial x^2} + \frac{\partial^2 u}{\partial y^2}\right) + v_t^V\frac{\partial^2 u}{\partial z^2}$$

and

$$\frac{\partial v}{\partial t} + u\frac{\partial v}{\partial x} + v\frac{\partial v}{\partial y} + w\frac{\partial v}{\partial z} + fu = -\frac{1}{\rho}\frac{\partial p}{\partial y} + v_t^H\left(\frac{\partial^2 v}{\partial x^2} + \frac{\partial^2 v}{\partial y^2}\right) + v_t^V\frac{\partial^2 v}{\partial z^2}$$

where the total time derivative has been split into the inertial term that describes the local acceleration and a second term describing the advective acceleration.

For the vertical component the hydrostatic approximation is used

$$\frac{\partial p}{\partial z} = -\rho g.$$

The conservation equations for heat and salt read

$$\frac{\partial T}{\partial t} + u\frac{\partial T}{\partial x} + v\frac{\partial T}{\partial y} + w\frac{\partial T}{\partial z} = v_T^H \left(\frac{\partial^2 T}{\partial x^2} + \frac{\partial^2 T}{\partial y^2} \right) + v_T^V \frac{\partial^2 T}{\partial z^2} + \frac{1}{\rho c_p}\frac{\partial Q_s}{\partial z}$$

and

$$\frac{\partial S}{\partial t} + u\frac{\partial S}{\partial x} + v\frac{\partial S}{\partial y} + w\frac{\partial S}{\partial z} = v_S^H \left(\frac{\partial^2 S}{\partial x^2} + \frac{\partial^2 S}{\partial y^2} \right) + v_S^V \frac{\partial^2 S}{\partial z^2} + (E - P)$$

with v_T^H, v_T^V and v_S^H, v_S^V the turbulent diffusivities (horizontal and vertical) for temperature and salinity. Finally the equation of state reads

$$\rho = f(T, S, p).$$

This completes the treatment of the hydrodynamic equations.

4.5. The Shallow Water Equations

The equations that are derived above can explain the dynamics of an arbitrary water body such as lagoons or coastal seas. However, sometimes one can simplify the equations furthermore in order to take account of the typical features found in shallow water bodies.

If the water body is very shallow, or if the water column can be considered well mixed along the water column, then all state variables may be taken constant over the depth of the water column. In shallow waters this is normally the case if there are no phenomena of stratification.

It is then easy to integrate the equations over the water column and derive the two-dimensional barotropic (vertically integrated) equations, also called *shallow water equations*. Please note that these equations may as well apply also to deeper basins, if the conditions of being well mixed and not stratified are fulfilled.

To derive the two-dimensional form of the continuity equation it is enough to integrate the three-dimensional continuity equation from the bottom to the surface and apply the kinematic boundary conditions at the surface

$$\frac{dw}{dt} = 0 \text{ at } z = \eta.$$

It is however more instructive to derive the continuity equation directly from first principles. Taking a control volume over the whole water column with rectangular area $dx\,dy$ the total inflow in x direction is $Hu(x, y)\,dy - Hu(x + dx, y)\,dy$ and in y direction is $Hv(x, y)\,dx - Hv(x, y + dy)\,dx$. Here the total water depth is called $H = h + \eta$ with h the undisturbed water depth. The total inflow must result in more water in the control volume and because of incompressibility the only way for water to accumulate is through the rise of the water level η. The rate of increase of the volume can therefore be written as $(\partial\eta/\partial t)\,dx\,dy$. If we divide this conservation equation by the area of the control volume $dx\,dy$ we have

$$\frac{\partial\eta}{\partial t} + \frac{Hu\,(x+dx,y) - Hu\,(x,y)}{dx} + \frac{Hv\,(x,y+dy) - Hv\,(x,y)}{dy} = 0$$

and in the limit

$$\frac{\partial\eta}{\partial t} + \frac{\partial Hu}{\partial x} + \frac{\partial Hv}{\partial y} = 0.$$

The horizontal pressure gradients may also be expressed in a simpler way. Since all variables are considered constant along the water column the density ρ that is mainly changing in the vertical direction and much less in the horizontal one may be considered constant all over and is denominated ρ_0. With this simplification the hydrostatic equation can be integrated from a depth $-h$ to the water surface η

$$p_0 - p\,(-h) = \int_{-h}^{\eta} \frac{\partial p}{\partial z} dz = -\int_{-h}^{\eta} g\rho_0\,dz = -g\rho_0\,(\eta+h).$$

Assuming the atmospheric pressure p_0 to be constant we can write the horizontal pressure gradient as

$$\frac{1}{\rho_0}\frac{\partial p}{\partial x} = g\frac{\partial\eta}{\partial x} \qquad \frac{1}{\rho_0}\frac{\partial p}{\partial y} = g\frac{\partial\eta}{\partial y}.$$

In this way the pressure can be completely eliminated from the equations and is substituted by the water level η.

The last term to be dealt with is the turbulent friction term. If the term $\nu_t\dfrac{\partial^2 u}{\partial z^2}$ is integrated over the whole water column we obtain

$$\nu_t \int_{b}^{s} \frac{\partial^2 u}{\partial z^2} dz = \nu_t \left(\frac{\partial u}{\partial z}\right)^s - \nu_t \left(\frac{\partial u}{\partial z}\right)^b$$

where the indices s and b stand for surface and bottom. These two terms are exactly the normalized stress (stress per density) in x direction that is exerted over the two interfaces of the surface and the bottom (see (9)):

$$v_t \int_b^s \frac{\partial^2 u}{\partial z^2} dz = \frac{1}{\rho_0} \left(\tau_s^x - \tau_b^x \right).$$

These values are boundary values that have to be imposed on the moving fluid (surface) or can be computed from the actual flow field (bottom friction). Their exact formulation will be given later.

If the vertical integrated equations are divided by the total depth H the 2D shallow water equations result

$$\frac{\partial u}{\partial t} + u \frac{\partial u}{\partial x} + v \frac{\partial u}{\partial y} - fv = -g \frac{\partial \eta}{\partial x} + \frac{1}{\rho_0 H} \left(\tau_x^s - \tau_x^b \right) + v_t^H \left(\frac{\partial^2 u}{\partial x^2} + \frac{\partial^2 u}{\partial y^2} \right)$$

$$\frac{\partial v}{\partial t} + u \frac{\partial v}{\partial x} + v \frac{\partial v}{\partial y} + fu = -g \frac{\partial \eta}{\partial y} + \frac{1}{\rho_0 H} \left(\tau_y^s - \tau_y^b \right) + v_t^H \left(\frac{\partial^2 v}{\partial x^2} + \frac{\partial^2 v}{\partial y^2} \right).$$

Together with the continuity equation

$$\frac{\partial \eta}{\partial t} + \frac{\partial Hu}{\partial x} + \frac{\partial Hv}{\partial y} = 0$$

they form a closed set of equations for the variables η, u, v provided that the boundary conditions for the stress and the values of v_t^H are specified. The values for the velocities u, v represent average values over the whole water column. The other state variables (p, w, T, S) have been eliminated from the equations.

Even if the shallow water equations are not depending anymore on temperature and salinity, there might still be a need to compute their evolution. Now, however, these equations are decoupled from the hydrodynamics and their solution may be obtained independently. Conservation equations for temperature and salinity may be also derived as above. After vertical integration and division through H of the conservation equations we have

$$\frac{\partial T}{\partial t} + u \frac{\partial T}{\partial x} + v \frac{\partial T}{\partial y} = v_T^H \left(\frac{\partial^2 T}{\partial x^2} + \frac{\partial^2 T}{\partial y^2} \right) + \frac{1}{\rho_0 c_p} \frac{Q_s}{H}$$

and

$$\frac{\partial S}{\partial t} + u \frac{\partial S}{\partial x} + v \frac{\partial S}{\partial y} = v_S^H \left(\frac{\partial^2 S}{\partial x^2} + \frac{\partial^2 S}{\partial y^2} \right) + (E - P)$$

where the fluxes at the surface and bottom have been set to zero. If these fluxes are important they can be easily included in the equations. The values T and S represent average values of the temperature and salinity over the water column.

It should be noted that these equations can be used to compute the transport and diffusion of temperature or salinity or any other conservative dissolved substance C by the velocity field $\vec{u}$. As explained above, they are not necessary for the solution of the hydrodynamic equations. This is due to the fact that the full hydrodynamic equations depend on T and S through the density ρ. In the case of the shallow water equations the density is constant ($\rho \equiv \rho_0$) and does not depend any more on T and S.

4.6. Boundary Processes

The equations that have been derived so far describe only the inner dynamics of a water body such as lagoons, lakes or the sea. They guarantee inner consistency of the variables and enforce the conservation of the physical properties, but they do not explain why the water moves. This movement of the water may be explained through the application of external forces, so-called forcings, and they fall in the field of boundary processes.

If we take for an example a lake where no external forcings act anymore, such as wind, heating, rain etc. In a first moment the lake will continue to move by inertia. But then frictional forces will slow down the water movement until no movement at all will be visible. The water masses have come to a standstill, there will be no gradients in the water level, T or S. The density and the pressure will be distributed according to the hydrostatic equation and all velocities will be zero. The final distribution will be a static one.

By the way, the same happens also when you climb out of the bath tube, creating whirls and currents. When you come back after half an hour, there is no water movement anymore. Without external forcing, no movement is initiated and can be sustained. This equilibrium condition is rather uninteresting for the scientist. The fluid is internally consistent, but no interesting features will result from this consistency.

From the mathematical point of view, the hydrodynamic equations represent partial differential equations. For the solution of these equations both initial and boundary conditions must be specified, otherwise the problem of finding a solution for these equations is not well posed.

Initial conditions are needed for all state variables, because the values of the variables must be well defined in order to solve the equations. In addition to these initial conditions also boundary conditions must be prescribed. For example at a material boundary we know that no water may cross the boundary. On other boundaries a flux of certain quantities must be prescribed.

Two different types of boundary conditions exist, one where the value of the state variable is directly specified (*Dirichlet* conditions) and one where the flux of the variable is imposed (*von Neuman* conditions). In case of the equation for temperature, the first type would correspond to specifying the value of the temperature at the water surface, whereas in the second type of boundary condition the heat flux (radiation) of the sun (and other processes like evaporation) would be specified at the surface. Both boundary conditions eventually change the water temperature at the surface and therefore also in the interior of the water body.

4.6.1. INITIAL CONDITIONS

As explained before, all state variables need to be given initial values, otherwise the mathematical problem is not well posed. Normally one is interested to solve the equations starting from a certain time (initial time) into the future. The values of the variables are then fixed for this time and the equations are solved from this time on.

For real world problems this requirement is sometimes a problem. Very often we do not have the complete set of initial conditions for all state variables. Actually, it would be an exceptional case if we knew the values of all state variables at a certain time. The opposite will be normally true: we have a very incomplete knowledge of our state variables at any time. If we take the example of a lake, we will be very lucky if we know all the water levels at every point in the lake, but it will be surely impossible to know also the velocity field of the entire lake.

Fortunately the types of equations we have to solve give us a way out of this problem. The primitive equations, as well as the shallow water equations, have frictional terms that damp out all oscillations that are present in a certain moment, if no other external forces act on the basin. Therefore, even if we do not know exactly (or not at all) the initial conditions of our problem, after some time the solution of our equation will converge to the solution that is compatible with the external forces applied. In a certain way, the solution has forgotten the initial conditions of the problem.

This kind of problem differs completely from other problems in physics, where the initial conditions are fundamental for the solution of the problem. In hydrodynamics we are actually not interested in the exact specification of initial conditions, because the system will forget the initial situation more or less quickly. If we take as an example again a lagoon subject to tides (boundary conditions, see later), then after some time the regular oscillations forced by the tide at the lagoon inlet will destroy all the memory that the system had at the initial time.

This feature is used especially when the equations are solved with numerical methods. In this case any initial condition can be used. However, the system will approach faster equilibrium if boundary conditions are used that are close to the real situation. After some time, depending on the strength of the frictional forces, the system has forgotten the initial conditions. The time it takes to reach this point is called the *spin up time* of the simulation. The initial energy, due to the initial conditions, has been removed from the system and the system is in equilibrium with its boundary conditions.

4.6.2. MATERIAL BOUNDARY CONDITIONS

On material boundaries there is normally no flux of any quantity going on. Temperature (heat) and salinity fluxes should be zero if the boundary is isolating. For the current velocities no mass should cross the material boundary and therefore the current velocity must be tangential to the boundary. Mathematically the condition is $\vec{u} \cdot \vec{n} = 0$ where $\vec{n}$ denotes the direction normal to the boundary.

For fluids with friction, however, another boundary condition has to be specified. In a flow with friction the single water particles adhere to the material boundary and right at the boundary there is no movement of the fluid, not even in the tangential direction. Clearly this phenomenon is observed only very close to the boundary. Mathematically we have therefore to impose $\vec{u} = 0$ all along the boundary.

If the equations have to be solved with numerical modeling, the problem becomes even more pressing. Because the frictional forces act only on the molecular level, only the water molecules close to the boundary will feel its influence. This will result in strong gradients close to the boundary. In order to resolve this boundary layer a numerical discretization must have high resolution in this area, otherwise its influence will be much too strong on the inner water body. Normally this high resolution cannot be afforded by the numerical models. A way out of this problem is to change the conditions at the material boundary from the former no-slip to a free-slip condition. This means that only the normal velocity of the currents is set to zero, while the tangential velocity is free to adjust to the flow field. In the case that there are not too many lateral boundaries, or that the bottom friction is dominating, this approximations is a good one.

4.6.3. OPEN BOUNDARIES CONDITIONS

In a certain way all boundaries in nature are material. If we speak about open boundary conditions we mean artificial boundaries inside the water body where the domain is ending. This is clearly a mathematical construction, but it is very useful. Otherwise we would be forced to search for solutions of the whole world ocean even if we are only interested in a small coastal lagoon.

Natural open boundary conditions would be to prescribe fluxes at the interface. Therefore the flux of heat and salinity or the flux of momentum could be specified. This would again correspond to a boundary condition of the von Neuman type. We can, however, also fix the value of our state values at the open boundary. This is very often done with the water level, but also temperature and salinity values may be fixed at the open boundary. Care must be taken if the scalar values T or S are fixed. In this case it is possible to prescribe these values only during inflow. During outflow the values at the open boundary are influenced by the values inside the water body that get advected from the inside.

4.6.4. CONDITIONS ON THE SEA SURFACE AND THE BOTTOM

Besides specifying boundary conditions at lateral boundaries (closed and open), there are other interfaces in the water body. The most important ones are the bottom and the water surface. Actually, these interfaces are extremely important due to the fact that in lagoons and coastal seas the horizontal dimensions are always much bigger than the vertical dimensions. Therefore, fluxes through the sea bottom and surface will, to a certain extent, dominate the fluxes through the lateral boundaries.

Through the sea surface all kind of fluxes can be observed. Solar radiation is changing the energy budget, as well as precipitation and evaporation changes the salinity budget. Rain is also changing the mass balance, as well as evaporation influences the heat budget. Other processes through the sea surface are long wave radiation and sensible heat flux that will change the heat budget close to the sea surface.

However, the most important exchange going on at the sea surface is the momentum transfer from the wind action to the sea. This momentum input sets up waves resulting in strong wind currents that change directly the circulation pattern in the surface layer. Even if part of this momentum transfer is not yet well understood, the wind stress acting across the sea surface may be parameterized through the following bulk formula

$$\tau_x^s = c_D \rho_a \, |u^w| \, u^w \qquad \tau_y^s = c_D \rho_a \, |u^w| \, v^w$$

where ρ_a is the density of air, u^w, v^w the components of the wind vector and $|u^w|$ the modulus of the wind speed. The parameter c_D is called the *drag coefficient* that has an empirical value between 1.5×10^{-3} and 3.2×10^{-3}. This drag coefficient can either be used as a constant or it may be parameterized itself through the wind velocity.

At the bottom of the water another type of boundary condition has to be imposed. Normally a good approximation of the bottom is to take it isolating. Therefore, heat and salinity fluxes can be set to zero. However, important momentum exchanges are going on through the bottom resulting in frictional forces acting across the water bottom. The bottom exerts a drag on the overlaying moving water column slowing its movement down. The physics is similar to the water surface, where momentum is exchanged. However, through the surface normally momentum is transferred into the water body, whereas through the bottom momentum is lost.

The bottom friction stress can be expressed in a similar way as the wind stress. A bulk formula of the form

$$\tau_x^b = c_B \rho_0 \,|u|\, u \qquad \tau_y^b = c_B \rho_0 \,|u|\, v$$

is normally used with ρ_0 the density of the water, u, v the components of the current velocity in the vicinity of the bottom and $|u|$ the modulus of the current velocity. The parameter c_B is called the *bottom friction coefficient* that has an empirical value of about 2.5×10^{-3}, similar to the wind drag coefficient. As with the wind drag coefficient, this bottom drag coefficient can be computed through other formulas such as the Chezy or Manning formula. It would however take us too far to discuss also the formulations of these equations.

References

Apel, J. R., 1987. Principles of Ocean Physics, Academic Press, London.
Batchelor, G. K., 1967. An Introduction to Fluid Mechanics, Cambridge University Press, Cambridge.
Defant, A., 1961. Physical Oceanography, Pergamon Press, New York.
Gill, A. E., 1982. Atmosphere-Ocean Dynamics, Academic Press, London.
Guyon, E., J.-P. Hulin, and L. Petit, 1991. Hydrodynamique Physique, CNRS Editions, Paris.
Holton, J. R., 1992. An Introduction to Dynamic Meteorology, Academic Press, London.
Lindzen, R., 1990. Dynamice in Atmospheric Physics, Cambridge University Press, Cambridge.
Pedlosky, J., 1994. Geophysical Fluid Dynamics, Springer-Verlag, Berlin.
Salby, M. L., 1996. Fundamentals of Atmospheric Physics, Academic Press, London.
UNESCO, 1981. The Practical Salinity Scale 1978 and the International Equation of State of Seawater 1980, UNESCO Technical Papers in Marine Science, No. 36, Paris.

5. MONITORING PHYSICAL PROCESSES

Boris V. Chubarenko
Atlantic Branch, P.P. Shirshov Institute of Oceanology, Russian Federation

5.1. Introduction

Clear understanding that the human year-by-year negatively affects natural environment brings up the idea of sustainable development (WCED, 1987), which includes reduction of anthropogenic impact down to a level assimilated by nature. To implement the last one practically, permanent efforts on development and step-by-step implementation of "good" decisions are needed at the global and local scales. A process of elaboration of such a decisions includes several obligatory steps: gathering of data, extracting an information and its analysis against selected criteria, and, finally, development of decisions and implementation of them (Figure 1).

Data collection as a first step includes precise measurements and analytical work to find out the answer to the question "What is going on?" in numbers. Then, these numerous amounts of data, only understandable mostly for experts, have to be generalized and converted into information, which contains processes characteristics, tendencies, trends showing dynamics in time and space for the studied system (e.g., hydrodynamic and hydrological regimes, dynamics in water quality state and, finally, status of elements of an ecosystem). The "information" has to include a description of current state of environment, clear comparison with the past and, very important, some predictions toward the future in terms of several scenarios. Such an information is an ideal basis for development of management decisions elaborated in accordance with some pre-defined criteria of sustainability, harmonization, rehabilitation, progress, etc. Next step is an implementation of the accepted decisions into actions to affect reality and change a state of environment.

The complexity increases along all these steps, as the process starts from someone who measure precisely some parameters and ends at practical actions implemented by institutions. The generalization of information and the rise of complexity in decisions and actions lead at the same time to increasing the uncertainty. As to avoid this uncertainty is impossible to secondary loop starting from data collection is essential to verify whether the actions done bring expected changes or what else is needed to achieve desirable criterion.

I. E. Gonenc et al. (eds.), Assessment of the Fate and Effects of Toxic Agents on Water Resources, 109–125.
© 2007 *Springer.*

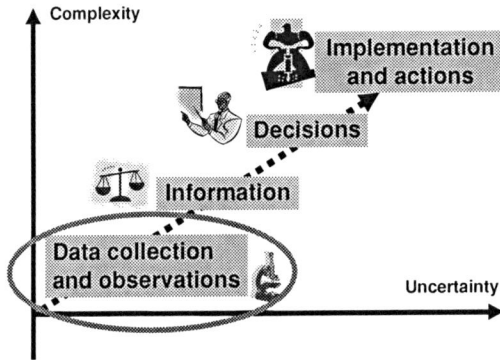

Figure 1. The consequence of steps toward the "clever" actions in water management starts from wide and precise data collection and measurements, then the data have to be compressed into information. Decisions will be concluded from the information and serve as a guideline for future actions. Each of these steps is fulfilled at the more and more high level of complexity, and it leads to decreasing of certainty

Monitoring belongs to the basic step of data collection, observations and measurements. It ends by valuable and representative data sets for numerous of parameters needed for preparation of information for decision makers. Monitoring is targeted toward the formulation of precise, clear, integrative and comprehensive information about the past, the present and the future of the water system.

Even monitoring designs vary in terms of spatial and temporal schedule, the processes of their development, however, is always the same. It begins from clearly defining the water-resource questions; outlining the information aspects are expectable to be made from the data and possible decisions those information will serve for; and, finally, identifying the data (or monitoring) needed to make the decision (Miller, 2005).

5.2. Definition and Objectives of Environmental Monitoring

This section contains an information concerning monitoring in general. It is fully applicable for a monitoring of physical processes as well as for others monitoring activities.

Monitoring is the application of fundamental scientific methods of observation of the environment. It is the assessment method of comprehensive determination of the current state of the environmental conditions. Monitoring measures are for description rather than prediction, however, monitoring data is used for various purposes including prediction scenarios/modelling (Andrulewicz and Chubarenko, 2004).

What are essential issues which exactly reveal a monitoring from any other types of observations? A following definition would be rather helpful, as it contains main features characteristic for a monitoring:

> "the process of repetitive observing for defined purposes, of one or more elements of the environment, according to prearranged schedules in space and in time and using comparable methodologies for environmental sensing and data collection" (ICES, 1988).

To understand better, what do the "defined purposes" mean, we may use an example of document "Preface to Monitoring" (Unites States Environmental Protection Agency) freely available in the INTERNET:

- *to learn about water environment;*
- *to create a baseline from which we can assess changes;*
- *to see if restoration efforts are working;*
- *to see if things are getting worse or staying the same;*
- *to meet national and international standards;*
- *to submit data for water managers, for public, ...*

As first three questions may be considered as general or scientific, the last three is directly related to management issues.

The purposes are the key issues which specify the main "vector" of monitoring activity.

> The purposes for undertaking monitoring vary, but, in any case, an information is collected for a <u>defined purposes</u>, and not simply because it <u>is available</u> (Andrulewicz and Chubarenko, 2004)

Some definitions of monitoring already contain the purposes inside the definition statement:

> Monitoring is (a) the repeated measurement of some parameters to assess the current status and changes over time of the parameters measured (U.S. Fish and Wildlife Service, USFWS), or (b) the periodic or continuous surveillance or testing to determine the level of compliance with statutory requirements and (or) pollutant levels in various media or in humans, animals, and other living things (U.S. Intergovernmental Task Force on Monitoring Water Quality, ITFM).

The following examples show some exact monitoring purposes (Miller, 2005) and illustrate the different general tasks which a monitoring serves to:

- *"To monitor the current status of the lagoon and its long-term variations"* obviously works upon environmental assessment.

BORIS V. CHUBARENKO

- *"Are regulatory requirements being met? Are concentrations or loads below those allowed in discharge permits?"*—these are the typical questions for "end of pipe" study.
- *"Is water quality varying and getting better or worse during certain times of the year?"* is the baseline question for development of the adaptive management scheme for a polluter.
- *"How do changes in land use or management practices affect water quality?"* brings the data for scenarios' study.

> None of these [*mentioned above*] questions is easy to answer, and each requires a different kind of monitoring—a specific set of data collected in certain places and at certain times. . . . Water-resource issues or questions determine monitoring objectives. And the objectives determine the monitoring design. No design, therefore, is "better" or "more successful" than another. Success is measured by whether the monitoring design addresses the specific objectives (Miller, 2005).

5.3. Multi-Purposefulness of Monitoring Activity

The "purposefulness" of any monitoring activity is a standard recommendation, but very often the reality is more complicated, and it is not clear to address monitoring to some specific purposes. To illustrate this we will refer to specific example of the City of Kaliningrad (the Baltic Sea, Russian Federation).

Since many years oxygen depletion zone accompanied by H_2S release develops in the segment of Pregolia River from the harbor up to the down town of Kaliningrad during a warm period (Figure 2). There are five (!) synchronous reasons, which bring own contribution to the phenomenon. The industrial sewage (i) from paper mill industry and municipal sewage (ii) are partly discharged into the Pregolia River through emergency outlets. The sewaged water is not efficiently flushed (iii) out in summer time because the river current is very slow. Weak river stream permits salt water intrusion (iv) from the Baltic Sea to penetrate easily deep upstream the river, and intrusion brings ions of chlorine, which react with sewages. Finally, seasonal warming (v) intensifies oxygen demands, and concentration of oxygen in the downstream segment of the river reduces practically up to zero. Remarkable is that time rate of the oxygen consumption is so high that low oxygen conditions permanently regenerate in this segment of the Pregolia River despite fresh waters constant coming. In result, the oxigen depletion zone steadily occupies the downstream of the river up to the city centre.

In above case all purposeful monitoring questions mentioned in previous paragraph have to be answered simultaneously. And, only the obvious baseline monitoring, i.e. a long-term collection of data with permanent time step at

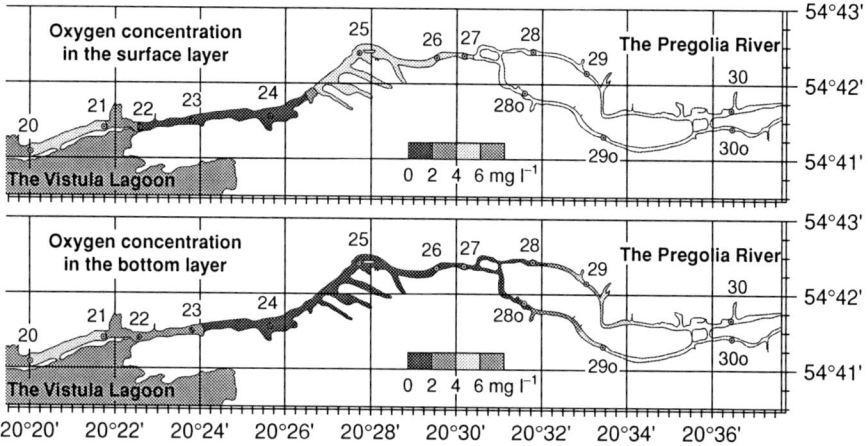

(a)

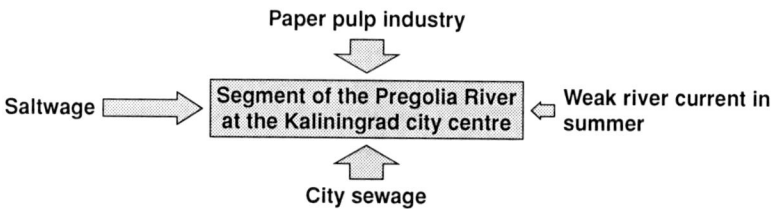

Paper pulp industry

Saltwage ⟹ **Segment of the Pregolia River at the Kaliningrad city centre** ⟸ Weak river current in summer

City sewage

(b)

Figure 2. The phenomenon of development of an oxygen depletion zone (a) in the low stream of the Pregolia River at the area of the City of Kaliningrad, Russian Federation (South-Eastern part of the Baltic Sea), is caused by unfortunate combination of natural factors and human impacts (b) as well as the seasonal warming in background. The existed monitoring scheme (monitoring points are numbered) allows to locate the phenomena, to reveal its temporal dynamics, resolve the influence factors and submit the data, which will be needed for development of scenarios for adaptive management of economic activity in the city. River segment 23–26 belongs to the port of Kaliningrad, historical centre of the City is between stations 26–28/28o

the stations regularly located in the study area (Figure 2), is the best for such multifactor situation.

As monitoring is a long-term activity, it should be planned on a system approach, not just for one purpose, even this purpose is of high current importance. Systematic monitoring guarantees that data collected will be really useful for many purposes, and monitoring scheme once established will be sustainable for new tasks.

Application of a system approach guides mostly a configuration of spatial scheme of monitoring, namely the positions of the monitoring stations. The sampling frequency is more flexible and exactly depends on purposes and available resources.

5.4. Types of Monitoring

Targeting to assessment of a quality of US Nation's waters (Miller, 2005) outlined two principal types of monitoring:

... "probabilistic" and "targeted" designs, which answer different sets of questions.
    Probabilistic: What percentage [of national waters] is in good condition? What percentage of streams is meeting their beneficial uses? Such questions require a broad-based probabilistic monitoring design, in which sites are chosen randomly and are distributed across a certain region. This type of monitoring provides a quantitative, statistically valid estimate of, for example, the number of impaired stream miles within a region or State.
    Targeted : Why are water-quality conditions happening and when? Do certain natural features, land uses, or human activities, and management actions affect the occurrence and movement of certain contaminants? Monitoring sites are therefore not selected randomly within a grid, because they represent certain human activities, environmental settings, or hydrologic conditions during different seasons or times of year.
    Although both of these designs can contribute to state-wide, regional, or national assessments, and improve understanding of the general or "ambient" water resource, they provide different types of information.

The above subdivisions of monitoring activities is in fact a view from the "top" of management system of a country, which possesses wide water resources and needs probabilistic monitoring as *a useful and cost-effective method for getting an unbiased, broad geographic snapshot of "whether there is a problem" and "how big the problem is." Targeted sampling brings an understanding of the causes of water-quality conditions. It establishes relations between water quality and the natural and human factors that affect water quality* (Miller, 2005).
    As for the single water pool, following (Miller, 2005), one may consider two main kinds of monitoring activities: "baseline" and "targeted" monitoring. First one is a basic routine monitoring, main questions for which could be the following:

What is the quality of waters? What is a long-term trend for monitoring parameters? (Miller, 2005).

For targeted monitoring another one gives the best fit:

Do certain natural features, land uses, or human activities, and management actions affect the occurrence and movement of certain contaminants? (Miller, 2005)

Sites for the baseline monitoring are selected as continuous grid coverage in the studied water pool to represent hydraulic and water quality parameters

variations during different seasons. Baseline monitoring helps to document—
is the water in good or poor condition, to prioritize both time stages and parts
of the water pool according to some water quality criteria.

In opposite, targeted monitoring focuses on understanding *the relations
between water quality conditions and the natural and human factors that
cause those conditions. Data helps to quantify relations between water quality
and the natural and human factors that affect water quality, to assess which
stressors (nutrients, sedimentation, . . .) are of most importance (Miller, 2005).*

Whether baseline and/or targeting monitoring activities are running on
many water bodies, the overview analysis of its results gives rise to probabilistic assessment, which fits to needs of probabilistic and targeting monitoring
in a scale of country:

> Probabilistic monitoring and assessments [*at the country scale*] help to docu-
> ment what is going well (how much of the resource is in good condition) and
> what is not (how much is in poor condition). The data collected help decision
> makers prioritize regions having the most degraded waters and assess which
> stressors—such as nutrients, sedimentation, and habitat disturbance—are of
> most importance in that region or State."
>
> Targeted monitoring and assessments [*at the country scale*] help decision
> makers to (1) identify streams, aquifers, and watersheds most vulnerable to
> contamination; (2) target management actions based on causes and sources of
> pollution; and (3) monitor and measure the effectiveness of those actions over
> time. *(Miller, 2005)*

Are baseline and targeted monitoring designs in contradiction or supplemen-
tary?

> Both types of monitoring are important, and therefore, should not be viewed as
> competitive or duplicative. In fact, these designs are so different that discussions
> should not focus on whether one design can substitute for another but on how
> to integrate the two in order to go beyond what each can provide individually?
> (Miller, 2005).

The single way to optimize a monitoring activity upon one water pool and
harmonize all types of monitoring is, as we emphasised at the end of previous
section, to organize monitoring on the basis of system principles, and in this
case its results will fit many tasks simultaneously.

5.5. Baseline and Targeted Monitoring From a View of System Approach

Main idea of system approach applied to natural systems of different scales
is to pick out the number of natural objects which together compose a system separated from neighbouring space by clear boundary, and are related

to each other "slightly closer" than to outside (or external) neighbours. The dynamics of the system in time is the very problem under study. Components of the system are in interactions between each other, and these interactions or relations exhibit themselves per se as internal processes between objects within the system. At the same moment, the system as a whole interacts with surroundings, and it become apparent in a form of specific impacts or driving forces from outside toward the component of the system. And finally, the system, being a part of a system of more large scale, takes part in many processes in the surrounding environment.

Baseline monitoring is addressed to monitor a dynamics of a system. So, any monitoring activity concerning internal objects as well as internal processes between them belongs to baseline monitoring. *Targeted monitoring,* being usually more specific and narrow, focuses on some internal objects of a system and compulsory takes into consideration the part of surroundings.

Lagoon as a system has obvious spatial boundaries related with four elements of environment (chapter 6 of present book Koutitonsky): lagoon catchment area (watershed), adjacent open area of a sea or ocean, an atmosphere and a bottom, and, the behaviour of a lagoon directly depends upon the impacts from them.

The *baseline monitoring* is addressed to a lagoon itself. Monitoring cites are inside the lagoon volume and data gathered there exhibit the state of a lagoon in terms of physical (and other) parameters. Monitoring of a lagoon coast (sediment dynamics, morphology) belongs also to baseline monitoring because data collected concerns directly the object of monitoring. The typical example of baseline monitoring is a routine monitoring activity on regular sampling or measurements in fixed location in a lagoon.

Any activity which belongs to *targeted monitoring* are to cover not only a lagoon but also some part of surroundings, because the targeted monitoring is responsible for collection of data about both a state of a lagoon in some respect and ambient conditions. The purpose is to reveal relationships between the entire state of a lagoon and causing factors, and to submit quantitative information about these relations. The example is a monitoring program which includes collection of samples (according fully agreed time schedule) both in a lagoon, in streams within the lagoon catchment and in adjacent marine area.

Figure 3 principally illustrates the scheme of relations between baseline and targeted monitoring for a lagoon.

5.6. Monitoring Parameters

State of any object in a system is described by several parameters. The processes or interactions are also quantified in terms of parameters which show

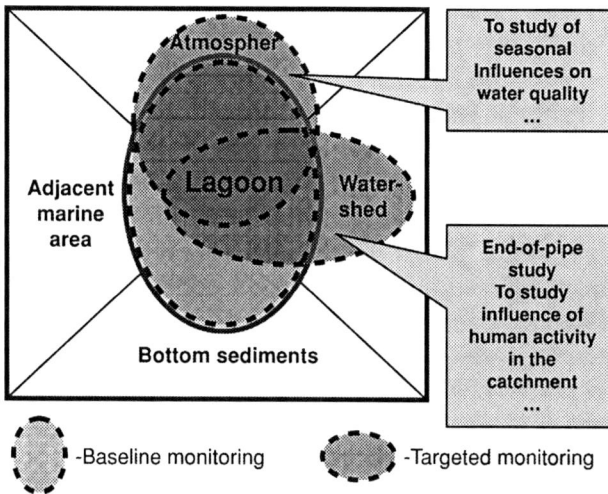

Figure 3. Lagoon and four segments of its surroundings: catchment, atmosphere, adjacent marine area and bottom. Baseline monitoring focuses on the system (a lagoon) or on an object of a system (e.g. lagoon water body) itself, and therefore monitoring cites are geographically located within the system boundaries. Every targeted monitoring includes not only monitoring of entire part of a system, but also some segment of system surroundings, therefore, it has a "cross-border" aspect. So, in principal, any monitoring activity may be considered as baseline or targeted in dependence of an identification of a system and its border

dynamics of different characteristics like processes intensity, rates, etc. Monitoring brings up the data about dynamics of a system as a whole (objects and processes together) through regular measurements of respective parameters in time and space.

There are many parameters which describe both an object state or its dynamics (e.g. water temperature), and, at the same moment, are indicators of processes characteristics (spatial and temporal variations of temperature indicate the differences in intensity of a process of heat exchange between the water and atmosphere). Therefore, when monitoring design is developing, a set of "informatively rich" key parameters should be considered to serve better for purposes of monitoring.

Traditional parameters of physical monitoring (as well as some others based on measurements of physical features) are:

- *Water level*, its dynamics shows the variations of quantity of water in the water body;
- *Currents*, it shows the intensity of water movement, and fluxes of water and water content both inside a lagoon volume and toward lagoon volume through boundaries;

- *Water temperature*, it is an important indicator of heat content and thermal stratification of water column;
- *Salinity,* it is an integral indicator on freshwater runoff, penetration of seawater, and finally of balance in water budget (remember fresh or hypersaline lagoons), it provides information of vertical stratification. Salinity is not measured, it is calculated from measurable conductivity. The more dissolved compounds (ions) in water, the greater the conductivity;
- *Dissolved oxygen* (DO), aquatic organisms require dissolved oxygen in the water column to survive (respiration requires oxygen). Freshwater and tidal saltwater standard is about of 6–10 milligrams per liter (ppm). Levels of DO less than 5 ppm are dangerous to fish health;
- *pH,* typical freshwater pH may range from 6.0 to 9.0, brackish and saltwater 7.5–8.5, and swamp water may be naturally low ranging 3–5. High pH in freshwaters may indicate high algal productivity;
- *Turbidity*, it is a measure of water clarity. Elevated turbidity can be a result of heavy storm water runoff, land-disturbing activity, algal blooms, or suspension of inorganic or organic matter. Usually measured by Secchi transparency, which is an integral characteristics of clarity of a surface waters.
- *Light attenuation*, it indicates the amount of light that fluxes down through the water column. Increasing turbidity and/or water color results in less light penetration. Important factor for organisms relying on photosynthesis;
- *Suspended matter*, it consists from fixed (terrestrial) and volatile (biotic) components;
- *Meteorological parameters* (wind, precipitation, evaporation, air pressure, temperature and humidity, cloudiness) are usually accompany measurements of all hydrophysical parameters.

All these parameters are usually measured during baseline as well as targeted monitoring program inside the lagoon as well as on its boundary. Monitoring activities differ because of different combinations of measured parameters and various spatial and temporal monitoring designs.

> The list of monitored hydrodynamic parameters usually depends on the monitoring goals, but generally includes: flows, salinity and temperature through the lagoon entrance; discharges and water temperature from all rivers and artificial outlets; level variation at the open lagoon entrance; level variation at some points remote from lagoon entrances; current, salinity and temperature vertical profiles at monitoring points inside the lagoon; spatial variation of salinity and temperature in the lagoon; wind wave height and spreading direction; parameters related to turbulent mixing; tidal characteristics of the adjacent marine area (Andrulewicz and Chubarenko, 2000).

5.7. Spatial Schemes of Monitoring

Lagoon as a system has a "boundary" which detaches a lagoon from catchment, atmosphere, bottom and adjacent marine area (Figure 4). Surroundings influence on a lagoon by means of fluxes of matter and energy through this boundary (Figure 4). Without regard to monitoring purposes one may subdivide the monitoring of a lagoon into two main activities: monitoring of state variables of an entire lagoon water body and monitoring of external factors or driving forces (or impacts or influences).

Baseline monitoring is a monitoring of the state of entire water body of a lagoon. A spatial net of monitoring stations covers all main compartments of the lagoon volume (e.g., Figure 5). Time schedule is usually tuned to resolve seasonal variations, and time interval is of fortnight or month.

Depending upon monitoring objects, locations of monitoring cites inside lagoon area may be different. For example, for transboundary Vistula Lagoon (Andrilewicz et al., 2004), there are two approaches to a monitoring (Figure 5), which are realized at national level in the northern, Russian, and southern, Polish, sides of the lagoon respectively. In the northern side monitoring stations are located along the central axis of the lagoon, along the line of maximum depths. The main purpose of monitoring is to collect data about general state of water quality in the lagoon. In contrary, at the Polish side, the main purpose is to search the water quality conditions just near the main local

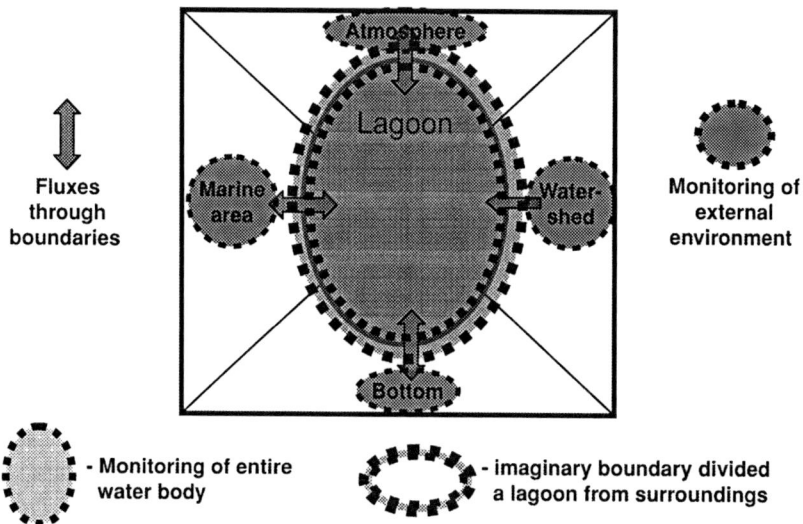

Figure 4. As concern the objects of monitoring one may divide any monitoring activities into two principal sections: a monitoring of water body and a monitoring of objects in external environment which affect a lagoon system

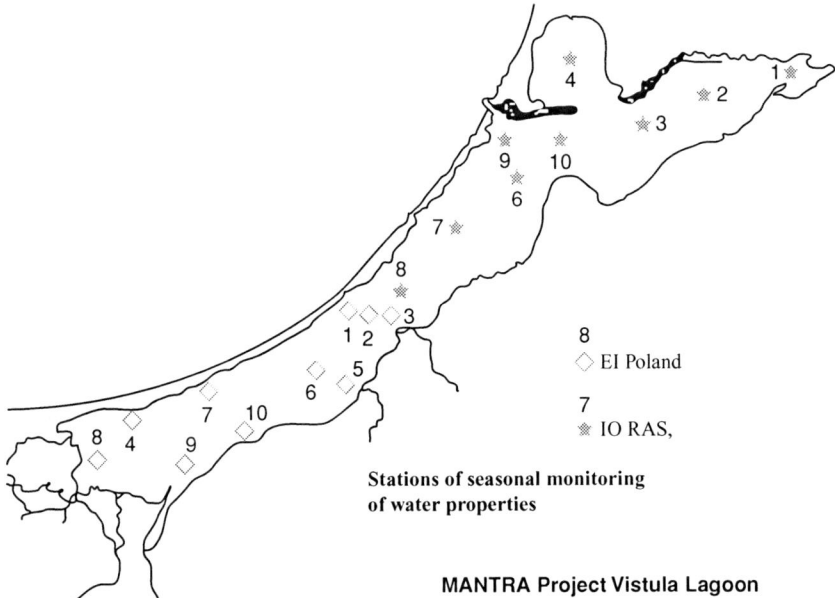

Figure 5. Spatial scheme of monitoring stations in the Vistula Lagoon (Poland-Russia, The Baltic Sea). Blue points are belong to Polish monitoring system and located along the coast showing sanitary conditions near settlements and rivers. The green points of Russian monitoring system are located along the central axis of the lagoon at the deepest places

settlements or river mouths, and, the monitoring is focused on sanitary needs more than on general baseline understanding of dynamics of water state in the lagoon.

Currents in a lagoon easy react to wind change, and at the lateral, the shallowest, parts of lagoon area they are very changeable. Water exchange with adjacent marine area is very intensive, and hydrodynamic situation varies frequently near a lagoon inlet. The most "conservative", if else this term is applicable for changeable lagoon environment, are the waters in a central lagoon part, which is equally remote from coast and inlet. These waters integrate changes which start and intensively develop on lagoon laterals. Therefore, an optimal way to organize a baseline monitoring is to assign monitoring cites far from lagoon coast, to locate them preferably in the central lagoon area and at the deepest places.

Targeted monitoring includes monitoring of external impacts, besides monitoring measurements within a lagoon volume. Monitoring sites are usually located in the lagoon and on the boundary, at the mouths of inflowing river streams, and in the inlet (or at the outward vicinity of inlet). Data collected is used to quantify the boundary conditions (see chapters 3, 4 of present book).

In lagoons, the major driving forces are usually wind stress, water level changes related to tides and wind action, density gradient of different origins and direct *atmospheric pressure*. Therefore, ..., following meteorological measurements are crucial: wind speed, wind direction, barotropic pressure, precipitation and evaporation, solar radiation, cloudiness, air temperature, humidity (Andrulewicz and Chubarenko, 2004).

The most critical hydrodynamic parameter is water exchange through lagoon inlet. There are different types of water flows in inlet, including steady flows, pulsing flows and backflow. More than one flow in different directions may occur in the water column at one time. These flows may be subject to considerable temporal variation and have a tremendous influence on lagoon hydrodynamics (Andrulewicz and Chubarenko, 2004).

If a lagoon bathymetry has narrowness and wideness, the principal recommendation is to have monitoring stations for measurement of water characteristics in the centre of wideness, and fluxes characteristics in the narrowness.

5.8. Types of Monitoring Measurements

Someone may select the following principal types of measurements used to use for monitoring activity in a lagoon.

Monitoring cruises by boat with expedition staff onboard is a common method of data and sample collection in a lagoon. The strength of this method is a possibility to gather different types of data at the same moment (hydrological, chemical, biological) through collecting of real water and sediment samples and making *in situ* measurements. The weakness is that, first, it is not possible to make such cruses frequently to cover the main important hydrological and hydrochemical cycles, and, second, because of time limitation, the area covered by measurements is limiting, and, third, samples are collected not simultaneously for the whole lagoon, especially, if the lagoon is of size of 20–100 km.

Continuous measurements in a lagoon are very effective to collect data about temporal or spatial variability. There are three principal types of application of this powerful approach: continuous measurements in one point, vertical profiling and horizontal profiling with help of moving boat.

Deployment of an automatic measuring station for a group of parameters (e.g., temperature, salinity, turbidity, level variation, current, pH, Eh, etc) will bring data for clear understanding of temporal dynamics of characteristics, and mutual correlation between them in wide range of temporal variations (from tidal cycle up to seasonal changes). The weakness is that all data collected relates to a very point of instrument deployment in the water volume, and the only currents may be registered along the vertical axis using ADCP. In case of the absence of stratification, when vertical differences are not important, deployment of a set of such stations brings nearly full representative data about variation of physical conditions in the lagoon.

Having onboard an automatic probe another type of continuous measurements is preferable. This is a continuous measuring of parameters from a boat moving along or across a lagoon area. These types of measurements bring clear understanding about spatial variability in the lagoon, about existed hydrological fronts and zones of well spatially mixed conditions. The weakness is that such continuous measurements could be made only at one depth level, usually at the surface.

As a lagoon or some of its parts very often exhibit vertical heterogeneity, vertical profiling by the instrument equipped by sensors for different parameters will bring the data about the vertical stratification of measured parameters and, in its turn, about the vertical mixing characteristics. The optimal monitoring strategy would be a vertical profiling at the stations, which are located in the key points of the lagoon area. For example, for the deep navigation channel at the northern part of the Vistula Lagoon the coupling of both strategies are used, namely, the vertical profiling at the fixed monitoring stations and instrument near surface towing with continuous recording of hydrophysical parameters between them.

Remote sensing measurements make a snap-short of the characteristics of the lagoon area. The modern technique of acoustic scanning the water area from the coastal stations (Badiey et al., 2003) will give the data about currents, water surface roughness, temperature and salinity. Satellite images gives an environmental information about whole lagoon water area (suspended sediments, concentration of nutrients, dissolved oxygen (FAO, 1999). The methodology applied in (FAO, 1999) can be easily extrapolated to other large lagoon areas. It can also serve to improve the understanding of circulation fronts, surface temperatures, geomorphological mapping and other aspects.

5.9. Temporal Schedule of Monitoring

The more frequent are the measurements the better informative they are. The wish is always to measure with the shortest time step as possible, but frequency of measurements depends mostly on resources available.

Baseline monitoring of a lagoon depends on time variability of the lagoon characteristics which are the subject of monitoring. Usually 1–3 surveys per month is a compromise to understand the seasonal variations of the lagoon characteristics. But, definitely more frequent measurements are needed to resolve the tidal cycle.

Frequency of measurements in any kind of targeted monitoring depends not only on the temporal scale of variations of internal parameters of the system, but also on variations of the external factors.

5.10. Monitoring and Short-Term Data Collection Dedicated to Modelling

Baseline monitoring covers an entire of a lagoon. Targeted monitoring covers both an entire of a lagoon and part of lagoon surroundings. Logically to consider *overall or complete or total monitoring* focused on a whole lagoon and all significant external factors. It is a desire of any scientist as this monitoring gives all needed data for understanding of system dynamics.

Such overall measuring activity is rather expensive even at the basic level of hydrological monitoring, and it is usually executed with specific purposes to collect data for modelling studies during relatively short period (several months or years).

> Monitoring is distinct from field data collection due to its long-term, continuous nature. Data collection efforts are sometimes referred to as short-term monitoring, but it is important to maintain a distinction with monitoring, because monitoring generally has different objectives than data collection (Andrulewicz and Chubarenko, 2004).

Monitoring brings data for modelling, for calibration and verification of a model, for current operational modelling activity. Modelling is used for verification and optimisation of monitoring design in space and time.

> Monitoring and modelling are two essential steps of water management practice. If the first mostly serves to environmental assessment, the second serves for environmental impact assessment and scenario analysis. Monitoring and modeling very often complement each other, as, for example, in case of feedback monitoring when both a short-term modeling forecast and adaptive monitoring are incorporated (Andrulewicz and Chubarenko, 2004).

Modeling needs specific amount of data depending on a stage of model design and implementation as well as on the level of model (hydraulics, water quality, transport, euthrophication, etc). As model is a mirror of a reality, it needs exact amount of data needed to describe the lagoon system dynamics.

> Hydrodynamic models are the basis for ecosystem modeling. Hydrodynamic modeling is the furthest developed type of modeling, but in the case of lagoons the implementation of models is a challenge due to the technical difficulties in

getting enough field data to calibrate the model for any type of application. The high variability of current patterns, its local peculiarities because of bathymetry variations, the complicated nature of water exchange between lagoons and the adjacent marine waters—all of these factors cause a dramatic increase of monitoring data amount needed for implementation of 2D and especially 3D hydrodynamic models (Andrulewicz and Chubarenko, 2004).

Monitoring is usually undertaken to reveal the long-term variations of the system, and data collected during monitoring are usually not enough to calibrate and validate a model, because modeling is not the purpose of standard monitoring. Therefore, a special type of data collection program for model calibration and verification is essential.

Very often such projects are called 'monitoring', but they are not actually monitoring, because they do not satisfy the time duration requirement. Such programs even being held according to a fixed prearranged spatial and temporal scheme as in the standard monitoring can be referred to as short-term data collection program for model implementation to emphasize that the duration of the program is short. In a practical sense it may vary from 2–3 months for calibration of the hydrodynamic module up to 1–3 years for calibration of either the advection-dispersion, water quality, or some biological modules (Andrulewicz and Chubarenko, 2004).

For model implementation and regular usage another type of modeling dedicated study, namely, model accompanied (attendant) current data supply *(Andrulewicz and Chubarenko, 2004)* has to be organized to provide the model with minimum data needed to run it at any given time for impact assessment or prediction. Mostly these are the data on driving forces for lagoon system like wind, water level variations, river discharge, etc; but also, some information on selected simulated variable or variables at least at one location in the lagoon are desirable to have a reference point for current model simulations.

Per se, modeling dedicated data collection is a short-term example of *a complete monitoring*, which bring together all data needed for assessment of system dynamics itself as well as its relations with surroundings.

References

Andrulewicz, E., and B. Chubarenko, 2004. Monitoring Program Design (Chapter 7), edited by I.E. Gonenc and J. Wolflin. Costal Lagoons: Ecosystem Processes and Modelling for Sustainable Use and Developments, CRC Press, Boca Raton, pp. 307–330.
Andrulewicz, E., B. Chubarenko, and I. Chubarenko, 2004. Case study: Vistula Lagoon (Poland/Russia)—Transboundary Management Problems and an Example of Modelling for Decision Making (in Chapter 9), edited by I.E. Gonenc and J. Wolflin. Costal Lagoons:

Ecosystem Processes and Modelling for Sustainable Use and Developments, CRC Press, Boca Raton, pp. 423–439.

Chubarenko, B.V., V.G. Koutitonski, R. Neves, and G.Umgiesser, 2004. Modelling concept (Chapter 6), edited by I.E. Gonenc and J. Wolflin. Costal Lagoons: Ecosystem Processes and Modelling for Sustainable Use and Developments, CRC Press, Boca Raton, pp. 231–306.

Miller, T.L., 2005. Monitoring in the 21st Century to Address our Nation's Water-Resource Questions, 25 February 2005, web-source.

Badiey, M., L. Lenain, K.-C. Wong, R. Heitsenrether, and A. Sundberg. Long-term Acoustic Monitoring of Environmental Parameters in Estuaries. http://www.udel.edu/dbos/assets/oceans_2003.pdf.

FAO Remote Sensing for Decision-makers Series, No. 14, 1999. Satellite imagery for aquaculture study and lagoon management—Pilot study in Morocco.

Gurel, M., 2000. Nutrient dynamics in coastal lagoons: Dalyan Lagoon case study, Ph.D. Thesis, Istanbul Technical University, Institute of Science and Technology, Istanbul, Turkey.

HELCOM, BSPF No. 27, 1988. Guidelines for the Baltic Monitoring Program for the Third Stage, Baltic Sea Environment Proceedings, No. 27.

HELCOM, EC 8/97, 1997. Manual for Marine Monitoring in the COMBINE Program of HELCOM.

OSPAR, 1996. Draft Guidelines on Contaminant-Specific Biological Effects Monitoring, Report of Ad Hoc Working Group on Monitoring, Oslo and Paris Commissions: Summary Record of the meeting of MON.

World Commission on Environment and Development (WCED), 1987. Our Common Future. Oxford University Press, Oxford, p. 43.

6. TIME SERIES ANALYSIS OF LAGOON VARIABLES

Vladimir G. Koutitonsky
Institut des science de la mer de Rimouski (ISMER), Canada

6.1. Introduction

Water masses, dissolved substances and particulate matters in coastal lagoons are set and maintained in motion by hydrodynamic forces such as sea level oscillations at the inlet, buoyancy inputs from river discharge and heat fluxes at the surface, winds, atmospheric pressure, gravity and earth rotation. Considering a lagoon as an input–output system, these forces are the system's inputs while the lagoon's response is the system's output. Given the physical nature of most forcing functions, it is customary to discuss oceanographic data analysis methods in terms of hydrodynamic variables such as sea levels, currents, etc, even though these methods apply equally well to time series of other ecosystem variables such as sunlight, chlorophyll-a, nutrients, etc. In this context, the objective of time series analysis is to isolate and describe the variability of the output and attribute this variability to one or more of the forcing input functions.

Two main classes of time series analysis methods are available: (i) *time domain analysis or TDA* and (ii) *frequency domain analysis or TDA*, the choice of one or the other ultimately depending on the study objectives, the number of series involved and the common duration of the time series sets. Time domain analysis normally involves correlation and regression analyses, empirical orthogonal functions (EOF) analysis, singular value decomposition analysis and others, while frequency domain analysis involves spectral analysis, harmonic analysis, wavelet analysis, frequency EOF analysis and others. Some of these methods are now briefly presented.

6.2. Lagoon Time Series

Most lagoon time series $x(t)$ will include a mean value $\bar{x}$ and stationary fluctuations $x'(t)$. In turn, these fluctuations probably include periodic fluctuations $x'_P(t)$ produced by outside tides and gravity waves, stochastic fluctuations $x'_S(t)$ produced by meteorological forces, a trend $x'_T(t)$ produced by large scale offshore forcing or seasonal buoyancy fluxes and stationary residual fluctuations of unknown sources $x'_R(t)$:

$$x(t) = \bar{x} + x'_T(t) + x'_P(t) + x'_S(t) + x'_R(t).$$

I. E. Gonenc et al. (eds.), Assessment of the Fate and Effects of Toxic Agents on Water Resources, 127–142.
© 2007 *Springer.*

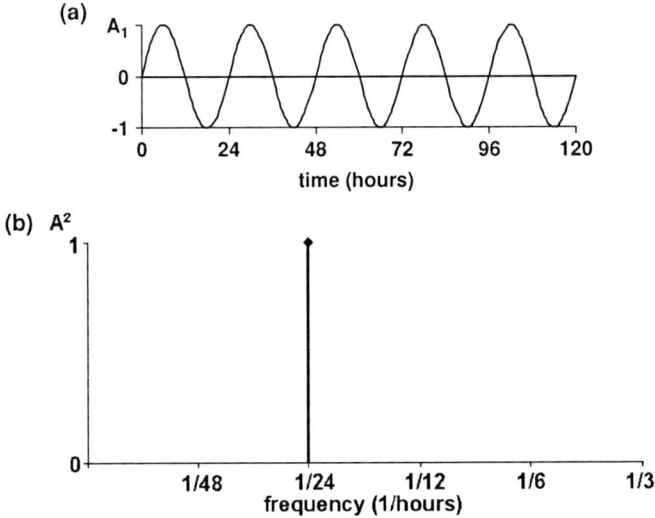

Figure 1. A periodic signal $x(t)$ represented in the time domain (a) and frequency domain (b)

The removal of the trend $\bar{x} + x'_T(t)$ from the series $x(t)$ will transform it into a stationary time series made up of fluctuations about a zero mean value. These fluctuations can be deterministic or stochastic in nature. Deterministic fluctuations can be accurately predicted by deterministic methods such as harmonic analysis. Examples of deterministic series are tidal fluctuations or daily light incidence durations. Stochastic fluctuations can also be predicted but only in terms of probabilities. Examples are winds, atmospheric pressure, and cloud cover time series. In any case, it is useful to consider lagoon time series as being periodic fluctuations, whether they are deterministic or stochastic. Periodic fluctuations can be described in the time domain by an intensity that changes with time t or in the frequency domain by its variance content that changes as a function of frequency f. A periodic time series presented in the time domain is shown in Figure 1(a). It has an amplitude $A = 1$ (arbitrary) units and a period $T = 24$ h. The same series is shown in the frequency domain in Figure 1(b). It has a power (variance) $X(f) = 1^2 = 1$ at only one frequency $f = 1/T = 1/24$.

6.3. Time Series Preprocessing

Some time series will require some preprocessing before being analyzed in the time domain or the frequency domain. For instance, data spikes must be removed from the series and data gaps must be filled by interpolation when the gap is not too long. It is then useful to estimate first order statistics such as the mean value:

$$\bar{x} = \sum_{n=1}^{N} x_n / N$$

and the variance:

$$\sigma^2 = \sum_{n=1}^{N} (x_n - \bar{x})^2 / (N - 1).$$

The standard deviation σ is the square root of the variance. It gives a measure of the "spread" of values within a time series. Normally, data points outside three standard deviations intervals about the mean are considered as noise and are replaced by interpolation.

Whenever series sampled at different time intervals are to be correlated, they must first be decimated or resampled at the same time step. However, decimating a series at a longer time step may introduce an artificial oscillation through a phenomenon known as *alaising*. Therefore, the series must first be smoothed before decimation in order to remove oscillations with periods smaller than the new sampling interval. A simple time-domain filter that will smooth out these oscillations is $\{[x_{n+1}^2 \cdot x_n] / [(n + 1)^2 \cdot n]\}$. For example, if it is desired to decimate a series x_j sampled at $\Delta t = 20$ min to an hourly series, such that $n = 60$ min $/20$ min $= 3$, the corresponding smoothing operator will be $[(x_4^2 x_3)/(4^2 \cdot 3)]$. In this case, the first and last 4 data points are lost and the smoothed series can then be decimated to hourly values without introducing artificial oscillations in the process.

Finally, it is also customary to remove the series trend in order to render it stationary. By definition, a trend is an incomplete oscillation over the discrete series record length T. A trend $\bar{x} + x'_T(t)$ can be estimated by fitting a third order polynomial to the series. When this trend is subtracted from the original series, the resulting series becomes stationary and ready for time domain or frequency domain analysis.

6.4. Time Domain Analysis

The correlation coefficient R_{xy} between two discrete time series x_n and y_n is considered as a measure of similarity between these two series. It is defined as

$$R_{xy} = \frac{\sum_{n=1}^{N} (x_n - \bar{x}) \cdot (y_n - \bar{y})}{\sqrt{\left[\sum_{n=1}^{N} (x_n - \bar{x})^2 \cdot \sum_{n=1}^{N} (y_n - \bar{y})^2\right]}}.$$

When $N > 30$, this coefficient has a normal distribution with a mean value $\mu = 0$ and a standard deviation $\sigma_r = 1/(N-2)^{1/2}$. In this case, the probability at the 0.975 level of a normal distribution is 1.96. In other words, close to 95% of the correlation coefficient estimates should be at ± 1.96 standard deviations from zero. A correlation coefficient located outside this band will be significant within a 95% confidence interval. The limitation in using the correlation coefficient is that it only provides a measure of similarity between data points measured at the same instant. However, in some cases, series y_n could be well correlated with series x_n when it is lagged in time relative to x_n by a few time steps. This is the case for example for sea level and atmospheric pressure time series. Sea level often responds to atmospheric pressure through the inverse barometric effect and in some cases this response lags atmospheric pressure by a few hours. This correlation will not show up in the correlation coefficient defined above. Such a "lagged" is why the notion of "lag" must be considered in time series correlation analysis. This brings up the concept of lagged-correlation analysis. In this case it is suggested to estimate the lagged covariance functions such as the auto-covariance between the same series and the cross covariance between two series. When normalized by their respective variances, these functions become the auto-correlation and cross-correlation functions, respectively. The auto-covariance at m lags is

$$C_{xx,m} = \frac{1}{N-m} \sum_{n=1+m}^{N} (x_n x_{n-m} - \bar{x}_n \bar{x}_{n-m}).$$

As a general rule, the number of lags m should exceed 10% of "N". The "auto-correlation" is simply the auto-covariance $C_{xx,m}$ divided the variance $\sigma^2 = C_{xx,0}$:

$$r_m = \frac{C_{xx,m}}{C_{xx,0}} = \frac{\dfrac{1}{N-m} \displaystyle\sum_{n=1+m}^{N} (x_n x_{n-m} - \bar{x}_n \bar{x}_m)}{\dfrac{1}{N-1} \displaystyle\sum_{n=1}^{N} (x_n - \bar{x})^2}.$$

The auto-correlogram is a plot of the auto-correlation as a function of lags. It shows an auto-correlation of 1 at zero lag and decreases thereafter but always remaining between -1 and $+1$. An example of the auto-correlogram between atmospheric pressure and sea level (station 2) in Boughara lagoon in Tunisia is shown in Figures 2(a) and (b), respectively.

The correlograms indicate that:

- The correlation coefficients are equal to 1.0 is at zero lag. This makes sense, since the series perfectly matches itself only at zero lag.

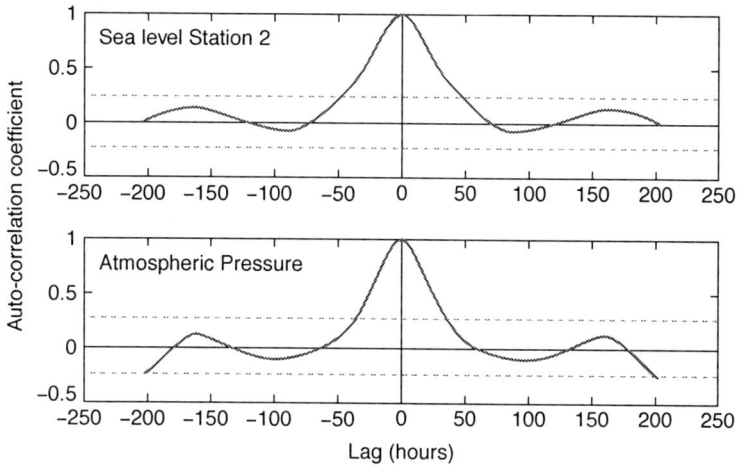

Figure 2. Auto-correlogram of sea level and atmospheric pressure

- The coefficients decrease with increasing lag until they reach zero at about 70 h for sea levels and 60 h for atmospheric pressure. This zero crossing is an important concept, since it characterizes the length of time that must elapse before the time series becomes uncorrelated, i.e., how long it takes to forget what has happened before.
- This first "zero" is known the *decorrelation time*. If one is analyzing a space series (e.g., a hydrographic section of regularly spaced stations), the first "zero" of the auto-correlation will be a measure of the spatial decorrelation length. This could be a measure of the eddy mixing scale: stations closer than this distance will tend to be correlated, and hence not independent of one another.
- At about 90 h, there is a negative correspondence. If the pressure is high now, it will likely be low about 3.75 days from now, and *vice versa*.
- As you approach 7.5 days, the covariance rises slightly to a small maximum. This is the indicator of periodicity. In this case, it is not significant as it remains within the 95% confidence intervals ±0.24

These auto-correlograms clearly show that consecutive observations are not independent from each other since their auto-correlation does not fall to zero right away. In this case, the assumption of independent samples is not valid and the series must be resampled at longer time intervals in order to obtain a series of N' independent observations, giving $N' - 2$ degrees of freedom. This new sampling interval is the integral time scale T^* of the series. It is defined as the integral of the auto-correlation coefficients over a time duration

$N'\Delta\tau$ of the time series after the start and is given by:

$$T^* = \frac{\Delta\tau}{2} \sum_{m=0}^{N'} \left[R_{xx,m} + R_{xx,m+1}\right], \quad \text{for} \quad N' \leq N - 1.$$

T^* is an indication of the time a data point in a series "remembers" previous data points. The number N' is difficult to specify. Normally, the summation is performed until the shape of T^* reaches a plateau. If not, the summation is performed up to the first zero crossing of the auto-correlation function. The integral time scale of the sea level series was found to be 32 h, such that the number of independent data points in the series decreases to $N^* = T/T^* = 2035/32 = 64$. The confidence intervals shown in Figure 2 were estimated using the N^* corresponding to each series.

The *cross-covariance* of two time series provides information about the relationship between two phenomena. For example, one might be interested in the influence of the wind stress on sea level set-up at one end of a lagoon. Of interest is not only the degree of correspondence between two records, but possible *time lags* between them. The cross-covariance function between two time series x_n and y_n of N simultaneous observations is

$$C_{xy,m} = \frac{1}{N} \sum_{n=1}^{N-m} (x_n - \bar{x})(y_{n+m} - \bar{y}), \quad [m = 0, 1, 2, \ldots, (N-1)]$$

$$C_{xy,m} = \frac{1}{N} \sum_{n=1-m}^{N} (x_n - \bar{x})(y_{n+m} - \bar{y}), \quad [m = -1, -2, \ldots, -(N-1)]$$

and the cross correlation coefficient between series x_n and y_n is

$$R_{xy\,m} = \frac{C_{xy\,m}}{\sqrt{C_{xx\,m} \cdot C_{yy\,m}}}.$$

One notes that the cross correlation is the general case of the auto-correlation. While the auto-covariance and auto-correlation functions are *symmetric* functions (value at lag m equals value at lag $-m$), the cross-covariance and cross-correlation functions are asymmetric functions (value at lag m not equal to value at lag $-m$). The asymmetry necessitates the two parts of the $C_{xy,m}$ equation. The first part applies to y_n lagging x_n, and the second part to x_n lagging y_n. The cross-correlation coefficient between the lagoon sea levels and atmospheric pressure should reveal a negative correlation between both series. Results of a lagged-correlation analysis between both series are presented in Figure 3.

The horizontal dashed red lines at $R_{xy} = \pm 0.24$ marks the 95% confidence intervals for the cross-correlation coefficient based on its estimated standard error. For the $\alpha = 0.05$ level and a two-tailed test, this confidence

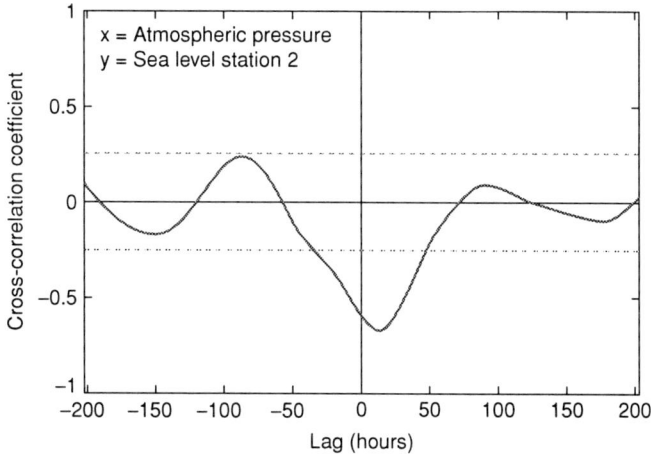

Figure 3. Cross-correlation coefficient between low passed sea level at station 2 and atmospheric pressure. The dotted lines are the 95% confidence intervals of the estimate

interval is $\pm 2/\sqrt{N^* - 3}$, where N^* is the effective number of independent data points separated by an integral time scale, in this case equal to 32 h. A cross-correlation estimate outside the confidence interval is "significant". The maximum correlation above is -0.82 and is found when sea level lags atmospheric pressure by 13 lags or, in this case, 13 h. Being outside the 95% confidence intervals, this is a significant lagged correlation.

6.5. Filters

A filter is a mathematical operator that eliminates oscillations of particular periods (or frequencies) from a time series while allowing other periods to pass. A frequency domain filter will act to pass, isolate or remove one or several frequency bands. For example, a low-pass filter can be applied to a sea level time series in order to eliminate tidal oscillations of frequencies higher than a cut off frequency f_c and allow lower frequencies of meteorological origin to pass. One such filter is based on the Fourier Transform concept. Let X_k be the Fourier transforms of series x_n:

$$X_k = \sum_{n=0}^{N-1} x_n \exp\left[-i\frac{2\pi kn}{N}\right]; \quad k = 0, 1, \ldots, N-1.$$

Oscillations of frequencies f_k higher than f_c can be removed by multiplying the transforms X_k by a low-pass filter operator W_k to obtain a new transform series X_k^*:

$$X_k^* = X_k \cdot W_k.$$

Here the filter operator W_k adopts one of the following values:

$$W_k = 1 \quad \text{for} \quad f_k < f_c,$$
$$W_k = 0 \quad \text{for} \quad f_k > f_c.$$

This filtered FT series X_k^* is then inverse transformed to generate the time domain low-passed series $x_{n\,\text{LP}}$.

$$x_{n\,\text{LP}} = \frac{1}{N} \sum_{k=0}^{N-1} X_k^* \exp\left[i\frac{2\pi kn}{N} \right]; \quad n = 0, 1, \ldots, N-1.$$

A cut-off frequency that removes tides effectively from a sea level time series is $(1/34)\ \text{h}^{-1}$.

6.6. Frequency Domain Analysis

Frequency domain analysis (FDA) is normally used to distribute the variance contained in time series into frequency bands, in an attempt to (i) detect dominant periodicities in the series and (ii) attribute them to similar frequency bands in one or several forcing functions. This is not possible in time domain analysis (TDA) since the portions of the variance produced by different forcing functions, acting at different frequencies, are all lumped into the overall series variance. There are several FDA methods for coastal time series, the major ones being: (1) spectral analysis performed on one or two series, where the variance is extracted at frequencies determined by the duration and the sampling interval of the time series, and (2) harmonic analysis performed on one series, where the variance is extracted at the particular frequencies of existing tidal harmonic constituents. Spectral analysis on one series yields the auto-spectral (variance) density of the series as a function of frequency. It is the Fourier transform (FT) in the frequency domain of the time domain auto-covariance function. Spectral analysis on two-series yields the cross-spectral (variance) density function, which is the FT of the cross-covariance function. When normalized by the product of the two auto-spectral densities, the cross-spectral density function yields the coherence and relative phase functions. The following sections briefly describe these methods and their application to sample series.

6.6.1. SPECTRAL ANALYSIS

One way of representing a time series in the frequency domain is to plot the contribution that each frequency makes to the total variance of the series. This representation has been given various names: the *variance spectrum* (in statistics), the *power spectrum* (originally in electrical engineering), or the *spectral density function*. A plot showing the discrete-time Fourier transform of a discrete finite stationary time series is called the *periodogram*. It shows the distribution of the variance per cycles of frequency, or the *variance density*, around each of the permissible Fourier series frequencies up to the Nyquist frequency $1/(2\Delta t)$.

The spectral variance density of a discrete time series x_n is called the *auto-spectral variance density* $G_{xx\ k}$. It is function of frequency k defined in terms of the Fourier transforms X_k as

$$G_{xx}(f_k) = \frac{2}{N\Delta t} \left[X^*(f_k) \cdot X(f_k) \right], \quad k = 0, 1, 2, \ldots, N/2$$

where X_k^* is the complex conjugate of the (complex) Fourier transform X_k. In simpler terms, it maybe thought off as the variance of a periodic wave of frequency k, fitted through the series x_n in a frequency band centered around k. The auto-spectral variance density can be estimated by expressing the Fourier transform X_k as a complex number:

$$X_k = X_{kr} + jX_{ki} \quad \text{and} \quad X_k^* = X_{kr} - jX_{ki}$$

and evaluating:

$$G_{xx\ k} = \frac{2}{N \cdot \Delta t} \left[X_{kr}^2 + X_{ki}^2 \right].$$

This demonstrates that the auto-spectral variance density function $G_{xx\ k}$ is a real number. The spectral density function is a Chi-squared function χ_n^2 as the variance function. Its mean and variance estimates being n and $2n$ respectively, the standard error associated to the variance density estimate is

$$\varepsilon = \frac{\sqrt{2n}}{n} = \sqrt{\frac{2}{n}}.$$

The number of degrees of liberty associated to a Chi-squared variable being $n = 2$, it can be seen that the error ε above equals 1, indicating that the error of the "raw" auto-spectral density estimate in the periodogram is as large as the estimate itself. It is therefore imperative to somehow increase the number of degrees of liberty n. This is achieved by smoothing the periodogram by either band averaging or segment averaging, or a combination of both. Band averaging consists of averaging the periodogram over m adjacent frequency bands on each side of the central frequency f_k of the estimate. The "smoothed" spectral estimate at f_k will now have $nd = 2m + 1$ degrees of freedom. Segment averaging consists of dividing the original series x_n in m segments, computing the auto-spectral density for each and averaging the m estimates at equal frequencies. The "smoothed" estimates in this case will also

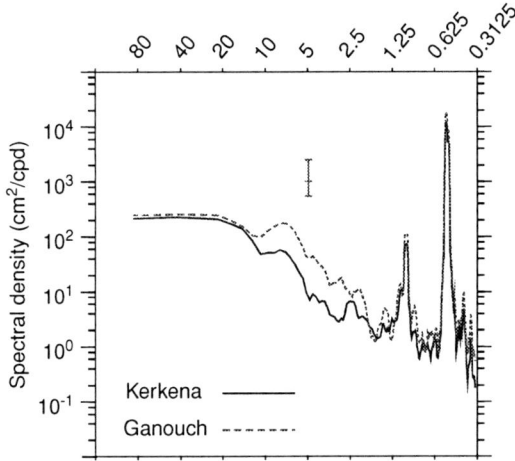

Figure 4. Auto-spectral variance density (cm^2/cpd) of sea level fluctuations at station 1 (Kerkena, blue) and station 2 (Ganouch, red) in the Gulf of Gabes, with 95% confidence intervals

have $nd = 2m$ degrees of freedom. Of course, several combinations of both approaches are possible, with time series segments overlapping or not, and so on. Normally, when the time series are relatively short, it is preferable to use the band-averaging approach if one is interested in the lower frequency fluctuations. If the interest is the high frequencies (e.g. inertial motion), seg-ment averaging may be appropriate. The error of the smoothed auto-spectral density estimate can then be computed as

$$\left[\frac{nd \cdot G_{xx\,k}}{\chi^2_{nd;\,\alpha/2}} \leq G_{xx\,k} \leq \frac{nd \cdot G_{xx\,k}}{\chi^2_{nd;\,1-\alpha/2}} \right].$$

For example, Figure 4 shows the auto-spectral variance densities of two sea level time series recorded around Boughara lagoon in the Gulf of Gabes. The 95% confidence intervals bar was computed by band averaging or smoothing the periodiogram over three frequency bands on each side of the frequency of the spectral estimate. This yields 14 degrees of freedom.

   Note that the first spectral estimate in the low frequency end is given at a period T, the fundamental period of the series (2035 hourly data points = 85 days), or at a frequency $1/T$. The last estimate in the high frequency end, not shown above, is given at a period of 2 h, or at the Nyquist frequency $1/2\Delta t$.

   These spectra are interpreted as follows. The highest levels of variance in both series are found in oscillations of semi-diurnal periodicities, proba-bly corresponding to the M2 tidal harmonic sea level component (see Sec-tion 2.6). Less variance is also centered at the diurnal period tidal components, probably close to the K1 harmonic component period. The semi-diurnal oscillations at Ganouch have slightly higher variance than those at Kerkena. The same is true for diurnal tides. Considerable energy is also found at lower frequencies, or longer periodicities, with more energy at Ganouch. However, given the 95% confidence intervals, no significant period (or frequency) of oscillation can be isolated in the lower frequency bands. These low frequency oscillations are generally produced by meteorological effects such as wind set-up, inverted barometers effects or other non-local forcing from the Mediterranean.

The cross-spectral variance density function is estimated as the autospectral density above except that the Fourier transforms now correspond to each of the two series being examined for similarities:

$$G_{xy\,k} = \frac{2}{N\,\Delta t}\left[X_k^* \cdot Y_k\right], \quad k = 0, 1, 2, \ldots, N/2.$$

The cross-spectral density reveals the frequencies at which a similarity exists between the two series. It is estimated by expressing the Fourier transform X_k as a complex numbers and evaluating:

$$G_{xx\,k} = \frac{2}{n \cdot \Delta t}\left[X_{rk}^2 + X_{ik}^2\right];$$

$$G_{yy\,k} = \frac{2}{n \cdot \Delta t}\left[Y_{rk}^2 + Y_{ik}^2\right];$$

and

$$G_{xy\,k} = C_{xy\,k} + iQ_{xy\,k} = \left|G_{xy\,k}\right|^2 e^{-i\theta_{xy\,k}}$$

where

$$C_{xy\,k} = \frac{2}{n \cdot \Delta t}\left[X_{rk} \cdot Y_{rk} + X_{ik} \cdot Y_{ik}\right]$$

and

$$Q_{xy\,k} = \frac{2}{n \cdot \Delta t}\left[X_{rk} \cdot Y_{ik} - X_i k \cdot Y_{rk}\right].$$

However, unlike the auto-spectral density, the cross-spectral density is a complex number that cannot be easily interpreted. As such, it can be expressed in terms of an amplitude and a phase. The amplitude is an equivalent of the cross-covariance function but at each frequency, and the phase is equivalent to the cross-covariance time lag. The *amplitude* of the cross-spectral density is

$$\left|G_{xy\,k}\right|^2 = C_{xy\,k}^2 + Q_{xy\,k}^2$$

and its *phase* is

$$\tan(\theta_{xy\ k}) = \frac{Q_{xy\ k}}{C_{xy\ k}}.$$

Since the cross-spectral density is often estimated between series with different units, e.g. sea levels and atmospheric pressure, or chlorophyll-a concentrations and light, its amplitude adopts strange units. For this reason, it is customary to normalize the amplitude by dividing it by the product of the auto-spectral densities of both series. This yields a number between zero and one, much like the correlation coefficient. It is the *coherency squared*, or coherence function, defined as

$$\gamma_{xy\ k}^2 = \frac{\left|G_{xy\ k}\right|^2}{G_{xx\ k} \cdot G_{yy\ k}}.$$

A coherence-phase analysis was performed between the low-passed atmospheric pressure time series and the low-passed sea level time series at station 2 (Ganouch). Results are shown in Figure 5.

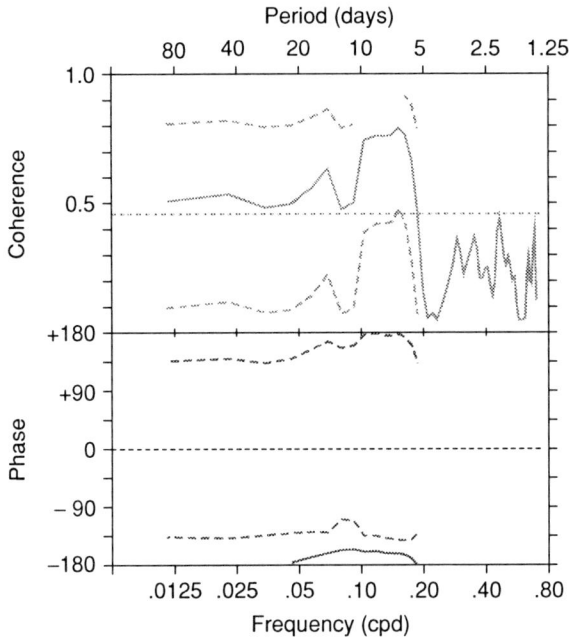

Figure 5. Coherence (top, blue) and relative phase (red, bottom) between low-frequency atmospheric pressure and Ganouch sea level time series. The coherence 95% confidence level (dotted) is at 0.46, while the 95% confidence intervals for coherence and relative phase are shown in dashed lines

Results indicate that the low-frequency sea level fluctuations at Ganouch respond to atmospheric as a lagged inverse barometer effect at periodicities longer than 6 days, and the response is coherent at periods between 6 and 10 days. The inverse barometer effect is evident from the out of phase (180 degrees) relation between both series, with pressure leading sea levels slightly between in the period band of 6–20 days. It is interesting to compare the above results to those obtained from the time domain cross-correlation analysis performed on the same series in Section 2 (Figure 2.12). Time domain analysis also revealed the inverse barometer effect but were not able to demonstrate significantly the return of the co-oscillation at periods between 6 and 10 days as shown here, although some non-significant evidence of this return was suggested by the TDA at periods around 7–8 days. This suggests that, in this case, FDA provides a better definition of the dynamic similarity between the series.

6.6.2. HARMONIC ANALYSIS

Harmonic analysis is used to extract deterministic signals and a major application is the study of tidal signals in the coastal oceans. Celestial mechanics has shown that the relative movements of the earth and the moon around themselves, around each other and around the sun are well known and do not change in time. Since ocean tides are produced by the gravitational attractions by the moon and the sun, the periods of the various tidal harmonic components can be predicted exactly. These deterministic signals will each have an amplitude and a relative phase that, in principle, change in space but not in time. The Table 1 below presents some of the tidal harmonic constituents found in most coastal ocean time series.

TABLE 1. List of tidal harmonic constituents, with corresponding periods and identifications

Harmonic constituent	Period (h)	Description
SSA	4383.00	Solar, semi-annual
MM	661.30	Lunar, monthly
MSF	354.48	Luni-solar, fortnightly
MF	327.90	Lunar, fortnightly
Q1	26.87	Lunar, elliptic diurnal
O1	25.82	Lunar principal, diurnal
P1	24.07	Solar principal, diurnal
K1	23.93	Luni-solar, diurnal
N2	12.66	Lunar elliptic, semi-diurnal
M2	12.42	Lunar principal, semi-diurnal
S2	12.00	Solar principal, semi-diurnal
K2	11.97	Luni-solar, semi-diurnal
M4	6.21	Lunar composite, quarter-diurnal
MS4	6.10	Luni-solar composite, quarter-diurnal

A sea level time series z_n measured at the mouth of a tidal lagoon can be represented by a mean value z_0, the sum of $j = 1, \ldots, J$ tidal periodic constituents of constant amplitudes A_j frequencies σ_j and relative phases ϕ_j, and non-tidal residual components as

$$z_j = z_0 + \sum_{j=1}^{N} A_j \cos\left(2\pi\left(\sigma_j n \Delta t - \phi_j\right)\right) + \text{non-tidal residuals.}$$

Here, σ_j is the phase of a tidal harmonic j at longitude $0°$ in the UT zone and ϕ_j is the phase lag of this harmonic to its phase at longitude $0°$. The objective of harmonic analysis is to find unique and optimal values for the A_j and ϕ_j estimates that best reproduce the tidal signal in z_n. The optimization criteria used in the method is the method of least squares.

Results from the application of harmonic analysis to the sea level time series at station 2 are presented in Table 2 Starting from the top, these results indicate that:

• The number of good data points is 2035 out of 2035;
• At a time step of 1 h, the period of observation is 84.79 days long;

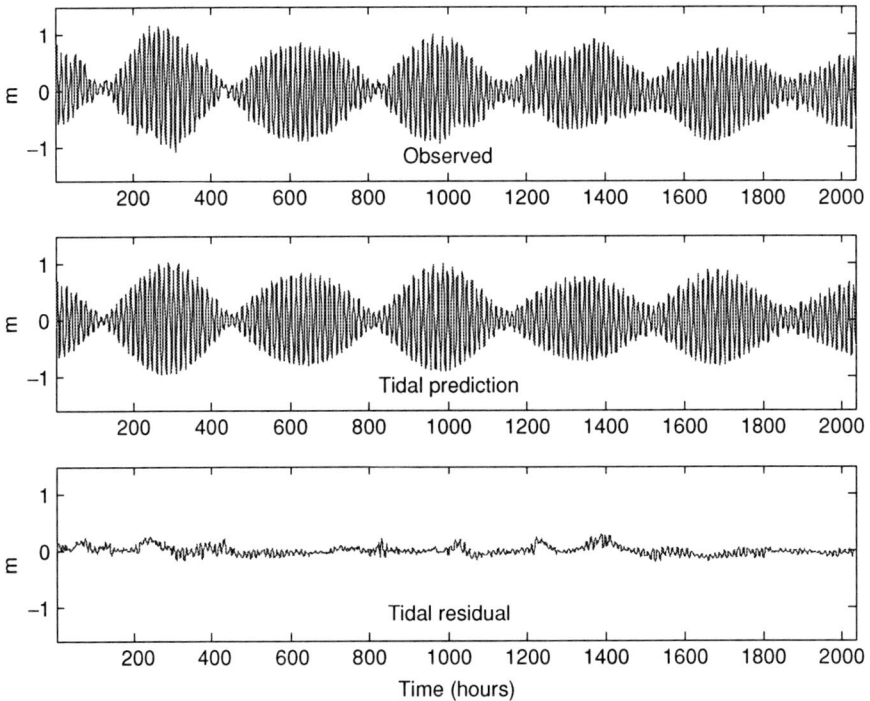

Figure 6. Observed hourly sea level time series at Ganouch (top), tides predicted from the harmonic analysis (center) and the tidal residuals (bottom)

TABLE 2. Harmonic analysis results for sea levels at station 2 (n_{obs} = 2035, n_{good} = 2035, record length (days) = 84.79; start time: 05-Apr-2003 13:00:00; $x0$ = −0.000965, x trend = 0; var (x) = 0.18555, var (xp) = 0.17913, var (xres) = 0.0064223; percent var predicted/var original = 96.5%).

Harmonic.	Frequency	Amplitude	Error	Phase	Error
MM	0.0015122	0.0365	0.037	260.32	62.96
MSF	0.0028219	0.0197	0.030	90.74	100.25
ALP1	0.0343966	0.0016	0.003	261.60	130.73
2Q1	0.0357064	0.0024	0.004	46.94	101.82
*Q1	0.0372185	0.0077	0.005	72.73	35.57
*O1	0.0387307	0.0094	0.004	93.47	27.99
NO1	0.0402686	0.0010	0.003	237.75	166.03
*P1	0.0415526	0.0114	0.005	22.67	24.52
*K1	0.0417807	0.0345	0.004	15.60	7.08
J1	0.0432929	0.0013	0.003	98.14	173.15
OO1	0.0448308	0.0025	0.003	79.55	80.63
UPS1	0.0463430	0.0007	0.002	322.78	183.33
EPS2	0.0761773	0.0089	0.016	109.51	143.44
MU2	0.0776895	0.0154	0.019	106.39	87.86
*N2	0.0789992	0.0796	0.021	66.54	14.03
*M2	0.0805114	0.5049	0.021	73.02	2.44
L2	0.0820236	0.0307	0.024	52.53	41.82
*S2	0.0833333	0.3092	0.023	98.00	4.50
*K2	0.0835615	0.0842	0.018	120.40	11.67
ETA2	0.0850736	0.0134	0.014	132.57	78.18
MO3	0.1192421	0.0021	0.002	129.66	66.61
*M3	0.1207671	0.0064	0.003	303.79	25.29
MK3	0.1222921	0.0021	0.003	292.17	75.83
SK3	0.1251141	0.0020	0.002	190.82	82.95
*MN4	0.1595106	0.0063	0.003	80.21	27.97
*M4	0.1610228	0.0087	0.003	105.32	21.85
SN4	0.1623326	0.0013	0.002	327.42	115.43
MS4	0.1638447	0.0009	0.002	128.58	136.95
S4	0.1666667	0.0031	0.003	19.62	62.82
2MK5	0.2028035	0.0001	0.001	116.59	210.87
2SK5	0.2084474	0.0005	0.001	209.29	137.23
2MN6	0.2400221	0.0002	0.001	130.95	203.32
M6	0.2415342	0.0004	0.001	255.98	155.39
2MS6	0.2443561	0.0017	0.001	263.51	48.01
*2SM6	0.2471781	0.0016	0.001	245.92	45.78
3MK7	0.2833149	0.0001	0.001	62.25	263.97
M8	0.3220456	0.0001	0.000	73.76	198.13

- The series starts at 13:00 h (UT) on April 5, 2003;
- The Rayleigh criteria used is 1;
- The mean is −0.000955 (practically zero) and there is no trend;
- The variance of the series is 0.18555 m^2;
- The variance explained is 0.17913 m^2;
- The residual variance is 0.0054223 m^2;
- The percent variance explained is 96.5 %;
- The number of significant estimates (names with *) is for 12 out of 37 harmonics;
- The dominant tidal constituent in the Gulf is the M2 semi-diurnal tidal constituent;
- It has an amplitude of 0.505 ± 0.02 m and a phase relative to UT of
- It has a phase relative to that at longitude zero (UT) of 73.02 ± 2.44 degrees;
- The next significant amplitudes are those of S2, N2, K1, P1, etc harmonic constituents.

Results in Table 2 can finally be used to reconstruct or predict the tidal signal present in the observations. Figure 6 shows the original (hourly) Ganouch sea level time series, the tidal signal predicted by the harmonic analysis and the tidal residual obtained by subtracting the prediction from the observation.

6.7. Summary

Time series analysis is a useful tool to describe the response of a lagoon to various forcing functions. Most series can be represented in the time domain or in the frequency domain. This has led to the development of corresponding time domain and frequency domain analysis methods. The time domain methods include correlation analysis, regression analysis and empirical orthogonal function analysis. Frequency domain methods include spectral analysis, coherency and phase analysis, frequency domain EOF analysis, harmonic analysis and wavelet analysis. Some of these methods have been described and examples of their application to real lagoon series were presented.

7. DISCRETIZATION TECHNIQUES

Georg Umgiesser
ISMAR-CNR, Venezia, Italy

7.1. Introduction

Computational fluid dynamics is concerned with the numerical solution of
the differential equations of fluid dynamics. The reason why this filed is so
important is because there are only few examples of flows where we can
find analytical solutions of the flow field. When applying these equations to
real world examples we have the problem that the flows become extremely
complex, impossible to be described by closed formulas.

For the resolution of these real world problems we have therefore to resort to numerical modeling. With this technique we are able to solve even
differential equations in complicated domains. However, care has to be taken
to guarantee that the solution obtained is actually approaching the solution of
the equation. Problems such as accuracy or stability have to be investigated.
This and other points are the topic of the following pages. Only a short introduction can be given to the complex field of computational fluid dynamics
that should, however, be enough to give an overview and an idea of the issues
connected to these discretization techniques.

7.2. The Basics

In the following the basic principles of discretization will be presented.

7.2.1. DISCRETIZATION

When an equation is discretised the temporal and spatial domain is divided into
discrete intervals and the variables are expressed on these discrete intervals.
For an differential equation that is only depending on the time t (an ordinary
differential equation), the time t is discretised into regular intervals Δt and
the dependent variable S is written on these intervals as $S(t) = S(n\Delta t) = S^n$
where we say that S is defined on time level n. We then express a derivative
in t in a convenient way, e.g.,

$$\frac{\partial S}{\partial t}(t) \to \frac{S(t + \Delta t) - S(t)}{\Delta t} = \frac{S^{n+1} - S^n}{\Delta t}.$$

I. E. Gonenc et al. (eds.), Assessment of the Fate and Effects of Toxic Agents on Water Resources, 143–170.
© 2007 *Springer.*

$$
\begin{aligned}
&\text{n=4} \quad \text{————————} \\
&\text{n=3} \quad \text{————————} \\
&\text{n=2} \quad \text{————————} \\
&\text{n=1} \quad \text{————————} \\
&\text{n=0} \quad \text{————————}
\end{aligned}
$$

Figure 1. Discretization of the time axis. Time levels are separated by the constant time step Δt

Now the derivative has been expressed as differences of the discrete values of S on the numerical grid given by the division of the time axis in regular intervals. Figure 1 shows an example of this division. An irregular division is also possible, but is out of the scope of this introduction and will not be treated here in this chapter. If there is also a dependence in x then also the spatial axis is divided into intervals Δx and the same procedure of expressing the variable on discrete spatial steps is carried out as for the temporal scale. We then have $S(t, x) = S(n\Delta t, j\Delta x) = S_j^n$ where the subscript gives the spatial and the superscript the temporal index of the discretization.

After all variables have been expressed on the discrete numerical grid the differential equation has been transformed into an algebraic equation, and the solution of the discrete values of the variable on the new time step, starting from the known values of the old time step, can be tackled with the usual methods of algebra and the solution of linear systems.

7.2.2. REQUIREMENTS

A priori some requirements must be fulfilled by the discrete equation and its solution to the differential equation. We will present them here in advance, even if they are discussed thoroughly only in later sections, when specific examples of equations are given.

The three fundamental assertions on which numerical modeling is based are

- Consistency.
- Accuracy.
- Stability.

What concerns consistency the original differential equation should be recovered in the limit when the temporal and spatial discretization step is going to zero. If we refer to the example given above, since we have

$$
\frac{\partial S}{\partial t}(t) = \lim_{\Delta t \to 0} \frac{S(t + \Delta t) - S(t)}{\Delta t}
$$

the representation of the time derivative as

$$\frac{\partial S}{\partial t}(t) = \left(\frac{\partial S}{\partial t}\right)^n \rightarrow \frac{S^{n+1} - S^n}{\Delta t}$$

is therefore a consistent discretization.

Accuracy is given by the truncation error of the discrete representation of the derivative. Since every variable can be expressed as a development of the variable into a Taylor series, the accuracy of the discretization is determined by the truncation error of the term. If we apply this to the derivative above, it can be shown (see later) that the discretization can be written as

$$\left(\frac{\partial S}{\partial t}\right)^n = \frac{S^{n+1} - S^n}{\Delta t} + O(\Delta t)$$

where $O(\Delta t)$ gives the truncation order of the discrete representation of the derivative, which is of order Δt. The accuracy of this discretization is therefore said to be of first order.

It will be shown later that basically one-side discretizations of derivatives always correspond to first-order accuracy, whereas central representations will give second-order accuracy. More details are given later.

The last point is even more crucial. Stability may be defined as the requirement that the computed solution should not diverge. However, stability has to be defined anew for every equation. If the solution to the equation is bounded, we may ask for a bounded numerical solution and stability would assure that the computed values will remain bounded too. If we know that a variable should be always greater than 0, stability requires that also the computed values are greater than 0. Finally, if the initial conditions are changed by an infinitesimal amount, we would hope that the computed solutions would be similar, if the original equation has the same property.

Stability of a discretization therefore depends both on the original equation and its discretization, and must be investigated separately for each equation. In the following section much care is devoted to this crucial topic.

7.3. The Decay Equation—A Simple Example

To gradually introduce the reader to the concept of discretization, a simple example is chosen that will, however, show much of the features that will be needed also later with the more complex equations. As a start an ordinary differential equation is chosen, which greatly simplifies the treatment of the discretization process.

7.3.1. BASIC DISCRETIZATION FORMULAS

The decay equation reads

$$\frac{\partial u}{\partial t} = -Ru \tag{1}$$

where R is the decay parameter, t is time and u is the state variable. It might stand for a culture of bacteria that are dying by a rate of R, a radioactive substance decaying or the current velocity in a channel that is slowing down due to the action of friction. An important feature of this equation is that we know its analytical solution, which is

$$u = u_0 e^{-Rt}$$

where u_0 is the initial concentration of the state variable. As can be seen, the decay follows an exponential decay.

In order to discretize the equation we have to express the derivative in (1) with a convenient difference equation. The most straightforward choice for this derivative is

$$\frac{\partial u}{\partial t}(t) \quad \rightarrow \quad \lim_{\Delta t \to 0} \frac{u(t + \Delta t) - u(t)}{\Delta t}$$

which uses the definition of the derivative. This type of discretization is clearly consistent, because for Δt going to 0 the difference representation will approach the derivative. Since we are moving on discrete time intervals that are Δt apart, we might as well label the state variable u with the number of the time level, and therefore we can write

$$u^n = u(n\Delta t) \qquad u^{n+1} = u((n+1)\Delta t) \tag{2}$$

where the superscript of n stands for the time level it represents. With these definitions we can rewrite the differential equation in a discrete form as follows

$$\frac{\partial u}{\partial t} = -Ru \qquad \rightarrow \qquad \frac{u^{n+1} - u^n}{\Delta t} = -Ru^n.$$

Please note that we had to decide on where to evaluate the right-hand side of the decay equations. One possibility, the one we have chosen, is to evaluate the term on the old time level. However, there are other possibilities that will lead to different schemes and will affect the characteristics of the solution.

Other solutions are summarized in the following list:

$$\frac{\partial u}{\partial t} = -Ru \quad \rightarrow \quad \frac{u^{n+1} - u^n}{\Delta t} = -Ru^n \qquad \text{forward in time,}$$

$$\frac{\partial u}{\partial t} = -Ru \quad \rightarrow \quad \frac{u^{n+1} - u^n}{\Delta t} = -Ru^{n+1} \qquad \text{backward in time,}$$

$$\frac{\partial u}{\partial t} = -Ru \quad \rightarrow \quad \frac{u^{n+1} - u^n}{\Delta t} = -R\frac{(u^n + u^{n+1})}{2} \qquad \text{centered in time,}$$

$$\frac{\partial u}{\partial t} = -Ru \quad \rightarrow \quad \frac{u^{n+1} - u^{n-1}}{2\Delta t} = -Ru^n \qquad \text{centered in time.}$$

$$(3)$$

The forward scheme is also called an explicit scheme, since we can explicitly compute the value of our concentration on the new time level. The backward scheme is also called an implicit scheme, because now the new value is only implicitly defined by the equation. In this simple case it is trivial to rewrite the implicit scheme to give us explicitly the value of the concentration on the new time level, but in more complex situations (partial differential equations) this is not any more possible and a linear system of equations has to be solved.

The last two discretizations are centered schemes, because the right-hand side is defined on a central point with respect to the time operator. The first one achieves this by averaging the right-hand side between the two levels, and the second one uses a third time level $n - 1$ in order to center the time difference. Even if they are both centered schemes, we will see shortly that their properties will be largely different.

7.3.2. ACCURACY

In order to study the accuracy of the schemes presented above we have to introduce the concept of Taylor series. A Taylor series is a development of a function in one point in terms of derivatives of the function nearby. The standard formula reads

$$u(t + \Delta t) = u(t) + \Delta t \frac{\partial u}{\partial t}(t) + \frac{\Delta t^2}{2}\frac{\partial^2 u}{\partial t^2}(t) + O\left(\Delta t^3\right).$$

In this formula the term we want to express through difference formulas is the first derivative which is the second term on the right-hand side. If we solve for this term, using $t = n\Delta t$ and also (2) we get

$$\frac{\partial u}{\partial t}(t) = \frac{u^{n+1} - u^n}{\Delta t} - \frac{\Delta t}{2}\frac{\partial^2 u}{\partial t^2}(t) - O(\Delta t^2) = \frac{u^{n+1} - u^n}{\Delta t} - O(\Delta t) \qquad (4)$$

The last two terms have been replaced simply by $O(\Delta t)$ which is the lower order term of both. As can be seen the forward scheme has been recovered through this procedure. It can be also seen that the forward scheme is of first order, e.g., it is linear in Δt. What this exactly means is that, if you let go your time step to 0, than the error you make when replacing $\dfrac{\partial u}{\partial t}$ with $\dfrac{u^{n+1} - u^n}{\Delta t}$ goes to zero in a linear way. Therefore we call the forward scheme a first-order accurate scheme, or simply a first-order scheme.

In the same way an expression for a backward scheme can be derived. We now expand $u(t - \Delta t) = u^{n-1}$ around t. The result is

$$u(t - \Delta t) = u(t) - \Delta t\frac{\partial u}{\partial t}(t) + \frac{\Delta t^2}{2}\frac{\partial^2 u}{\partial t^2}(t) - O\left(\Delta t^3\right)$$

which gives the formula

$$\frac{\partial u}{\partial t}(t) = \frac{u^n - u^{n-1}}{\Delta t} + \frac{\Delta t}{2}\frac{\partial^2 u}{\partial t^2}(t) - O(\Delta t^2) = \frac{u^n - u^{n-1}}{\Delta t} + O(\Delta t). \qquad (5)$$

Please note that this seems different from the formula given in equation (2b). However, both are backward formulas since the equation here developed $\frac{\partial u}{\partial t}$ around time level n and the one in equation (2b) is referred to time level $n + 1$.

We can now easily derive a formula for a centered scheme. To this purpose we subtract (5) from (4). This results in

$$u^{n+1} - u^{n-1} = 2\Delta t\frac{\partial u}{\partial t}(t) + O\left(\Delta t^3\right)$$

or after solving for $\dfrac{\partial u}{\partial t}$

$$\frac{\partial u}{\partial t}(t) = \frac{u^{n+1} - u^{n-1}}{2\Delta t} + O\left(\Delta t^2\right).$$

Now we see that the centered representation is of second order. Therefore, a centered scheme, all other properties being equal, is normally preferred to first-order schemes like the backward and forward scheme, because its accuracy is superior to the others.

7.3.3. STABILITY

Having dealt with consistency and accuracy, we now have to study the stability properties of our discretizations of the decay equation. Stability of a discretization depends crucially on the equation that is to be investigated. Since we are dealing here with the decay equation, we really would like to impose on our numerical solution that the value computed in the new time level is not bigger than the one on the actual time level. Here bigger must be seen as bigger in modulus. Therefore, we could impose that $|u^{n+1}| \leq |u^n|$. This would satisfy the stability requirements in the mathematical sense. However, looking at the physical solution of the decay equation it is clear that, if the solution is positive at one point, it should stay positive forever. Therefore, physical stability requires that $0 \leq u^{n+1} \leq u^n$. If we define the amplification factor $\lambda = u^{n+1}/u^n$ then we see that for physical stability $0 \leq \lambda \leq 1$ and for mathematical stability $|\lambda| \leq 1$ or $-1 \leq \lambda \leq 1$.

If we start from the forward scheme (2a) and solve for u^{n+1} we find

$$u^{n+1} = (1 - \alpha)u^n \quad \text{with} \quad \alpha = \Delta t R$$

and therefore $\lambda = 1 - \alpha$. If we impose physical stability we see that, with the definition of α, we have $\alpha = \Delta t R \leq 1$ or $\Delta t \leq 1/R$. We therefore see that the stability criterion transforms into a limitation for the allowable time step. The bigger the decay parameter R is, the smaller the time step.

We want to look further into the physical meaning of this restriction. If we compare the solution of the decay equation $u = u_0 e^{-Rt} = u_0 e^{-t/\tau}$ where $\tau = 1/R$ is a time scale that characterizes the physical process of decay, with the stability criterion, then we see that, for stability, we should have $\Delta t \leq \tau$. Formulating this differently, we can say that, for physical stability, the time step of discretization should be always less then the time scale that characterizes the process of decay. In the case of mathematical stability we see that the criterion is somewhat less restrictive, asking only for $\Delta t \leq 2\tau$.

Typical solutions of this decay equation can be found in Figure 2. As can be seen, the solutions that satisfy the physical stability criterion are always positive and have lower values with respect to the analytical solution. However, mathematically stable solutions show values that oscillate more or less strongly between positive and negative values. In the limiting case (marginal stability) of $\lambda = 0$ the solution drops in one time step to zero and then stays there, whereas for $\lambda = -1$ (marginal mathematical stability) the solution oscillates between the two values $\pm u_0$ forever and does not decay to smaller values.

Instead of starting from the forward discretization and deriving the explicit formula for the value u^{n+1} we can also start from the backward formula given before that defines the value of u^{n+1} implicitly

$$\frac{u^{n+1} - u^n}{\Delta t} = -Ru^{n+1}.$$

This is called an implicit discretization of the decay equation, because the solution at the new time level is defined only implicitly. A solution of these

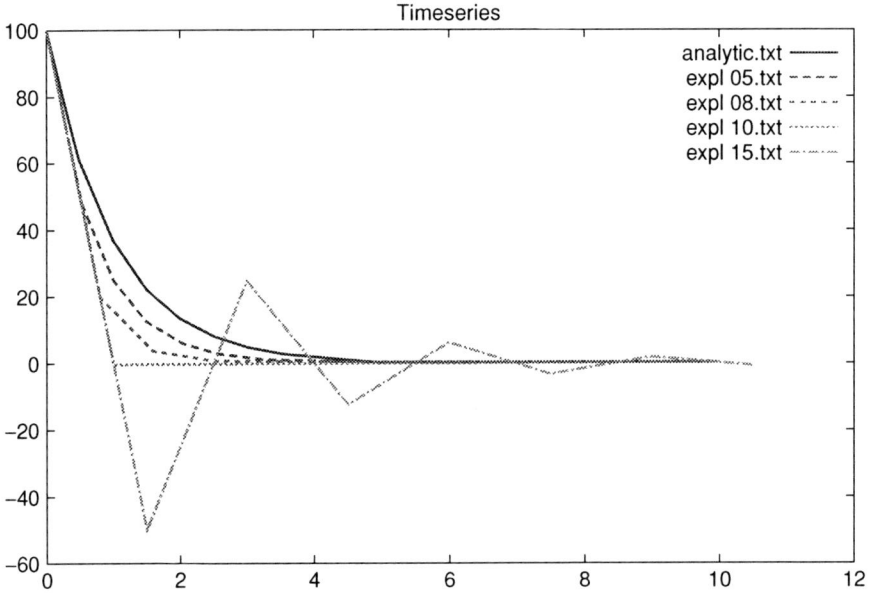

Figure 2. Explicit solutions of the decay equation. The parameter α varies between 0.5 and 1.5

implicitly defined discretizations normally requires the solution of a system of equations. However, in case of the decay equation the, it can still be easily solved to give

$$u^{n+1} = \frac{1}{(1 + \Delta t R)} u^n.$$

This leads to $\lambda = 1/(1 + \Delta t R)$ for the amplification factor, which is, for all values of the time step $0 \leq \lambda \leq 1$. Therefore, in this case, physical stability is guaranteed for every choice of the time step. Therefore the implicit scheme is said to be an unconditionally stable scheme, a nice property where the choice of the time step is dictated only by accuracy requirements and not by the need of fulfilling a stability criterion. On the other hand, the forward scheme treated above, is called an explicit scheme, where the solution of the new time step is directly available even in more complicated cases as the decay equation. In this case the discretization leads to conditionally stable scheme, where the time step must fulfill some criterion and cannot be chosen arbitrarily.

Figure 3 summarizes these findings. Explicit solutions can be found below the analytical solution, whereas the solutions of the implicit discretization are

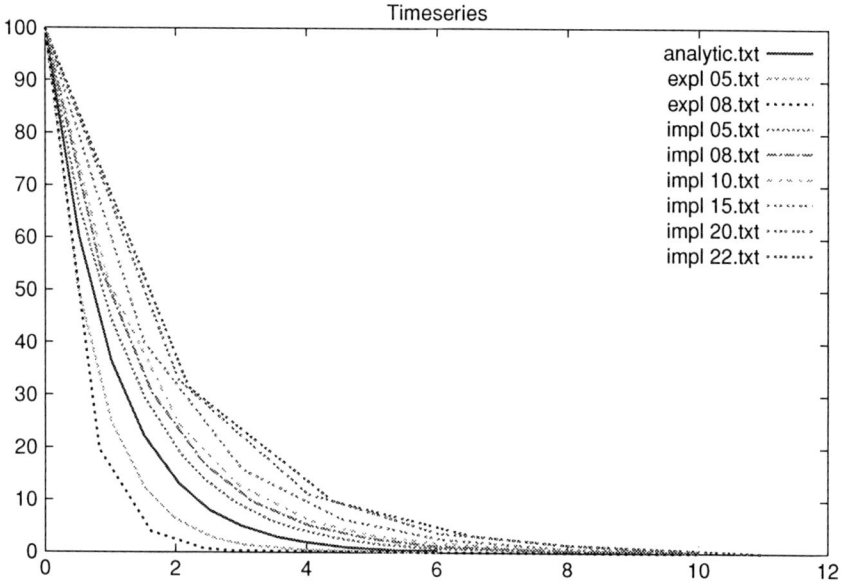

Figure 3. Explicit and implicit solutions of the decay equation. The parameter α varies between 0.5 and 0.8 for the explicit and between 0.5 and 2.2 for the implicit case

all above it. This is also the reason why the implicit discretization is stable for all time steps, because it can never fall below the analytical solution.

7.4. Inertial Motion—A System of Equations

We will now apply what we have learned to a slightly more complicated topic: a system of equations, but still with only the time variable as the independent variable. A typical example of this kind of equations is the inertial motion of a water particle that moves under the only influence of the Coriolis force, without any pressure forces present.

The equations that govern inertial motion are

$$\frac{du}{dt} - fv = 0$$

$$\frac{dv}{dt} + fu = 0$$

where u and v are the current velocity and f is the Coriolis parameter. Solutions of this equation are circular paths that turn around without any damping of the current speed.

7.4.1. MATRICES AND EIGENVALUES

When we apply again the explicit scheme and a forward time stepping scheme we can discretize above equations as

$$\frac{u^{n+1} - u^n}{\Delta t} = fv^n,$$

$$\frac{v^{n+1} - v^n}{\Delta t} = -fu^n.$$

Solving for the variables at the new time level, and putting it in matrix form we have

$$\begin{pmatrix} u \\ v \end{pmatrix}^{n+1} = \begin{pmatrix} 1 & +\Delta t f \\ -\Delta t f & 1 \end{pmatrix} \begin{pmatrix} u \\ v \end{pmatrix}^n.$$

This is a matrix equation that advances the values of the variables from time step n to time step $n + 1$. Before we can say anything about the stability of this system of equations, we have to have a little deeper look at the properties of matrices.

From linear algebra we know that a linear system can be transformed into a more convenient form by a unitarian matrix. A unitarian matrix is one that does not change the norm of a vector: $\|x\| = \|Ux\|$. If the matrix system is given by $x^{n+1} = Ax^n$ than by multiplying with the unitarian matrix we can transform this into $Ux^{n+1} = UAU^{-1}Ux^n$. If we now define a new matrix $D = UAU^{-1}$ and vector $y = Ux$ we can rewrite the above system as

$$y^{n+1} = Dy^n \text{ with } D = \begin{pmatrix} \lambda_1 & 0 \\ 0 & \lambda_2 \end{pmatrix}.$$

The matrix A has now been transformed into a diagonal matrix D. This is a big advantage, because now the matrix equation separates into two single equations $y_1^{n+1} = \lambda_1 y_1^n$ and $y_2^{n+1} = \lambda_2 y_2^n$. From the last section we know that the scheme is stable if both amplification factors $|\lambda| \leq 1$. But these amplification factors are exactly the eigenvalues of the matrix A. Therefore, the problem of finding a stability criterion for the matrix equation has been reduced to finding the eigenvalues of the original matrix. If we can assure that these eigenvalues are smaller or equal to 1 we have solved the problem.

From linear algebra we know that in order to find the eigenvalues of a matrix we have to proceed as following: The defining equation of the eigenvalues $Ax = \lambda x$ is transformed to $(A - \lambda I)x = 0$ (I is the unit matrix), and the condition that this is true (for x not the null vector) is that the matrix $(A - \lambda I)$ is singular, or that its determinant is 0: $|A - \lambda I| = 0$.

The values of λ that fulfill this conditions are called the eigenvalues of the system.

7.4.2. VARIOUS SCHEMES

We can now apply the theory of the last paragraph to the stability analysis of the discretised equation that governs inertial motion. We therefore have to find the eigenvalues of the matrix that describes the advance of the solution of the old to the new time step:

$$\begin{pmatrix} u \\ v \end{pmatrix}^{n+1} = \begin{pmatrix} 1 & +\Delta t f \\ -\Delta t f & 1 \end{pmatrix} \begin{pmatrix} u \\ v \end{pmatrix}^{n} \Rightarrow \begin{vmatrix} 1 - \lambda & +\Delta t f \\ -\Delta t f & 1 - \lambda \end{vmatrix} = 0.$$

The determinant can be solved easily to give $(1 - \lambda)^2 + (\Delta t f)^2 = 0$ or $\lambda = 1 \pm i\Delta t f$. Here i is the imaginary number. Since we are only interested in the module of λ we have to compute $|\lambda|^2 = (1 + i\Delta t f)(1 - i\Delta t f) = 1 + (\Delta t f)^2 > 1$. As we can see, for all values of the time step and $f > 0$ this scheme is unconditionally unstable.

What does this mean? Well, it simply means that for whatever small value of the time step we use, the solution will always diverge. It is therefore not possible to use this scheme in our computation of inertial motion.

In the treatment of the decay equation we have seen that implicit schemes had the property that they were unconditionally stable for any time step. We can therefore try to apply an implicit scheme to the equation of inertial motion. If we do this we obtain

$$\frac{u^{n+1} - u^n}{\Delta t} = f v^{n+1}$$

$$\frac{v^{n+1} - v^n}{\Delta t} = -f u^{n+1}$$

and after simplifying

$$\begin{pmatrix} 1 & -\Delta t f \\ +\Delta t f & 1 \end{pmatrix} \begin{pmatrix} u \\ v \end{pmatrix}^{n+1} = \begin{pmatrix} u \\ v \end{pmatrix}^{n}.$$

If we now invert the matrix and solve for the amplification factor we obtain $|\lambda|^2 = \frac{1}{1+(\Delta t f)^2} < 1$. It can be seen that this value is always smaller than 1. Therefore, no matter what value the time step assumes, the scheme is always stable. This is called an unconditionally stable scheme. However, the scheme is always damping, i.e., the modulus of current velocities is decreasing. Clearly this is an artifact of the numerical scheme, because we know that in the physical solution the modulus of the velocity should be always constant. Through the application of the numerical scheme we have therefore also introduced some numerical damping.

However, we can do better. If we use a scheme that is centered in time, we might be able to combine the properties of the explicit and implicit scheme.

We can discretize the equation as

$$\frac{u^{n+1} - u^n}{\Delta t} = f \frac{v^n + v^{n+1}}{2}$$
$$\frac{v^{n+1} - v^n}{\Delta t} = -f \frac{u^n + u^{n+1}}{2}$$

where we have actually centered the right-hand side at the time level $n + 1/2$, in between the time derivative, by taking an average between the old and the new time step. Without going into details, we can again solve easily for the values of the new time level by inverting a 2×2 matrix, and can then determine the amplification factor, just as we have done before with the other schemes. The result is surprising, because we get $|\lambda|^2 = 1$, which means that the scheme is unconditionally stable for any time step. Moreover, now the amplification factor is exactly 1 which is what we would expect from the physical solution. Therefore, the numerical solution does not alter the characteristics of the physical one, a very desirable property.

7.5. The Diffusion Equation—The Spatial Dimension

We now turn to a new type of equation, the diffusion equation, which is given by

$$\frac{\partial S}{\partial t} = -K \frac{\partial^2 S}{\partial x^2}$$

where S is the variable and K the diffusion parameter. This equation differs from the preceding decay equation due to the fact that now we have a real partial differential equation because of the derivative with respect to the space coordinate x. The treatment of these must be described first.

7.5.1. FIRST AND SECOND SPATIAL DERIVATIVE

We make reference to Figure 4 where the space coordinate is shown. This space coordinate is discretised into regular pieces with a distance of Δx, the

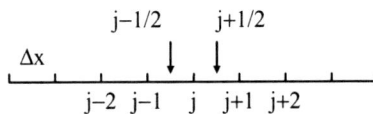

Figure 4. Spatial discretization for the diffusion equation

spatial step. The values for the variable can be given in this discrete points as $S(j\Delta x)$, or for short also as $S_j = S(j\Delta x)$.

A first-order derivative can be constructed in the same way as it has been done in the case of the time derivative. Therefore we can write

$$\left(\frac{\partial S}{\partial x}\right)_j \rightarrow \begin{cases} \dfrac{S_{j+1} - S_j}{\Delta x} & \text{forward in space} \\[2ex] \dfrac{S_j - S_{j-1}}{\Delta x} & \text{backward in space} \\[2ex] \dfrac{S_{j+1} - S_{j-1}}{2\Delta x} & \text{centered in space} \end{cases}$$

where the order of accuracy can be deduced from what has been said for the time discretization: first order for the forward and backward scheme, and second order for the centered scheme.

However, in the case of the diffusion equation, we need to define a second-order derivative. In this case it is advantageous to define the first-order derivatives in points half way between the discretised points, even if no values for S is available. If this is done we can write

$$\left(\frac{\partial^2 S}{\partial x^2}\right)_j = \frac{\partial}{\partial x}\left(\frac{\partial S}{\partial x}\right) = \frac{\left(\dfrac{\partial S}{\partial x}\right)_{j+1/2} - \left(\dfrac{\partial S}{\partial x}\right)_{j-1/2}}{\Delta x}$$

$$= \frac{\dfrac{S_{j+1} - S_j}{\Delta x} - \dfrac{S_j - S_{j-1}}{\Delta x}}{\Delta x} = \frac{S_{j+1} - 2S_j + S_{j-1}}{\Delta x^2}$$

This is the standard discretization of a second-order derivative. Since it is a centered scheme, it is also of second-order accuracy.

7.5.2. EXPLICIT AND IMPLICIT SCHEMES

Once the diffusion equation has been discretized in the spatial domain, we still have to apply a time stepping scheme, as has been done before with the decay equation. Here we will show two possibilities, the explicit forward scheme, and the implicit backward scheme. If the time operators are applied, the equations read

$$\frac{S_j^{n+1} - S_j^n}{\Delta t} = K\left(\frac{S_{j+1} - 2S_j + S_{j-1}}{\Delta x^2}\right)^n \qquad \text{explicit}$$

$$\frac{S_j^{n+1} - S_j^n}{\Delta t} = K\left(\frac{S_{j+1} - 2S_j + S_{j-1}}{\Delta x^2}\right)^{n+1} \qquad \text{implicit}$$

The difference between the two schemes is simply that in the explicit scheme the right-hand side is taken at the old time level, whereas in the implicit scheme it is taken at the new time level. With the decay equation there was really no big difference between the two schemes, because both could easily be solved for the variable at the new time step. However, now the story is different. The explicit scheme can still be solved easily for S_j^{n+1}, but this is not possible anymore for the implicit scheme. In this case there are three unknowns in the equation $(S_j^{n+1}, S_{j+1}^{n+1}, S_{j-1}^{n+1})$ and therefore a system of equations has to be solved. In this case the solution of the implicit scheme is much more complicated than in the explicit case.

7.5.3. STABILITY

In the case of the decay equation we were arguing with physical arguments what stability was meaning. We will do this now also for the diffusion equation. Later on we will see how this problem of stability can be solved in mathematically satisfying way, but for now we will again apply physical reasoning to deduce a stability criterion.

Starting from the explicit scheme above we have, after multiplying with Δt

$$S_j^{n+1} - S_j^n = \frac{\Delta t K}{\Delta x^2} \left(S_{j+1} - 2S_j + S_{j-1} \right)^n$$

and with the definition $\alpha = \dfrac{2\Delta t K}{\Delta x^2}$ and $\tilde{S}_j = \dfrac{S_{j+1} + S_{j-1}}{2}$ this becomes

$$S_j^{n+1} = (1 - \alpha)S_j^n + \alpha \left(\frac{S_{j+1}^n + S_{j-1}^n}{2} \right) = (1 - \alpha)S_j^n + \alpha \tilde{S}_j^n. \tag{6}$$

This formula can be interpreted as a simple linear interpolation between the central point j and the arithmetic mean of the points surrounding the central point. Since diffusion is a process that is trying to smooth differences, we know that the new value of S at point j should certainly not be higher than the maximum or lower than the minimum at point j and its surrounding points $j + 1$ and $j - 1$. Therefore we must ask for the formula above that it should be a real interpolation, and not an extrapolation, which is the same as asking for $0 \leq \alpha \leq 1$. The fact that α must be positive simply says that the diffusion parameter must be positive and negative diffusion is an unphysical process. Therefore, in order to have a stable scheme, the stability criterion for the explicit diffusion scheme is given by $\alpha \leq 1$:

$$\frac{2\Delta t K}{\Delta x^2} \leq 1 \quad \text{or as a criterion for the time step } \Delta t \leq \frac{\Delta x^2}{2K}.$$

7.5.4. THE VON NEUMAN STABILITY ANALYSIS

The stability analysis given for the diffusion equation above is unsatisfactory for two reasons: first because it needs some reasoning like the maximum principle that might be difficult to apply in the case of higher dimensions or more complicated equations, and second because it only works for the explicit case. We cannot repeat the same procedure also for the implicit case. We will therefore introduce a new way of analyzing the stability of equations, which is the von Neuman stability analysis. This will be also helpful in case of the transport equation. The key to understanding the von Neuman stability analysis is that every periodic function can be developed in a Fourier series $S(x) = \sum S_k e^{ikx}$ where the complex exponential functions represent basically sine and cosine functions and the S_k are the (complex) amplitudes of the sinus waves. In the case of the discrete functions used in the discretization process we can write $S_j = S(j\Delta x) = \sum S_k e^{ikj\Delta x}$. The advantage of this approach is that now the sinusoidal functions to be considered are much simpler and can be handled more easily. Now stability means that for every single component of S the scheme should be stable. If we can prove this, then it is clear that also the sum of all partial waves will be stable.

Therefore it is enough to test stability only for one component of the special exponential functions given above. If we use $S_j = S_k e^{ikj\Delta x}$ and insert this into equation (6) we get

$$S_k^{n+1} e^{ikj\Delta x} = (1-\alpha)S_k^n e^{ikj\Delta x} + \frac{\alpha}{2}S_k^n \left(e^{ik(j+1)\Delta x} + e^{ik(j-1)\Delta x}\right)$$

where we have used $S_{j+1} = S_k e^{ik(j+1)\Delta x}$ and $S_{j-1} = S_k e^{ik(j-1)\Delta x}$. If we divide by the common factor $e^{ikj\Delta x}$ we arrive at

$$\frac{S_k^{n+1}}{S_k^n} = (1-\alpha) + \frac{\alpha}{2}\left(e^{ik\Delta x} + e^{-ik\Delta x}\right).$$

Now the expression $\frac{S_k^{n+1}}{S_k^n}$ is simply the ratio between the wave amplitude of the new and old time step, which we have already encountered in the decay equation and identified with the amplification factor. This fraction must be smaller than 1, because otherwise our wave would start to amplify. This must be true for all values of k because, if not, there would be at least one of the partial waves that would grow continuously and would eventually lead to the blow up of the computation. Therefore we have

$$\lambda = \frac{S_k^{n+1}}{S_k^1} = (1-\alpha) + \frac{\alpha}{2}\left(e^{ik\Delta x} + e^{-ik\Delta x}\right) = 1 + \alpha\left(\cos(k\Delta x) - 1\right)$$

where we have also used the Euler formulas

$$e^{ix} = \cos x + i \sin x \qquad \text{and} \qquad e^{-ix} = \cos x - i \sin x$$

$$\sin x = \frac{e^{ix} - e^{-ix}}{2i} \qquad \text{and} \qquad \cos x = \frac{e^{ix} + e^{-ix}}{2}.$$

Since the factor in parenthesis is always negative, the requirement $|\lambda| \leq 1$ translates to $-1 \leq \lambda$ an therefore $\alpha \leq 1$, because the cosine is always greater than -1. This is the same as the stability criterion that has been derived before based only on the characteristics of the diffusion equation.

However, now in exactly the same way we can also derive a stability criterion for the implicit scheme. In this case we have

$$S_k^{n+1} e^{ikj\Delta x} = S_k^n e^{ikj\Delta x} + \frac{\alpha}{2} S_k^{n+1} \left(e^{ik(j+1)\Delta x} - 2 e^{ikj\Delta x} + e^{ik(j-1)\Delta x} \right)$$

where the only difference is now that we have substituted S_k^n with S_k^{n+1} in the right-hand side. Solving this for $\frac{S_k^{n+1}}{S_k^n}$ we arrive at

$$\left(1 + \alpha - \alpha \frac{e^{ik\Delta x} + e^{-ik\Delta x}}{2} \right) S_k^{n+1} = S_k^n$$

or

$$\lambda = \frac{S_k^{n+1}}{S_k^1} = \frac{1}{(1 + \alpha(1 - \cos(k\Delta x)))}.$$

It is now easy to see that this ratio can never be greater than 1, because the value in the denominator is always greater than 1. Therefore, for all values of α this implicit scheme is stable. We also say that the implicit scheme of the diffusion equation is unconditionally stable. This is clearly a nice property that we do not have to worry about the length of the time step as was the case of the explicit scheme. On the other side, this advantage has been achieved through a higher computational cost. In the case of the explicit scheme we can directly solve for the unknown variable at the new time step. In the case of the implicit scheme, this is not possible. Now we have to solve a linear system in order to proceed to the new time step. On the other side, we can make the time step arbitrarily large, at least what stability is concerned.

7.6. The Transport Equation

We now treat a new type of differential equation: the transport equation. Many physical processes we can observe in nature obey this process, where a quantity is transported by a fluid from one place to another. In its simplest

$$\frac{\partial S}{\partial t} + u\frac{\partial S}{\partial x} = 0 \qquad u > 0 \text{ is velocity of transport}$$

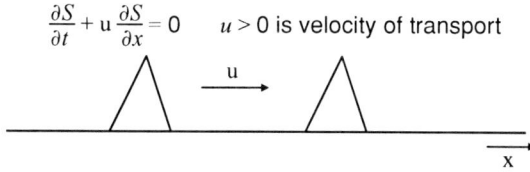

Figure 5. Graphical solution for the transport equation

form the transport equation reads

$$\frac{\partial S}{\partial t} + u\frac{\partial S}{\partial x} = 0$$

where u is the velocity of transport, such as the current velocity of the fluid. With no other terms the solution of the transport equation in case of constant velocity is simply a shift of S along the space axis. This is shown in Figure 5 where a simple peak function is advected to the right by the constant current velocity.

But even if the analytical solution is easy to understand, that does not mean that it is also easy to solve the transport equation. In fact, we will see that solving the transport equation is quite a complex task. For the whole section we will assume that the current velocity is constant.

7.6.1. THE FTCS SCHEME

As a first try we will apply the simplest scheme possible. This scheme uses the usual forward time and central space discretization, much like the explicit scheme for the diffusion equation. The scheme is called FTCS (Forward in Time, Central in Space) and it reads

$$\frac{S_j^{n+1} - S_j^n}{\Delta t} + u\left(\frac{S_{j+1} - S_{j-1}}{2\Delta x}\right)^n = 0$$

This scheme is forward in time, because the spatial derivative is taken on the old time level, and it is a central scheme because it is centered on the node j. With the definition $\alpha = \frac{u\Delta t}{\Delta x}$ the scheme can be rewritten as

$$S_j^{n+1} = S_j^n - \frac{\alpha}{2}\left(S_{j+1}^n - S_{j-1}^n\right)$$

To explore the properties of this scheme we will again apply the von Neuman stability analysis. We therefore insert the special functions $S_j = S_k\, e^{ikj\Delta x}$ in the above discretization, just as we have done in the case of the diffusion equation, and we obtain

$$S_k^{n+1} = S_k^n - \frac{\alpha}{2} S_k^n \left(e^{ik\Delta x} - e^{-ik\Delta x}\right)$$

or, after applying one of the Euler equations and dividing through S_k^n

$$\lambda = \frac{S_k^{n+1}}{S_k^n} = 1 - i\alpha \sin(k\Delta x).$$

The right-hand side of this equation is complex, but we are only interested in the modulus of $|\lambda|$, which we know must be at most 1 in order that the scheme can be considered stable. The modulus of a complex number can be computed by multiplying its value by its conjugated value, which can be obtained by multiplying the imaginary part by -1. If $z = a + ib$ then $|z|^2 = (a + ib) \cdot (a - ib)$ which is a real value. If we apply this to the amplification factor of the FTCS scheme we finally end up with

$$|\lambda|^2 = 1 + \alpha^2 \sin^2(k\Delta x) > 1.$$

As can be seen, the amplification factor of the FTCS scheme is always greater than 1. This means that, no matter what time step we are going to use, the scheme always amplifies all waves present in the solution, and the solution will diverge. This is an example of an unconditionally unstable scheme.

The reason why we were using the central spatial differentiation simply was the hope that by using a second-order discretization we would not only have a more accurate scheme, but also in some way a stable scheme. This was, however, not true, and we will have to resort to some other type of schemes.

7.6.2. THE FORWARD AND BACKWARD SCHEME

We will now try to apply two other schemes we have already encountered before, the forward and the backward scheme. Even if these two discretizations are only second order, it might be possible that they consent a stable computation.

We recall that the forward and backward schemes, applied to the spatial derivative, read respectively

$$\frac{S_j^{n+1} - S_j^n}{\Delta t} + u \left(\frac{S_{j+1} - S_j}{\Delta x}\right)^n = 0$$

$$\frac{S_j^{n+1} - S_j^n}{\Delta t} + u \left(\frac{S_j - S_{j-1}}{\Delta x}\right)^n = 0$$

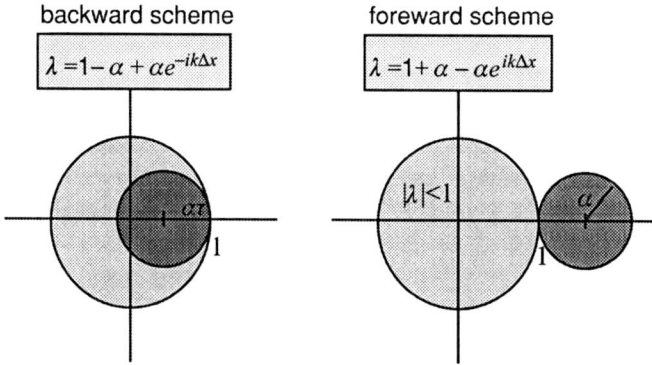

Figure 6. Solutions of the transport equation in the complex plane

where the only difference between the two is the fact that the spatial derivative is taken once forward and once backward starting from j. As before, we can introduce $\alpha = \frac{u\Delta t}{\Delta x}$ and end up with $\lambda = 1 + \alpha - \alpha\,e^{ik\Delta x}$ for the forward scheme and $\lambda = 1 - \alpha + \alpha\,e^{-ik\Delta x}$ for the backward scheme. The values for the amplification factor is complex again, and to find out if the modulus is smaller than 1 we have to multiply its value with the conjugated one. However, it is easier to represent the value of λ in the complex plane. Figure 6 shows the resulting values. Since all the values of the complex exponential function $e^{ik\Delta x}$ lie on a unit circle around the origin, the values of $\lambda = 1 + \alpha - \alpha\,e^{ik\Delta x}$ lie on a circle with radius α around the point $1 + \alpha$ that is situated on the real axis. The values of λ that are stable are the ones smaller than 1 and are therefore situated inside the unit circle. If we have a look at Figure 6 we can see that the values for λ of the forward scheme lie completely outside the unit circle, and therefore also the forward scheme is unconditionally unstable. However, if we take the backward scheme, we can see (Figure 6) that the values for $\lambda = 1 - \alpha + \alpha\,e^{-ik\Delta x}$ lie on a circle with radius α around the value $1 - \alpha$. This circle lies completely inside the unit circle, provided that the radius of the circle is smaller than 1. Therefore, we find as a stability criterion for the backward scheme $\alpha = \frac{u\Delta t}{\Delta x} \leq 1$.

It is interesting to note that of the three straight forward schemes studied only one proved actually to be useful and could be made stable: the backward scheme. The two other schemes were not useful because they were unconditionally stable. In order that the backward scheme was stable α had to be smaller than 1. In case of the transport equation, α is called the Courant number and the stability criterion just says that the Courant number must be smaller or equal to 1.

Rewriting the stability criterion, we can also write $\frac{\Delta x}{\Delta t} \geq u$. The physical meaning of this equation is simple. The ratio $\Delta x / \Delta t$ is the numerical velocity of propagation of the signal computed on the discrete mesh. This numerical velocity must be always greater than the physical velocity of propagation of the signal. Otherwise, the scheme becomes unstable. In the following section this will be explained more thoroughly.

Please note that for $\alpha < 1$ the numerical solution shows a decay of the amplitude of the wave, a property that the real solution does not have, because the amplitude should always stay constant. It would clearly be desirable to have a scheme that could assure $\lambda = 1$ so no numerical damping would result. But please note that for $\alpha = 1$ the numerical solution exactly reproduces the solution of the physical problem. It would therefore seem that we should always work with a Courant number of 1. However, we normally are not in the ideal situation where the propagation velocity is constant, and therefore we will always have areas where we have to accept a Courant number less than 1, resulting in numerical damping.

7.6.3. THE PROPAGATION CONE

If we plot the computational grid with the spatial axis along the x-axis and the time domain along the y-axis, we obtain Figure 7. Starting from the origin, we have evidenced all points that are influenced by the origin through the backward scheme by a light color, and all points that can influence the origin with a darker color. For example, only the points $j = 0$ and $j = 1$ on time level 1 may be influenced by the point $j = 0$ on time level 0. If we follow this up into the future, we obtain the light colored points. If, however, we lock back for the points that might influence $j=0$ on time level 0, we can identify points $j = 0$ and $j = -1$ from time level -1. If this is done for the points on the earlier time levels, the darker points are produced.

We can see that the figure that is formed by these points resembles closely a cone. In fact, this cone is called the numerical propagation cone for the

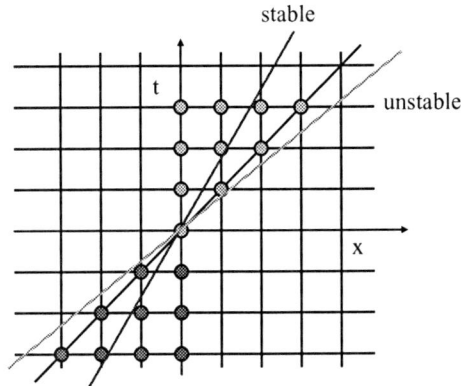

Figure 7. Propagation cone for the backward scheme of the transport equation. Physical propagation velocities corresponding to stable and unstable solutions have been inserted

backward scheme. All points inside the cone are either points that sooner or later will be influenced by the point at the origin (future cone), or are points that will influence it (past cone). All points outside the cone will neither be influenced by nor influence the point situated on the origin.

Figure 7 also shows some lines that represent the physical propagation of a signal with velocity u. Of these lines, the steepest one corresponds to a propagation velocity which is smaller than the other ones, because it covers less space in the same amount of time. This line is characterized by the fact that its velocity is given by $u < \frac{\Delta x}{\Delta t}$ where the (numerical) velocity $\Delta x / \Delta t$ is given by the line that runs along the border of the propagation cone. Therefore, this line completely lies inside the propagation cone and because of the Courant criterion we can conclude that this scheme is stable. On the contrary, the other line represents a signal that is propagating with velocity $u > \frac{\Delta x}{\Delta t}$, and therefore the numerical scheme is too slow to propagate the signal on the numerical grid. This scheme has a Courant number greater than 1 and is therefore unstable. As we can see, in order to be stable, the line representing the physical velocity must be always contained inside the numerical propagation cone.

It becomes now clear why, for example, the forward scheme was unstable. In Figure 8 the propagation cone for the forward scheme has been plotted. With a physical propagation of the signal to the right, the numerical cone is directed towards the other side and the line representing the physical propagation is always outside the propagation cone. Please note that, no matter how small the time step, this scheme will always be unstable. But in case of the backward scheme, if we lower the time step, we will be able to find a value for the time step when the propagation line of the signal will be eventually inside the propagation cone. This procedure corresponds to lower the time step in order to bring the Courant number below 1.

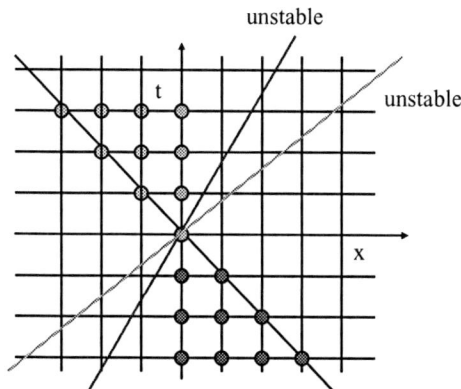

Figure 8. Propagation cone for the forward scheme of the transport equation. Physical propagation velocities have been inserted. The scheme is unstable for all velocities >0

If we have a situation where the velocity of propagation is negative, all arguments have to be reversed. Now it is the forward scheme that guarantees stability, and the backward scheme is unconditionally stable. For this reason, the scheme to be applied is called *upwind* scheme, because the spatial derivative is discretised always in the direction opposed to the propagation. This is the same direction an observer would face if he would stand and look against the wind, looking up-wind.

7.6.4. THE LEAP-FROG SCHEME

As we have seen, the central difference scheme in space (FTCS) has shown to be unconditionally unstable. There is however a way to stabilize the system, without having to resort to numerical diffusion. If we also center the time derivative then we can derive a discretization scheme that is actually stable.

During the discussion of the decay equation two second-order schemes in time have been presented. One centered the right-hand side bye taking the average of the old and the new time step, the other one instead used three time levels, taking the right-hand side at the central time level.

This second way of centering the time derivative is applied here. We introduce a time level $n - 1$ and center the time derivative around level n. The scheme that we obtain reads

$$\frac{S_j^{n+1} - S_j^{n-1}}{2\Delta t} + u \left(\frac{S_{j+1} - S_{j-1}}{2\Delta x} \right)^n = 0$$

where now both derivatives have been centered around time level n and point j. This scheme has been given the name of *leap-frog* scheme. The name is due to the fact that the time derivative jumps over the time level n in a frog like manner. If we introduce α as before we can write this scheme as $S_k^{n+1} - S_k^{n-1} = -S_k^n \alpha 2\,\mathrm{i}\sin(k\Delta x)$ and if we remember the definition of $\lambda = \frac{S_k^{n+1}}{S_k^n} = \frac{S_k^n}{S_k^{n-1}}$ we obtain

$$\lambda - \frac{1}{\lambda} = -\alpha 2\,\mathrm{i}\sin(k\Delta x).$$

This is a quadratic equation in λ and the solution is easy to obtain. We will omit the details here and give only the final result. For values of α greater than 1 the leap-frog scheme, as all the other schemes here, is unstable. However, for $\alpha \leq 1$ the scheme not only is stable, but its amplification factor is $|\lambda| = 1$. This not only means that the scheme is stable, but that the scheme is non-damping, a property that is a characteristic of the transport equation. This very desirable property makes the leap-frog scheme one of the most used numerical schemes in meteorology and oceanography, because it assures the propagation of waves

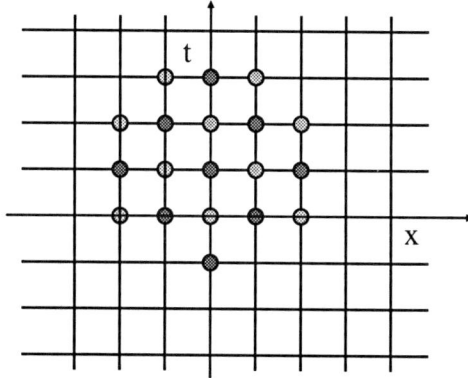

Figure 9. Checkerboard instability for the leap-frog scheme

without numerical damping. Further analysis that cannot be presented here shows that, unfortunately, the scheme shows numerical dispersion, i.e., waves of different wavelength are advected at speeds different from each other. Therefore, the form of the advected function is not constant but changing in time, even if all partial waves are advected without de-amplification.

Another problem with the leap-frog scheme is shown in Figure 9. Here all points starting or involving point (n, j) are shown in light color, and points involving $(n + 1, j)$ or $(n, j + 1)$ are shown in dark color. As can be seen, the two sets of points are clearly separated in space and time. This means that the two sub-grids do not influence each other. Therefore, two solutions may develop in time that do not exchange information between each other and may even diverge. This is the well-known *checkerboard* instability of the leap-frog scheme. In fact, the leap-frog scheme is never used alone but always together with some other scheme that once in a while brings the two solutions in sync with each other.

7.6.5. FINAL THOUGHTS

In this section that dealt with the numerical discretization of the transport equation we have investigated some schemes. For the first time we have seen numerical schemes that, even if accurate and consistent, were not usable because unconditionally unstable. The upwind scheme was stable for Courant numbers smaller than 1, but was also damping, except for the case of Courant equal to 1. The accuracy of this scheme was only of first order.

There was the leap-frog scheme that showed an accuracy of second order, and, in case of Courant number smaller than 1, was also stable and not damping. One could be tempted to say that this scheme would be the preferred scheme for the solution of the transport equation. However, other problems

(numerical dispersion, checkerboard instability) do set a limit to the usability
of this scheme. As for all numerical schemes there is no ideal solution in all
cases, but the advantages must be weighted with the disadvantages and only
then a choice of the numerical scheme can be carried out.

7.7. Gravity Waves

As a last application we will deal shortly with an application to the equation
governing the propagation of gravity waves. These waves, in the simplest case
of linearized equations in one spatial direction can be written as

$$\frac{\partial u}{\partial t} + g\frac{\partial \eta}{\partial x} = 0 \qquad \frac{\partial \eta}{\partial t} + h\frac{\partial u}{\partial x} = 0$$

where u is the velocity of the fluid, η the level elevation of the fluid surface,
h the fluid depth and g the gravitational acceleration. If u is eliminated from
the two equations, a single equation in η results:

$$\frac{\partial^2 \eta}{\partial t^2} = gh\frac{\partial^2 \eta}{\partial x^2}.$$

This is the wave equation with a wave propagation speed of $c = \pm\sqrt{gh}$,
the well-known speed of the gravity waves. The solutions are similar to the
transport equation, i.e., the water level is advected with no change in form
along the x-axis. However, now the propagation is in both directions of the
x-axis.

7.7.1. STAGGERED SCHEMES

For the numerical solution of the above equations it is convenient not to define
both variables at the same point. Instead, the grids are shifted by $\Delta x/2$ in order
that the variables η are taken at full indices $(j, j + 1, \ldots)$ but u is taken at
point in between that correspond to indices $(j - 1/2, j + 1/2, \ldots)$. Figure 10
shows the resulting grid.

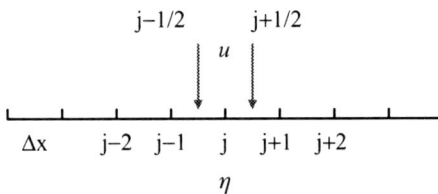

Figure 10. Staggered grid for the resolution of the gravity wave equation

This arrangement allows for a simple centering of the first derivatives in space that are appearing at the right-hand side of the equations. Once discretized, they read

$$\frac{\eta_j^{n+1} - \eta_j^n}{\Delta t} = -h \left[\frac{u_{j+1/2} - u_{j-1/2}}{\Delta x}\right]^{m_1}$$

$$\frac{u_{j+1/2}^{n+1} - u_{j+1/2}^n}{\Delta t} = -g \left[\frac{\eta_{j+1} - \eta_j}{\Delta x}\right]^{m_2}.$$

As can be seen, the equation for η has been centered at point j, around which also the spatial derivative is taken, and the equation for u is centered around point $j + 1/2$. The dependence on the time step is contained in the indices m_1 and m_2. Possible values can be between n and $n + 1$.

These equations form a system of equations that can be solved with the usual techniques of linear algebra. If $m_1 = m_2 = n$ the scheme is a fully explicit scheme and the solution at the new time step can be given immediately. On the other side, for $m_1 = m_2 = n + 1$ the system is fully implicit, and a system of equations has to be solved, as already explained with the diffusion equation. Finally, some ways exist to center the gravity wave discretization. One is the choice $m_1 = n$ and $m_2 = n + 1$, the other one is $m_1 = m_2 = n + 1/2$. This second scheme is called a semi-implicit scheme, and it is achieved by averaging the discretizations of time step n and time step $n + 1$.

7.7.2. STABILITY OF SCHEMES

In principle, the stability analysis can be done exactly as has been shown before, i.e., we develop the variables η and u in a Fourier series (complex exponential function), insert it in the discrete equations and obtain a system of equations in the complex amplitudes η_k and u_k. The eigenvalues of the amplification matrix give the amplification factor and, as we already know, the module of these eigenvalues, in order to have a stable scheme, must be at most 1.

We will not develop the mathematical details of this procedure, which is, by the way, quite straight forward. Here we give only the results in terms of the stability criterion and the amplification that results from these schemes.

It is interesting to note that the completely explicit scheme ($m_1 = m_2 = n$) is unconditionally unstable. This situation corresponds therefore to the FTCS scheme seen in the transport equation and the discretization is not usable. On the other side, the completely implicit scheme ($m_1 = m_2 = n + 1$) is unconditionally stable (no stability criterion). Therefore, at least in theory, we could use it with an arbitrary long time step. The scheme is, however, damping and the waves tend to decrease in amplitude.

The two schemes which have been time centered show interesting characteristics. The first one with $m_1 = n$ and $m_2 = n + 1$ is stable for Courant numbers smaller or equal than 1, but once the stability criterion is fulfilled, the amplification factor is equal to 1, i.e., the waves are propagating without damping, a very nice property. Please also note that it is not necessary to solve a linear system when solving the equations in this case. This is because we can first solve the first equation obtaining η at the new time level directly (explicitly) because the right-hand side is taken at the old time step. After this step we can plug in the new value of η into the right-hand side of the second equation and we can again solve directly the second equation. This is a nice property that makes this type of scheme very economic in terms of computer time.

Finally, for the semi-implicit scheme ($m_1 = m_2 = n + 1/2$) we have to solve a linear system. In change we get an unconditionally stable scheme with no limitation in the time step. Moreover, the amplification factor is again equal to 1, so also this scheme is not damping the gravity waves. Clearly, we have to make a choice between the former scheme, where we could solve explicitly our equation, but with limited time step, and the last scheme where we are not any more limited by the time step but have to solve a linear system at every time step. The choice of what scheme to use will be dictated by the type of application that one has to carry out.

7.7.3. THE SHALLOW WATER EQUATIONS

The last part of this chapter wants to give a real world example: the shallow water equations. These equations have been derived earlier in this book and govern the water circulation in shallow basins and lagoons. When faced with the problem how to tackle the problem of discretizing the various terms, we can now use what we have learned before.

The shallow water equations for the conservation of momentum read

$$\frac{\partial u}{\partial t} + u\frac{\partial u}{\partial x} + v\frac{\partial u}{\partial y} - fv = -g\frac{\partial \eta}{\partial x} + K\left(\frac{\partial^2 u}{\partial x^2} + \frac{\partial^2 u}{\partial y^2}\right) + Ru$$

$$\frac{\partial v}{\partial t} + u\frac{\partial v}{\partial x} + v\frac{\partial v}{\partial y} + fu = -g\frac{\partial \eta}{\partial y} + K\left(\frac{\partial^2 v}{\partial x^2} + \frac{\partial^2 v}{\partial y^2}\right) + Rv$$

where all symbols have already been explained before. The first term is the inertial term, the next two terms give the non-linear advective terms, and the last one on the left-hand side is the Coriolis acceleration. On the right-hand side we find the barotropic pressure gradient (water level gradient), the diffusion of momentum and the bottom friction, where R is still dependent on the velocity and the water depth.

From the discussion before we already know that the friction term (decay equation) and probably the diffusion term is best to be taken implicitly, due to the stability of this discretization. On the other hand, the pressure gradient (gravity waves) and the Coriolis term (inertial movement) should be discretized in a energy conserving way, i.e., with the semi-implicit algorithm. Finally from the treatment of the transport equation we know that the advective terms should be discretized with an upwind scheme. As can be seen, even if we have derived all the algorithms on their own, there is really no reason why we should not be able to use the stability criterion of each of these schemes and derive a stability criterion of the whole equation. In our case the only stability criterion left is the one of the advective transport, where the propagation velocity is given by the current. This is normally a much weaker constraint than, e.g., the velocity of the gravity waves.

7.8. Conclusions

In this chapter we have tackled the problem of how to transform a differential equation into a discrete equation that can then be solved by means of numerical modeling. The equations are discretised according to the need of satisfying constraints in accuracy and mostly stability.

Every term shows different characteristics and must be dealt with separately. The sections above have shown how to find out if a chosen scheme is stable, and under what conditions. After this, the scheme may be applied to the equation and combined with other schemes to solve more complex equations such as the shallow water equation.

Many different important points have not been treated, due to space constraints and the complexity that were out of the scope of this book. But the text might have given you an introduction into the beauty of computational fluid dynamics and oceanography, and may tempt you to dive more into this subject.

References

Ames, W. F., 1977. Numerical Methods for Partial Differential Equations, 2nd ed., Academic Press, New York.

Anderson, John D., Jr., 1995. Computational Fluid Dynamics, McGraw-Hill, New York.

Fletcher, C. A., 1988. Computational Techniques for Fluid Dynamics, Vol. I: Fundamental and General Techniques, Springer-Verlag, Berlin.

Fletcher, C. A., 1988. Computational Techniques for Fluid Dynamics, Vol. II: Specific Techniques for Different Flow Categories, Springer-Verlag, Berlin.

Leveque, Randall J., 2002. Finite Volume Methods for Hyperbolic Problems, Cambridge University Press, Cambridge.

Press, William H., Flannery, Brian P., Teukolsky, Saul A. and Vetterling, William T., 1986. Numerical Recipes: The Art of Scientific Computing, Cambridge University Press, Cambridge.

Richtmyer, R. D. and Morton, K. W., 1967. Difference Methods for Initial Value Problems, 2nd ed., Wiley-Interscience, New York.

Roache, P. J., 1976. Computational Fluid Dynamics, Hermosa, Alberquerque.

Stoer, J. and Bulirsch, R., 1980. Introduction to Numerical Analysis, Springer-Verlag, New York.

8. NUMERICAL MODELS AS DECISION SUPPORT TOOLS IN COASTAL AREAS

Ramiro Neves
Instituto Superior Técnico Av. Rovisco Pais, 1 1049-001 Lisboa, Portugal

8.1. Decision Support Tools

Whenever policy decisions are to be made affecting the natural resource Water—arguably the most precious natural resource—an implacable scrutiny is to be expected. Legal demands are huge as the large number of EU directives targeting water testifies, from which stand out the Nitrates Directive, the Urban Wastewater Directive, the Drinking Water Directive, the Bathing Water Directive and the Water Framework Directive. Media and public opinion at large are continuously exerting a strong pressure over these policies. Decisions need to be thoroughly supported and documented and this is where computers come into play. Modelling tools, in the form of Decision Support Tools, are extensively used both to detect and select the "best" solution and to prove that the best solution was chosen.

Nowadays modelling tools are used as Decision Support Tools. In fact models' results are acceptable justification to major decisions, e.g. the location and configuration of ports, sewage systems and ecological reserves.

Increasingly, computer models are the corner stone in ecological impact studies, being the single most important factor to support policy decisions.

For a model to qualify as a Decision Support Tool it must be able to produce results that describe a reference situation—usually representing conditions for a generic year—and hypothetical scenarios. Most often it is the comparison among a set of results that enables decision makers to make good decisions.

What is a decision support tool?

- It is a model or set of models;
- It produces quantified results;
- The results must be available in time for decisions to be made;
- There is a trend to be run by non-specialists in numerical methods. The operator should be able to pre-process and post-process using a GUI (Graphical User Interface). Usually it is operated by someone that, in one hand, has a deep knowledge of the processes being modelled but, on the other hand, does not know how to build a model;

I. E. Gonenc et al. (eds.), Assessment of the Fate and Effects of Toxic Agents on Water Resources, 171–195.
© 2007 *Springer.*

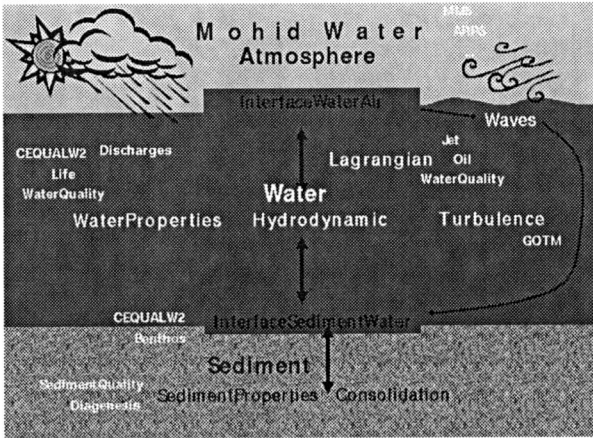

Figure 1. MOHID graphical representation

- A Decision Support Tool integrates several domains covering hydrodynamics, transport and diffusion phenomena and some chemical and biochemical cycles;
- The results are not the decisions in themselves. A decision maker is mandatory.

MOHID Water grew from a 2D hydrodynamic modelling tool in the nineteen eighties to the Decision Support Tool it is today. Figure 1 shows a graphical representation of MOHID's structure. The hydrodynamic module (component model) is the foundation of the entire modelling system. Hydrodynamic fields are used by other components as inputs, e.g., pollutants drift and dilute due to water currents, density, temperature, etc.

Lets suppose that a deep water sewage system is to be built. Inevitably a Decision Support Tool will be used, first of all to provide an accurate characterization of the ecosystem prior to the intended intervention—reference situation. Afterwards several locations and configurations might be simulated, covering as wide a range of natural conditions and ecological stress as possible—scenarios. Some sensitivity analysis should be expected. The comparison between each scenario and the reference situation illustrates the respective technical solution's impact on the ecosystem.

This set of ecological impacts must be evaluated against quantified and objective criteria.

8.2. Brief History of Modelling Tools

Hydrodynamic modelling was initiated in the early 1960s, with the birth of computation, a decade where the first temporal discretization methods for flows with hydrostatic pressure were published (Leendertse, 1967; Heaps, 1969) and developed for two-dimensional vertically integrated models. In the 1970s, the number of applications was multiplied and extensive research on numerical methods was carried out, namely on forms to minimize numerical diffusion introduced from solving advection terms (e.g. Spalding, 1972; Leonard, 1979). Three-dimensional models necessary to simulate oceanic circulation had a high development in the 1980s, benefiting from the increase in computing capacity and in the breakthroughs in turbulence modelling based on work since the 1970s which had in Rodi (1972) one of its main pioneers. In the 1990s, hydrodynamic models were consolidated and several models with great visibility started to emerge, e.g. POM (Blumberg and Mellor, 1987), MOM (Pacanowskid et al., 1991) but also from European schools, e.g. GHER model (Nihould et al., 1989). Benefiting from technological advances, including both hardware and software (e.g. compilers, data management, graphical computation), from the second half of the 1990s, the dawn of integrated models, coupling modules developed by several authors, was witnessed. Turbulence modelling packages like GOTM (Burchard et al., 1999) constitute one of the first examples of this integration, but coupling GOTM to other models constitutes a second level integration example.

Together with the development of hydrodynamic models, ecological models were also developed. Among the pioneer models one can mention WASP developed at EPA (Di Toro et al., 1983) and BOEDE model developed at NIOZ (Ruardij and Baretta, 1982). These models were developed in boxes and in former times used a time step of one day, being the short term variability of flow (e.g. tidal) accounted using diffusion coefficients. Ecological models have improved a lot during the 1980s and 1990s, benefiting from the scientific and technological progress and have been coupled to physical (hydrodynamic) models thus generating the present integrated models.

Current research on modelling is oriented towards operational modelling, integrating different disciplines and assimilating as much field data as possible, with especial emphasis for remote sensing.

Modelling at UTL followed the world trends and benefited from high investments on computing systems in the 1980s. The development of MO-HID system (http://www.MOHID.com) was initiated at that time (Neves, 19855) as a 2D hydrodynamic model and was subsequently developed for becoming an integrated modelling system for tidal flow in estuaries and progressively generalized to waves (Silva, 1991), water quality (Portela, 1996), three-dimensional flows (Santos, 1995), new numerical methods (Martins, 2000), extended set of different open boundary conditions (Leitão, 2003) and finally to be reorganized in an integrated perspective in order to accommodate alternative modules for different processes (Braunschweig et al., 2004). The model evolution enabled to couple alternative modules to compute biogeo-chemical and water quality processes (Trancoso et al., 2005; Saraiva et al., 2006; Mateus, 2006), the broadening to flow through porous media (Galvão et al., 2004), model water flow in a river basin (Braunschweig and Nevas, 2006), and ocean circulation (Leitão et al., 2006).

This model is a working tool of the environmental modelling group of MARETEC research centre, having been used in more than 30 research projects, 50% of which with European funds and currently has around 500 registered users in its online website.

8.3. MOHID

With the growing model complexity, it was necessary to reorganize the MOHID model. In 1998 the whole code was submitted to a complete rear-rangement, using new FORTRAN features and also the capacities of modern computers. The main goal of this rearrangement was to make the MOHID more robust, reliable, protect it against involuntary programming errors and make it scalable. An object oriented philosophy based on Decyk's framework was put to in place (Decyk et al., 1997). The whole model is programmed in ANSI FORTRAN 95.

The philosophy of this new version of MOHID (Miranda et al., 2000) allows it to be applied to one-, two- or three-dimensional problems. MOHID makes intensive use of FORTRAN modules, corresponding as far as possible to logical entities, being it:

- physical domains, e.g. water column, benthos, air;
- interfaces, e.g. air/water, water/sediments, domain/sub-domain;
- phenomena, e.g. turbulence;
- numerical methods, e.g. Lagrangian and Eulerian approaches;
- models that make an intensive use of hydrodynamic results, e.g. pollutants dispersion, oil spills, biogeochemical cycles.

Presently MOHID is composed of more than 40 modules which complete over 150 thousand code lines. Each module is responsible to manage a certain kind of information. The main modules are the modules listed in Table 1.

Another important feature of MOHID is the possibility to run nested mod-els. This feature enables the user to study local areas, obtaining the boundary conditions from the parent model. Computer power is the unique limitation to the number of nested models.

TABLE 1. MOHID's main modules

Module name	Module description
Model	Manages the information flux between the hydrodynamic module and the two transport modules and the communication between nested models.
Hydrodynamic	Full 3D dimensional baroclinic hydrodynamic free surface model. Computes the water level, velocities and water fluxes.
Water Properties (Eulerian Transport)	Eulerian transport model. Manages the evolution of the water properties (temperature, salinity, oxygen, etc) using an Eulerian approach.
Lagrangian	Lagrangian transport model. Manages the evolution of the same properties as the water properties module using a Lagrangian approach. Can also be used to simulate oil dispersion.
Water Quality	Zero-dimensional water quality model. Simulates the oxygen, nitrogen and phosphorus cycle. Used by the Eulerian and the Lagrangian transport modules. Based on a model initially developed by EPA (Bowie et al., 1985).
Oil Dispersion	Oil dispersion module. Simulates the oil spreading due thickness gradients and internal oil processes like evaporation, emulsification, dispersion, dissolution and sedimentation.
Turbulence	One-dimensional turbulence model. Uses the formulation from the GOTM model.
Geometry	Stores and updates the information about the finite volumes.
Surface	Boundary conditions at the top of the water column.
Bottom	Boundary conditions at the bottom of the water column.
Open Boundary	Boundary conditions at the frontier with the open sea.
Discharges	River or Anthropogenic Water Discharges
Hydrodynamic File	Auxiliary module to store the hydrodynamic solution in an external file for posterior usage.

8.4. MOHID: A Modular System

8.4.1. MODEL MODULE

8.4.1.1. *Introduction*
Module Model is MOHID's topmost module and has two main responsibilities:

- Hydrodynamic and the transport modules execution coordination and; Figure 2 illustrates these relations.
- Parent-son communication management (nested models).

8.4.1.2. *Single Model*
A single model execution coordination consists of the global model time actualization and hydrodynamic and transport modules update. Transport

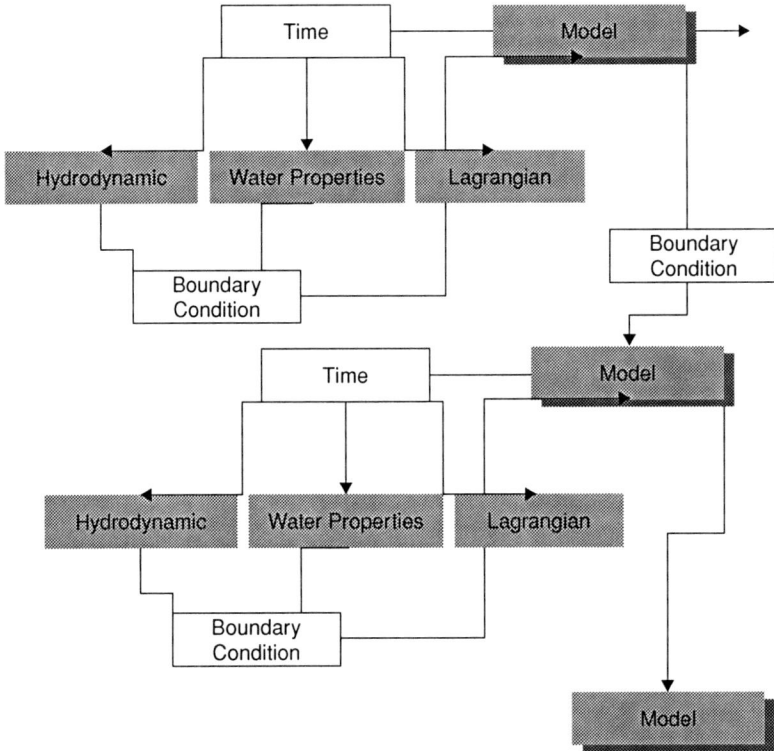

Figure 2. Information flux among nested models

modules' time steps may differ from hydrodynamic module's time step (it is mandatory that transport modules' time steps are multiples of hydrodynamic's time step).

8.4.1.3. *Nested Models*

Information flux coordination among nested models includes their synchronization because nested models may run with different time steps. Nested models coordination is done in a hierarchical way. Every model can have one or more nested child models which, recursively, can have one or more child models. Information flow is one way, consisting on boundary conditions being passed from parent to son(s).

8.4.2. BATHYMETRY MODULE

The Bathymetry module is one of the bottom modules of the MOHID water modelling system. It reads bathymetry data from the input file and publishes this data to all client modules. Bathymetric data can be stored in any regular

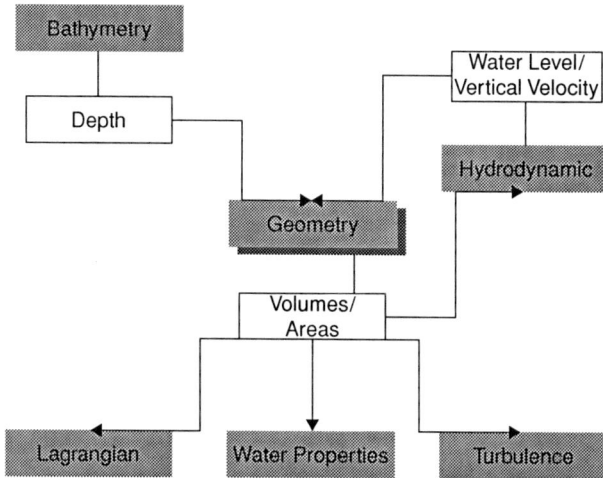

Figure 3. Information flux among the Geometry Module and other modules

grid, with independent variable spacing along the X and Y directions. For every grid point the depth of this point must be given. The horizontal coordinates can be supplied in variety of coordinates systems; the most commonly used are metric and geographic coordinates.

8.4.3. GEOMETRY MODULE

The Geometry Module computes finite volume's lateral areas and volumes, based upon surface elevation and bathymetric data. This information is updated as needed, and made available to other modules. Figure 4 represents the information flux among geometry module and other modules.

8.4.3.1. *Finite Volume*

MOHID uses a finite volume approach (Chippada et al., 1998; Martins et al., 1999, 2000) to discretize equations. In this approach the discrete form of the governing equations is applied macroscopically to a cell control volume. A general conservation law for a scalar U, with sources Q in a control volume Ω is then written as

$$\partial_t \int_\Omega U \, d\Omega + \oint_S \vec{F} \, d\vec{S} = \int_\Omega Q \, d\Omega,$$

where F are the fluxes of the scalar through the surface S embedding the volume. After discretizing this expression in a cell control volume Ω_j where

Au: I prov capti Figu 11, 1 20 ar cite I 5–9, 15, 1 in se the t

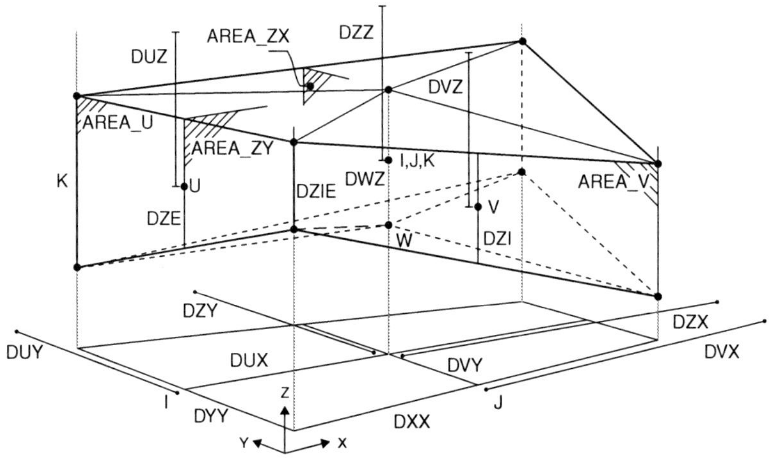

Figure 4. Finite volume element of MOHID model

U_j is defined:

$$\partial_t (U_j \Omega_j) + \sum_{\text{faces}} \vec{F} \cdot \vec{S} = Q_j \Omega_j.$$

This way the procedure for solving the equations is independent of cell geometry. Cells can have any shape with only some constraints—the computational mesh must be regular—because only fluxes among cell faces are required (see Montero (1999) or Martins (2000)). Therefore, a complete separation between physical variables and geometry is achieved (Hirsch, 1988). As volumes can vary during a run, geometry is updated in every time step after computing flow properties. Moreover, spatial coordinates are independent, meaning that different geometry types can be chosen for each dimension, e.g., Cartesian or curvilinear coordinates can be used in the horizontal dimensions and a generic vertical coordinate with several sub-domains can be used in the vertical. This general vertical coordinate allows minimizing errors if compared with some classical vertical coordinates (Cartesian, sigma, isopycnal) as pointed in (Martins et al., 2000).

8.4.3.2. *Vertical Coordinates*
The Geometry module can divide the water column in different vertical coordinates: Sigma, Cartesian, Lagrangian (based on Sigma or based on Cartesian), "Fixed Spacing" and Harmonic. A water column subdivision into different domains is also possible. Sigma and Cartesian sub-domains are often used. The Cartesian coordinate can be used with or without "shaved cells". Lagrangian coordinates move both top and bottom faces with the vertical flow velocity.

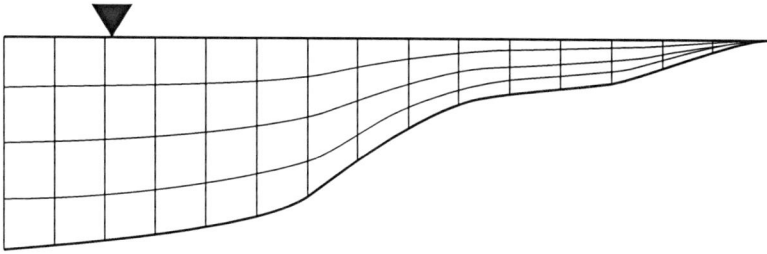

Figure 5. Sigma domain with 4 Layers

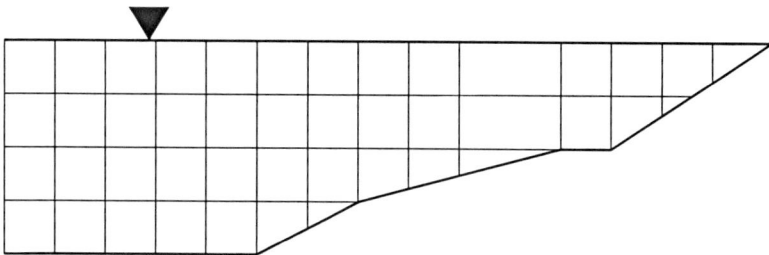

Figure 6. Cartesian domain with 4 Layers (shaved cells)

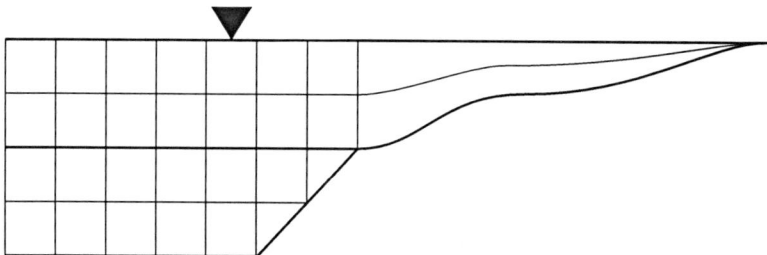

Figure 7. Water column sub-division in a Cartesian domain (inferior) and a Sigma domain (superior)

"Fixed Spacing" coordinates allow the user to study flows close to the domain bottom and Harmonic coordinates work like Cartesian coordinates, just that the horizontal faces close to the surface expand and collapse depending on the variation of the surface elevation. This Harmonic coordinates system was implemented to simulate reservoirs.

8.4.4. HYDRODYNAMIC MODULE

In this section MOHID's hydrodynamic module is described. The information flux of the hydrodynamic module, relative to the other modules of MOHID, is shown in Figure 10.

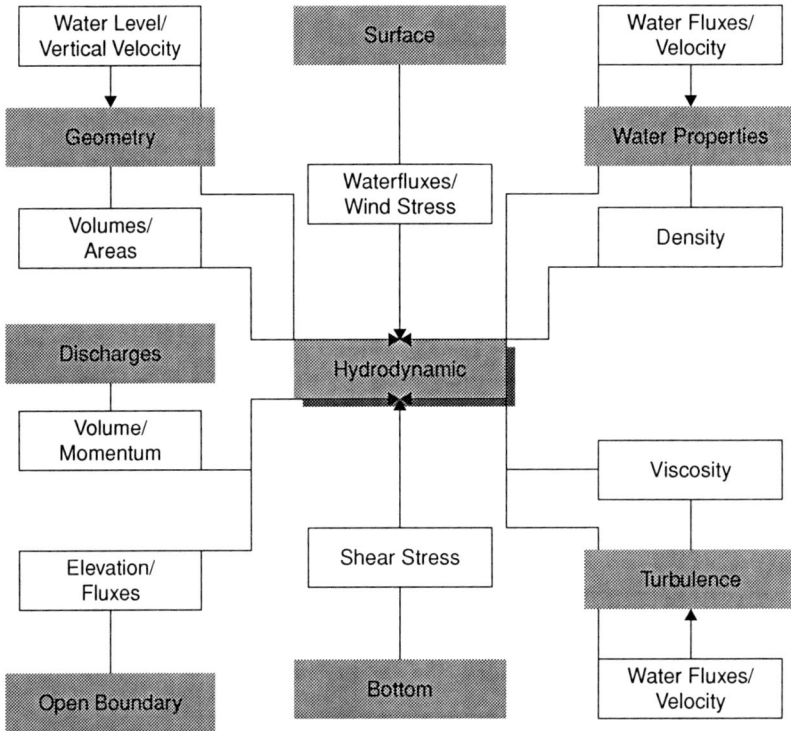

Figure 8. Information flux among the Hydrodynamic Module and other modules

The model solves the three-dimensional incompressible flow primitive equations. Hydrostatic equilibrium is assumed as well as Boussinesq and Reynolds approximations. The density is obtained from salinity and temperature fields, which are transported by the water properties module.

8.4.4.1. *Open Boundary Conditions*

Open boundaries arise from the necessity of confining the domain to the study area. Variables values must be introduced in such a way that information about what is happening outside the domain is guaranteed to enter the domain, so that the solution inside the domain is not corrupted. Waves generated inside the domain should be allowed to go out. There exists no perfect open boundary condition and the most suitable would depend on the domain and the phenomena being modelled. A recent review paper comparing open boundary conditions in test cases can be found in Palma and Matano (1999) and in Blayo (2005). Some different open boundaries are already introduced in MOHID 3D (Santos, 1995; Montero, 1999) and some others like FRS (Flow Relaxation Scheme), radiation processes (Flather, 1987; Orlansky, 1991) and the viscosity sponge layer.

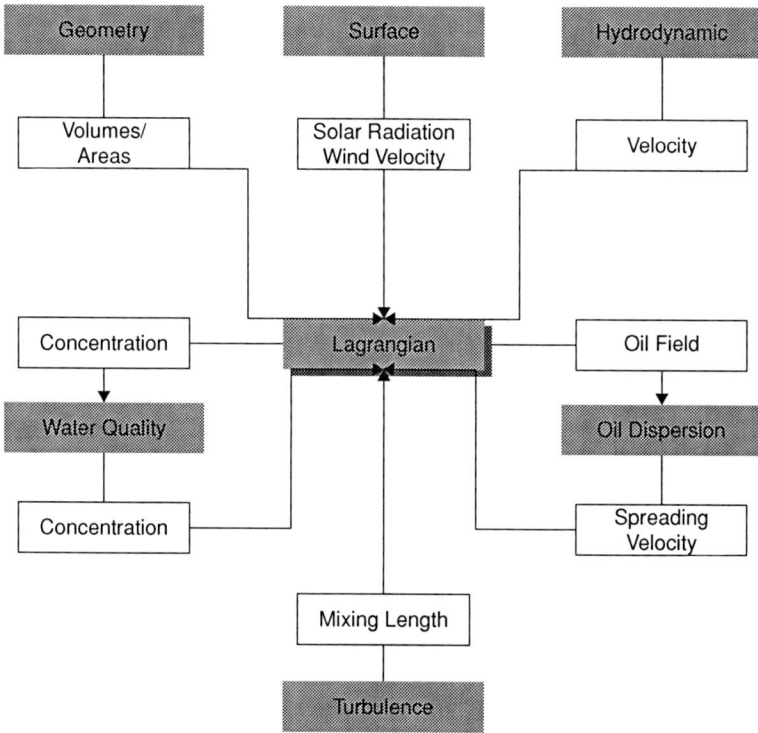

Figure 9. Information flux among the Lagrangian module and other modules

8.4.4.2. *Moving Boundaries*

Moving boundaries are closed boundaries that change position in time. If there are intertidal areas in the domain some points are periodically covered and uncovered, depending on tidal elevation. A stable algorithm is required for modelling these zones and their effect on hydrodynamics of estuaries. A detailed exposition of the algorithms used in MOHID can be found in Martins et al. (1999) and Martins (1999).

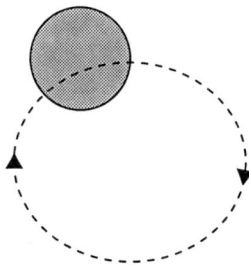

Figure 10. Random movement forced by an eddy larger than the particle

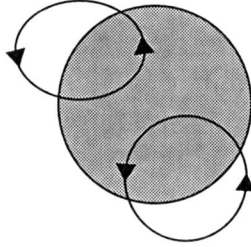

Figure 11. Random movement forced by an eddy smaller than the particle

8.4.5. LAGRANGIAN MODULE

Lagrangian transport models are very useful to simulate localized processes with sharp gradients (submarine outfalls, sediment erosion due to dredging works, hydrodynamic calibration, oil dispersion, etc).

MOHID's Lagrangian module uses the concept of tracer. The most important property of a tracer is its position (x, y, z). For a physicist a tracer can be a water mass, for a geologist it can be a sediment particle or a group of sediment particles and for a chemist it can be a molecule or a group of molecules. A biologist can spot phytoplankton cells in a tracer (at the bottom of the food chain) as well as a shark (at the top of the food chain), which means that a model of this kind can simulate a wide spectrum of processes.

Tracers movement can be influenced by the velocity field from the hydrodynamic module, by the wind from the surface module, by the spreading velocity from oil dispersion module and by random velocity.

At the present stage the model is able to simulate oil dispersion, water quality evolution and sediment transport. To simulate oil dispersion the Lagrangian module interacts with the oil dispersion module. To simulate water quality evolution in time the Lagrangian module is a client of the water quality module. Sediment transport can be associated directly to the tracers using the concept of settling velocity.

Figure 12 represents the information flux among the Lagrangian module and other modules of MOHID.
Another feature of the Lagrangian transport model is its ability to calculate residence times. This can be very useful when studying the exchange of water masses in bays or estuaries.

8.4.5.1. *Tracer Concept*
Like referred above, the MOHID's Lagrangian module uses the concept of tracer. Tracers are characterized by three spatial coordinates, volume and a list of properties (each with a given concentration). Properties can be the same ones described in the water properties module or coliform bacteria. Each tracer has associated a time to perform the random movement.

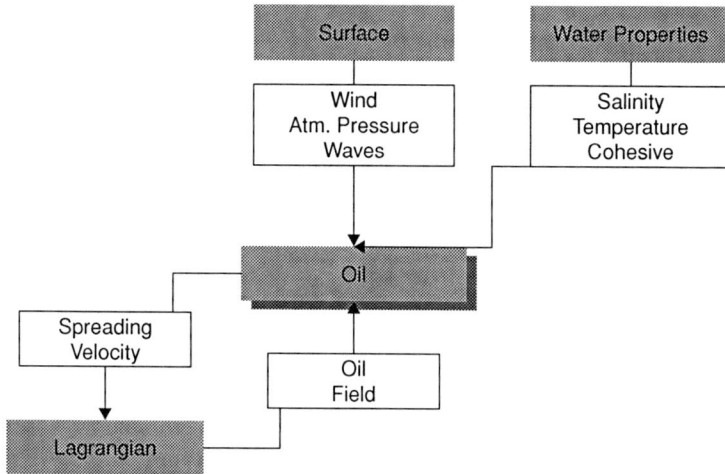

Figure 12. Information flux between the oil module and other modules

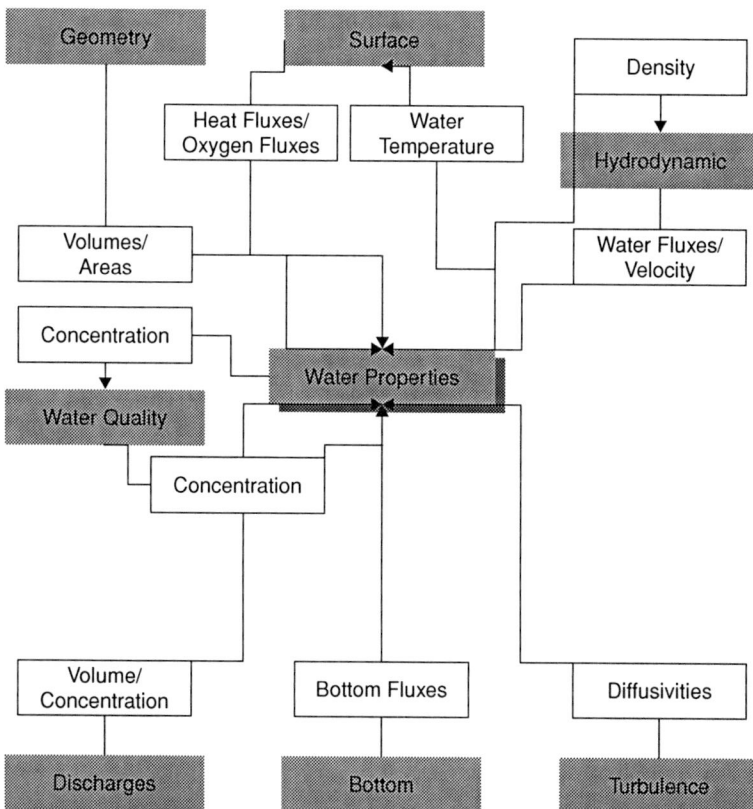

Figure 13. Information flux among the Water Properties Module and other modules

The tracers are "born" at origins. Tracers which belong to the same origin have the same list of properties and use the same parameters for random walk, coliform decay, etc. Origins can differ in the way they emit tracers. There are three different ways to define origins in space:

- a "Point Origins" emits tracers at a given point;
- a "Box Origins" emits tracers over a given area.

There are two different ways in which origins can emit tracers in time:

- a "Accident Origins" emit tracers in a circular area around a point;
- a "Continuous Origins" emits tracers during a period of time;
- a "Instantaneous Origins" emits tracers at one instant.

Origins can be grouped together in Groups. Origins which belong to the same group are grouped together in the output file, so it is easier to analyse results.

8.4.5.2. *Tracer Movement*
Usually the mean velocity is the major factor influencing particles movement. Spatial coordinates are given by the definition of velocity:

$$\frac{dx_i}{dt} = u_i(x_i, t)$$

where u stands for mean velocity and x for particle position.

The Lagrangian module allows several tracers trajectory computations for each hydrodynamic time step.

8.4.5.3. *Turbulent Diffusion*
Turbulent transport is responsible for dispersion. The effect of eddies over particles depends on the ratio between eddies and particle size. Eddies bigger than the particles make them move at random as explained in Figure 14. Eddies smaller than the particles cause entrainment of matter into the particle, increasing its volume and its mass according to the environment concentration, as shown in Figure 16.

Mass decay rate. The decay rate of coliform bacteria, which are can associated to tracers, is computed by the following equation:

$$\frac{dC}{dt} = -\frac{\ln 10}{T_{90}}C$$

where C represents the concentration, and T_{90} the time interval for 90% of the coliform bacteria to die.

A backward in time method is used to solve the above equation numerically, preventing a negative number of coliform bacteria.

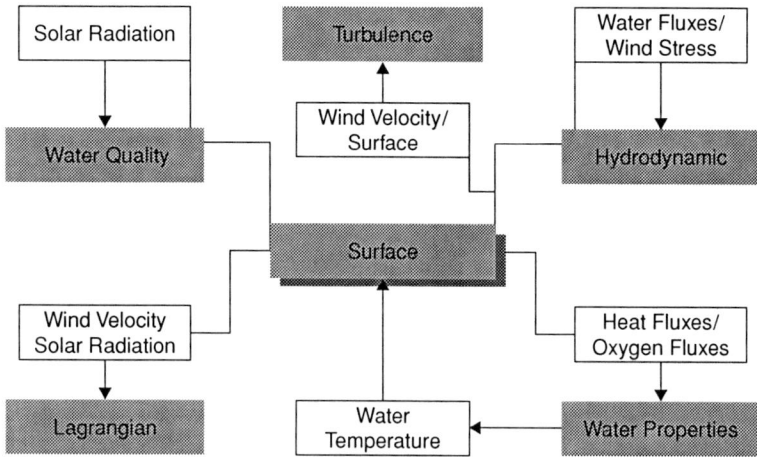

Figure 14. Information flux among the Surface Module and other modules

8.4.5.4. *Monitoring Boxes*

The Lagrangian module permits to monitor particles distribution inside "monitoring boxes". This feature is very useful to compute the residence time of water inside these monitoring boxes and the origins of the water present inside each box at each moment. The Lagrangian module "monitors" the boxes the following way:

- In every instant the volume of each box b, InstBoxVol(b) is calculated:

$$\text{InstBoxVol(b)} = \int (h + Z)\, dx\, dy.$$

- In every instant the water origin "o" inside each monitoring box "b" is identified and the water volume from each origin is stored in the variable InstVolumeByOrigin(b, o):

$$\text{InstVolumeByOrigin}(b, o) = \sum_o \text{Vol}_j^b.$$

- *In the case of instantaneous emissions in boxes*, these contributions are integrated over time, given the integrated contribution over time, IntgVolumeByOrigin(b, o)

$$\text{IntgVolumeByOrigin}(b, o) = \int \text{InstVolumeByOrigin}(b, o)\, dt.$$

A residence time measure for tracers emitted in box "o" in monitoring box "b" is given by

$$\text{ResidenceTimePerBox}(b, o) = \text{IntgVolumeByOrigin}(b, o)/\text{IntialVol(o)}.$$

Adding the values for all monitoring boxes inside the estuary one gets the residence time inside the whole system of the water emitted into box "o":

$$\text{ResidenceTime}(o) = \sum_b \text{ResidenceTimePerBox}(b, o).$$

These values also permit to compute how each monitoring box is influenced by each emitting box:

$$\text{InfluenceOverBox}(b, o) = \text{IntgVolumeByOrigin}(b, o)/\text{InitialVol}(b).$$

In case of a continuous emission, the residence time can be computed as:

$$\text{ResidenceTimePerBox}(b, o) = \text{InstVolumeByOrigin}(b, o)/\text{DischargeRate}(o).$$

8.4.6. OIL MODULE

The prediction and simulation of oil spills trajectory and weathering are essential to the development of pollution response and contingency plans, as well as to the evaluation of environmental impact assessments.

In order to predict the behaviour of oil spilled in coastal zones, an oil weathering model was developed, which predicts the evolution and behaviour of processes (transport, spreading and behaviour) and properties. Some pollution response methods are also integrated in the model.

8.4.6.1. *Implementation*

Oil density and viscosity, and many different processes are included in oil module, such as oil spreading, evaporation, dispersion, sedimentation, dissolution, emulsification, oil beaching and removal techniques.

Different alternative methods were coded for the prediction of some processes like oil spreading, evaporation, dispersion, sedimentation and emulsification. Therefore, when using the model, there is more than one way of simulating the same process, depending, for example, on the characteristics of the computational mesh or on the magnitude of the spill.

The oil weathering module (OWM) uses mainly the 3D hydrodynamics and 3D Lagrangian transport modules. The hydrodynamic module simulates the velocity field necessary for the Lagrangian module to calculate oil trajectories. These oil trajectories are computed assuming that oil can be idealized as a large number of particles that independently move in water. Water properties and atmospheric conditions are introduced in the Lagrangian module and used by the oil module for determination of oil processes and properties. Excepting spreading and oil-beaching, all weathering processes and properties are assumed uniform for all tracers, like water properties and atmospheric conditions, which are considered equal to these environmental conditions determined in accident origin.

As it was already mentioned, the movement of oil tracers can be influenced by the velocity field from the hydrodynamic module, by the wind from the surface module, by the spreading velocity from the oil module and by random velocity.

Oil temperature is assumed equal to water temperature, neglecting solar radiation or any other energy transfer process that may influence oil temperature.

8.4.6.2. *Oil-Beaching*
When oil reaches a coastal zone, it might become beached. This model estimates the amount of beached oil when the model user predefines a beaching probability (or different beaching probabilities for different coastal zones).

8.4.6.3. *Removal Techniques*
Some removal techniques like chemical dispersion or mechanical cleanup are also included in model.

8.4.7. WATER PROPERTIES MODULE

The water properties module coordinates the evolution of the water properties in the water column, using an Eulerian approach. This coordination includes the transport due to advective and diffusive fluxes, water discharges from rivers or anthropogenic sources, exchange with the bottom (sediment fluxes) and the surface (heat fluxes and oxygen fluxes), sedimentation of particulated matter and internal sinks and sources (water quality).

In its present state MOHID can simulate 24 different water properties: temperature, salinity, phytoplankton, zooplankton, particulate organic phosphorus, refractory dissolved organic phosphorus, non-refractory dissolved organic phosphorus, inorganic phosphorus, particulate organic nitrogen, refractory organic nitrogen, non-refractory organic nitrogen, ammonia, nitrate, nitrite, biological oxygen demand, oxygen, cohesive sediments, ciliate bacteria, particulate arsenic, dissolved arsenic, larvae and fecal coliforms. Any new property can be easily added, due to the object-orientated programming used within the MOHID model.

In the water quality module, nitrogen, oxygen and phosphorus cycle can simulate the terms of sink and sources. Figure 19 represents the information flux of the water properties module.

8.4.8. WATER QUALITY MODULE

Efforts towards ecological modelling are being made in most countries where water quality management is a major concern. Fransz et al. (1991) notices that most new generation models tend to become much more biologically and chemically diversified than earlier models, as it is now largely

recognized that there is no way to simulate in sufficient detail the ecosystem behaviour without an in-depth treatment of the full cycle of organic
matter.

These processes are not strange to the preoccupations caused by the eutrophication and its various manifestations. Although there is general consensus that the inputs of nutrients to the sea must be reduced there is so far no
firm scientific basis to decide upon the extent of such reductions.

An appropriate way of addressing the problem of eutrophication and of
testing nutrient reduction scenarios is to simulate the phenomenon with numerical models. It is probably correct to assume that any ecological model
with a sufficiently complex internal structure and the multiple relationships
that are found at the lower trophic levels will come close to an answer, provided
the right time scale is applied.

The ecological model included in MOHID is adapted from EPA (1985)
and pertain to the category of ecosystem simulations models, i.e., sets of
conservation equations describing as adequately as possible the working and
the interrelationships of real ecosystem components. It is not correct to say
that the model describes the lower trophic levels with great accuracy. In fact the
microbial loop that plays a determinant role in water systems in the recycling
processes of organic waste is very simplified in MOHID.

Lower trophic levels appear in nearly all marine ecosystem simulation
models since there is at least a compartment "phytoplankton" required to
compute the organic matter cycle. Some early models applied in the North
Sea were one-compartment models, especially endeavouring to simulate phytoplankton growth, in relation with the physical environment and with grazing pressure (treated as a forcing variable). Both the influence of the Lotka-
Volterra equations—developed in the 1920s—and that of findings in the field
of plant physiology (photosynthesis-light relationship) were discernible. It
was not long before limiting nutrient and herbivorous zooplankton were incorporated as well, as state variables in simulation models (Fransz et al.,
1991)

8.4.9. SURFACE MODULE

The surface module stores boundary conditions at the water column surface.
These boundary conditions can be divided in two types. One type of boundary conditions which are given directly by the user, usually meteorological
data (wind velocity, air temperature, dew point, evaporation, cloud cover)
and boundary conditions calculated by the model from the meteorological
data/conditions of the water column (wind stress, solar radiation, latent heat,
infra-red radiation, sensible heat, oxygen flux). The information flux between
the surface module and other modules is shown in Figure 21.

8.4.9.1. *Wind*

Wind stress is calculated according to a quadratic friction law:

$$\vec{\tau w} = C_D \rho_a \vec{W} \left| \vec{W} \right|$$

where C_D is a drag coefficient that is function of the wind speed, ρ_a is air density and W is the wind speed at a height of 10 m over the sea surface.

The drag coefficient is computed according to Large and Pond (1981):

$$(W < 10\,\text{m/s})$$

$$C_D = 4.4\,\text{e}^{-4} + 6.5\,\text{e}^{-5}\,\vec{W} \left| \vec{W} \right| (10\,\text{m/s} < W < 26\,\text{m/s}).$$

8.4.9.2. *Heat Fluxes*

Heat fluxes at the surface can be separated into five distinctive fluxes: solar short-wave radiation, atmospheric long-wave radiation, water long-wave radiation, sensible heat flux and latent heat flux. These fluxes can be grouped into two ways: in (i) radiative fluxes (first three fluxes) and (ii) non-radiative fluxes (last two fluxes) or in (iii) fluxes independent of the water temperature (first two fluxes) and in (iv) fluxes dependent of the water temperature (last three fluxes).

8.4.9.3. *Solar Radiation*

Solar radiation is an important ecological parameter, and is often the key driving force in ecological processes (Brock, 1981). The solar radiation flux of short wavelength is computed by:

$$Q = Q_0 A t (1 - 0.65 C_n^2)(1 - R_s)$$

where Q_0 is the solar radiation flux on top atmosphere (W m^2), A_t the coefficient for atmospheric transmission, C_n the cloud cover percentage and R_s stands for albedo (0.055). The solar radiation flux on top atmosphere can be expressed as

$$Q_0 = \frac{I_0}{r^2} \text{senz}$$

where I_0 stands for the solar constant which is the energy received per unit time, at Earth's mean distance from the Sun, outside the atmosphere, a standard value is 1353 W m^{-2} (Brock, 1981), r stands for the radius vector and z stands for the solar high.

8.4.10. BOTTOM MODULE

The bottom module computes boundary conditions at the bottom of the water column. It computes shear stress as a boundary condition to the hydrodynamic and turbulence modules. It is also responsible for computing fluxes at the water-sediment interface, managing boundary conditions to both the water column properties and the sediment column properties.

Both in the water column or in the sediment column, properties can be either dissolved or particulate. The evolution of dissolved properties depends greatly on the water fluxes, both in the water column and in the sediment interstitial water. Particulate properties evolution in the water column depends also on the water fluxes and on settling velocity. Once deposited in the bottom they can either stay there or be resuspended back to the water column. If they stay there for a determined period of time, they can become part of the sediment compartment by consolidation.

8.4.11. FREE VERTICAL MOVEMENT MODULE

The free vertical movement module computes particulate properties vertical fluxes. It is normally used to compute settling velocity for cohesive sediment or particulate organic matter transport.

8.4.12. HYDRODYNAMIC FILE MODULE

In this section the hydrodynamic file module of the model MOHID is described. This module can be seen as an auxiliary module, which permits the MOHID user to integrate the hydrodynamic solution in space and time and store this solution in a file. This file can be later used to simulate longer periods, like water quality simulation which needs simulation times for at least one year.

References

Abbott, M.B., A. Damsgaardand, and G.S. Rodenhuis, 1973. System 21, Jupiter, a design system for two dimensional nearly horizontal flows, J. Hydr. Res., 1, 1–28.

Allen, C.M., 1982. Numerical simulation of contaminant dispersion in estuary flows, Proc. R. Soc. London. A, 381, 179–194.

Arakawa, A., and V.R. Lamb, 1977. Computational design of the basic dynamical processes of the UCLA General Circulation Model. Methods Comput. Phys., 17, 174–264.

Arhonditsis, G., G. Tsirtsis, M.O. Angelidis, and M. Karydis, 2000. Quantification of the effects of nonpoint nutrient sources to coastal marine eutrophication: Application to a semi-enclosed gulf in the Mediterranean Sea, Ecol. Modelling, 129, 209–227.

Backhaus, J., 1985. A three dimensional model for the simulation of shelf sea dynamics, Dt. Hydrogr. Z., 38, 165–187.

Blumberg, A.F., and G.L. Mellor, 1987. A description of a three-dimensional coastal ocean circulation model, Three-Dimensional Coastal Ocean Models, edited by N. Heaps., Vol. 4, American Geophysical Union, 208 pp.

BOEDE Publ., 1982. en Versl. No. 2, Texel.

Bowie, G.L., W.B. Mills, D.B. Porcella, C.L. Cambell, J.R. Pagendorf, G.L. Rupp, K.M. Johnson, P.W. Chan, S.A. Gherini, and C.E. Chamberlin, 1985. Rates, Constants and Kinetic Formulations in Surface Water Quality Modeling, U. S. Environmental Protection Agency.

Braunschweig, F., 2001. Generalização de um modelo de circulação costeira para albufeiras, MSc. Thesis, Instituto Superior Técnico, Technical University of Lisbon.

Braunschweig, F., and Neves, R., 2006. Catchment modelling using the finite volume approach, Relatório final do projecto http://www.tempQsim.net, Instituto Superior Técnico.

Braunschweig, F., P. Chambel, L. Fernandes, P. Pina, and R. Neves, 2004. The object-oriented design of the integrated modelling system MOHID, Computational Methods in Water Resources International Conference, Chapel Hill, North Carolina, USA.

Brock, T.D., 1981. Calculating solar radiation for ecological studies, Ecological Modelling

Buchanan, I., and N. Hurford, 1988. Methods for predicting the physical changes in oil spilt at sea, Oil Chem. Pollut., 4 (4), 311–328.

Burchard, H., K. Bolding, and M.R. Villarreal, 1999. GOTM—a general ocean turbulence model, Theory, applications and test cases, Tech. Rep. EUR 18745 EN, European Commission.

Cabeçadas, L., 1993. Ecologia do fitoplâncton do Estuário do Sado para uma estratégia de conservação, Estudos de Biologia e Conservação da Natureza Vol. 10, SNPRCN, Lisboa, 50 pp.

Cancino, L., and R. Neves, 1999. Hydrodynamic and sediment suspension modelling in estuarine systems. Part II: Application to the Western Scheldt and Gironde estuaries, J. Marine Syst., 22, 117–131.

Chippada, S., C. Dawson, and M. Wheeler, 1998. Agodonov-type finite volume method for the system of shallow water equations, Comput. Methods Appl. Mech. Eng., 151 (01), 105–130.

Coelho, H., A. Santos, T.L. Rosa, and R. Neves, 1994. Modelling the wind driven flow off Iberian Peninsula, GAIA, 8, 71–78.

Costa, M.V., 1991. A Three-Dimensional Eulerian–Lagrangian Method for Predicting Plume Dispersion in Natural Waters—Diplôme d'Etudes Approfondies Européen en Modélisation de l'Environnement Marin—ERASMUS.

Decyk, V.K., C.D. Norton, and B.K. Szymanski, 1997. Expressing Object-Oriented Concepts in Fortran 90, ACM Fortran Forum, Vol. 16.

Delvigne, G.A.L., and C.E. Sweeney, 1998. Natural dispersion of oil, Oil Chem. Pollut., 4, 281–310.

Di Toro, D.M., J.J. Fitzpatrick, and R.V. Thomann, 1983. Water Quality Analysis, Simulation Program (WASP) and Model Verification Program (MVP) Documentation. Hydroscience, Inc. Westwood, NY, USEPA Contract No. 68-01-3872.

Duarte, M., and M. Henriques, 1991. Caracterização físico-química das águas do Estuário do Rio Sado, INETI DEII 14/91.

Eilers, P.H.C., J.C.H. Peeters, 1988. A model for the relationship between light intensity and the rate of photosynthesis in phytoplankton, Ecol. Modelling, 42, 113–133.

EPA, 1985. Rates, constants, and kinetics formulations in surface water quality modeling, 2nd edn, United States Environmental Protection Agency, Report EPA/600/3-85/040.

ERM, 2000. Criteria for the Definition of Eutrophication in Marine/Coastal Waters, Final Report of European Commission Contract number B4-3040/98/000705/MAR/D1.

Falkowski, P.G., and C.D. Wirick, 1981. A simulation model of the effects of vertical mixing on primary productivity, Mar. Biol., 65, 69–75.

Fay, J.A., 1969. The spread of oil slicks on a calm sea, Oil on the Sea, Plenum Press, NY, pp. 53–63.

Fingas, Mervin, 1998. The evaporation of oil spills: Development and implementation of new prediction methodology. Marine Environmental Modelling Seminar'98, Lillehammer, Norway.

Fletcher, C.A.J., 1991. Computational Techniques for Fluid Dynamics, Vol. I, 2nd edn, Springer Series in Computational Physics, Springer Verlag, New York, 401 pp.

Flores, H., A. Andreatta, G. Llona, and I. Saavedra, 1998. Measurements of oil spill spreading in a wave tank using digital image processing. Oil and Hydrocarbon Spills, Modeling, Analysis and Control, WIT Press, Southampton, UK, pp. 165–173.

Fransz, H.G., J. P. Mommaerts, and G. Radach, 1991. Ecological modelling of the North Sea, Netherlands J. Sea Res., 28 (1/2), 67–140.

Galvão, P., P. Chambel-Leitao, R. Neves, and P. Leitao, 2004. A different approach to the modified Picard method for water flow in variably saturated media, Computational Methods in Water Resources, Part 1, Developments in Water Science, Vol. 55, Elsevier.

Heaps, N.S, 1969. A two-dimensional numerical sea model, Philosophy Transactions Royal D.B.

Hirsch, C., 1988. Numerical computation of internal and external flows. Vol I: Fundamentals of numerical discretization. Wiley Series in Numerical Methods in Engineering, John Wiley and Sons, Chichester, 515 pp.

Huang, J.C., and F.C. Monastero, 1982. Review of the state-of-the-art of oil spill simulation models. Final Report submitted to the American Petroleum Institute.

Humborg, C., K. Fennel, M. Pastuszak, and W. Fennel, 2000. A box model approach for a long-term assessment of estuarine eutrophication, Szczecin Lagoon, southern Baltic, J. Marine Syst., 25, 387–403.

James, I.D., 1987. A general three-dimensional eddy-resolving model for stratified seas, in Three-dimensional models of marine and estuarine dynamics, edited by J.C. Nihoul and B.M. Jamart, Elsevier Oceanography Series 45, Amsterdam, pp. 1–33.

Krone, R.B., 1962. Flume studies of the transport in estuarine shoaling processes, Hydr. Eng. Lab., Univ. of Berkeley, California, USA.

Leendertse, J.J., 1967. Aspects of a computational model for long-period water-wave propagation, Rand Corporation, Santa Monica, California, RM-5294-PR, 165 pp.

Leendertsee, J.J., and S.K. Liu, 1978. A three-dimensional turbulent energy model for non-homogeneous estuaries and coastal sea systems, Hydrodynamics of Estuaries and Fjords, edited by J.C.J. Nihoul, Elsevier Publ. Co., Amsterdam, pp. 387–405.

Leitão, 2003. Integração de Escalas e de Processos na Modelação ao Ambiente Marinho, Universidade Técnica de Lisboa, Instituto Superior Técnico. Tese de Doutoramento (in Portuguese).

Leitão, P.C., 1996. Modelo de Dispersão Lagrangeano Tridimensional. Ms. Sc. Thesis, Universidade Técnica de Lisboa, Instituto Superior Técnico.

Leitão, P., H. Coelho, A. Santos, and R. Neves, et al., 2006. Modelling the main features of the Algarve coastal circulation during July 2004: A downscalling approach, J. Atmos. Ocean Sci. (submitted).

Leonard, B.P., 1979. A stable and accurate convective modelling procedure based on quadratic upstream interpolation, Comput. Methods Appl. Mech. Eng., 19, 59–98.

Lobo, G., J. Almeida, N. Carvalhais, and S. Costa, 2000. Gestão Ambiental do Estuário do Sado. (em preparação).

Mackay D., I.A. Buistt, R. Mascarenhas, and S. Paterson, 1980. Oil spill processes and models, Environment Canada Manuscript Report No. EE-8, Ottawa, Ontario.

Martins, F., 1999. Modelação Matemática Tridimensional de Escoamentos Costeiros e Estuarinos usando uma Abordagem de Coordenada Vertical Genérica, Ph.D. Thesis, Universidade Técnica de Lisboa, Instituto Superior Tecnico.

Martins, F., P. Leitão, A. Silva, and R. Neves, 2000. 3D modeling in the Sado estuary using a new generic vertical discretization approach, Oceanologica Acta (submitted).

Martins, M., and M.J.L. Dufner, 1982. Estudo da qualidade da água. Resultados referentes às observações sinópticas em 1980, Estudo Ambiental do Estuário do Tejo (2ªsérie), no. 14, Comissão Nacional do Ambiente, Lisboa, pp. 1–212.

Mateus, M., 2006. A Process-Oriented Biogeochemical Model for Marine Ecosystems Development. Numerical Study and Application. Universidade Técnica de Lisboa, Instituto Superior Técnico. Tese de Doutoramento (submitted).

Miranda, R., 1999. Nitrogen Biogeochemical Cycle Modeling in the North Atlantic Ocean. Tese de Mestrado, Universidade Técnica de Lisboa, Instituto Superior Técnico.

Miranda, R., F. Braunschweig, P. Leitão, R. Neves, F. Martins, and A. Santos, 2000. MOHID 2000, A Costal integrated object oriened model, Hydraulic Engineering Software VIII, WIT Press.

Monteiro, A.J., 1995. Dispersão de Efluentes Através de Exutores Submarinos. Uma contribuição para a modelação matemática. Universidade Técnica de Lisboa, Instituto Superior Técnico.

Montero, P., 1999. Estudio de la hidrodinámica de la Ría de Vigo mediante un modelo de volúmenes finitos (Study of the hydrodynamics of the Ría de Vigo by means of a finite volume model), Ph.D. Dissertation, Universidad de Santiago de Compostela (in Spanish).

Montero, P., M. Gómez-Gesteira, J.J. Taboada, M. Ruiz-Villarreal., A.P. Santos, R.J.J. Neves, R. Prego, and V. Pérez-Villar, 1999. On residual circulation of Vigo Ría using a 3D baroclinic model, Boletín Instituto Español de Oceanografía, no. 15, SUPLEMENTO-1.

Mooney, M., 1951. The viscosity of a concentrated suspension of spherical particles, J. Colloidal Sci., 10, 162–170.

Nakata, K., F. Horiguchi, and M. Yamamuro, 2000. Model study of Lakes Shinji and Nakaumi— a coupled coastal lagoon system, J. Marine Syst., 26, 145–169.

Napolitano, E., T. Oguz, P. Malanotte-Rizzoli, A. Yilmaz, and E. Sansone, 2000. Simulation of biological production in the Rhodes and Ionian basins of the eastern Mediterranean, J. Marine Syst., 24, 277–298.

Neumann, T., 2000. Towards a 3D-ecosysytem model of the Baltic Sea, J. Marine Syst., 25, 405–419.

Neves, R.J.J., 1985. Étude Experimentale et Modélisation des Circulations Trasitoire et Résiduelle dans l'Estuaire du Sado, Ph.D. Thesis, Univ. Liège, 371 pp. (in French).

Neves, R., H. Coelho, P. Leitão, H. Martins, and A. Santos, 1998. A numerical investigation of the slope current along the western European margin, edited by V. Burgano, G. Karatzas, A. Payatakas, C. Brebbia, W. Gray, and G. Pinder, 1998. Comput. Methods Water Resources XII (2), 369–376.

Nihoul, J.C.J., E. Deleersnijder, and S. Djenidi, 1989. Modelling the general circulation of shelf seas by 3D k-epsilon models, Earth Sci. Rev., 26, 163–189.

NOAA, 1994. ADIOS[TM] (Automated Data Inquiry for Oil Spills) User's Manual, Hazardous Materials Response and Assessment Division, NOAA, Seattle, Prepared for the U.S. Coast Guard Research and Development Center, Groton Connecticut, 50 pp.

NOAA, 2000. ADIOS™ (Automated Data Inquiry for Oil Spills) Version 2.0, Hazardous Materials Response and Assessment Division, NOAA, Seattle. Prepared for the U.S. Coast Guard Research and Development Center, Groton Connecticut.

Pacanowski, R.C., K.W. Dixon, and A. Rosati, 1991. GFDL Modular Ocean Model, Users Guide Version 1.0, GFDL Tech. Rep., 2, 46 pp.

Palma, E., and R.P. Matano, 1998. On the implementation of passive open boundary conditions for a general circulation model: The barotropic mode, J. Geophys. Res., 103, 1319–1342.

Parsons, T.R., M. Takahashi, and B. Hargrave, 1984. Biological Oceanographic Processes, 3rd edn, Pergamon Press, Oxford, 330 pp.

Partheniades, E., 1965. Erosion and deposition of cohesive soils, J. Hydr. Div., ASCE, 91, No. HY1, pp. 105–139.

Payne, J.R., B.E. Kirstein, J.R. Clayton, C. Clary, R. Redding, D. McNabb, and G. Farmer, 1987. Integration of Suspended Particulate Matter and Oil Transportation Study, Final Report, Report to Minerals Management Service, MMS 87-0083.

Pérez-Villar, V., 1999. Ordenación Integral del Espacio Maritimo-Terrestre de Gali-cia: Modelización informática (Integrated Management of the Galician Maritime-Terrestrial Space: Numerical Modelling), Final report by the Grupo de Física Non Lineal, Consellería de Pesca, Marisqueo e Acuicultura. Xunta de Galicia.

Pina, P.M.N., 2001. An Integrated Approach to Study the Tagus Estuary Water Quality, Tese de Mestrado, Universidade Técnica de Lisboa, Instituto Superior Técnico.

Platt, T., C.L. Galeggos, and W.G. Harrison, 1980. Photoinhibition of photosynthesis in natural assemblages of marine phytoplankton, J. Mar. Res., 38, 687–701.

Portela, L., 1996. Modelação matemática de processos hidrodinâmicos e de qualidade da água no Estuário do Tejo. Dissertação para obtenção do grau de Doutor em engenharia do Ambiente.Instituto Superior Técnico, Universidade Técnica de Lisboa, 240 pp.

Portela, L.I., 1996. Mathematical modelling of hydrodynamic processes and water quality in Tagus estuary, Ph.D. Thesis, Instituto Sup. Técnico, Tech. Univ. of Lisbon (in Portuguese).

Proctor, R., R.A. Flather, and A.J. Elliot, 1994. Modelling tides and surface drift in the Arabian Gulf—application to the Gulf oil spill, Continental Shelf Res., 14, 531–545.

Rasmussen, D., 1985. Oil Spill Modelling—A tool for cleanup operations, Proc. 1985 Oil Spill Conference, American Petroleum Institute, pp. 243–249.

Reed, M., 1989. The physical fates component of the natural resource damage assessment model system, Oil and Chem. Pollut., 5, 99–123.

Rivera, P.C., 1997. Hydrodynamics, sediment tranport and light extinction off Cape Bolinao, Philippines, Ph.D. Dissertation, A.A.Balkema/Rotterdam/Brookfield.

Rodi, W., 1972. The prediction of free turbulent boundary layers by use of a two-equation model of turbulence, Ph.D. Thesis, Imperial College, University of London, UK.

Ruardij, P., and J.W. Baretta, The EmsDollart Ecosystem Modelling Workshop.

Santos, A.J., 1995. Modelo Hidrodinâmico Tridimensional de Circulação Oceânica e Estuarina, Ph.D. Thesis, Universidade Técnica de Lisboa, Instituto Superior Técnico.

Saraiva, S., P. Pina, F. Martins, M. Santos, F. Braunschweig, and R. Neves, 2006. EU-Water Framework: Dealing with nutrients loads in Portuguese estuaries, Hydrobiologia (accepted).

Silva, A.J.R., 1991. Modelação Matemática Não Linear de Ondas de Superfície e de Correntes Litorais, Tese apresentada para obtenção do grau de Doutor em Engenharia Mecânica, IST, Lisboa (in Portuguese).

Somlyódy, L., and L. Koncsos, 1991. Influence of sediment resuspension on the light conditions and algal growth in lake Balaton, Ecological Modelling, 57, 173–192.

Spalding, 1972. A novel finite difference formulation for differential expressions involving both first and second derivatives, Int. J. Numer. Methods Eng., 4, 551–559.

Steele, J.H., 1962. Environmental control of photosynthesis in the sea, Limnol. Oceanogr., 7, 137–150.

Stiver, W., and D. Mackay, 1984. Evaporation rate of spills of hydrocarbons and petroleum mixtures, Environ. Sci. Technol., 18 (11), 834–840.

Taboada, J.J., 1999. Aplicación de modelos numéricos al estudio de la hidrodinámica y del flujo de partículas en el Mar Mediterráneo (Application of numerical models for the study of hydro-dynamics and particle fluxes in the Mediterranean Sea), Ph.D. Dissertation, Universidad de Santiago de Compostela (in Spanish).

Taboada, J.J., M. Ruíz-Villarreal, M. Gómez-Gesteira, P. Montero, A.P. Santos, V. Pérez-Villar, and R. Prego, 2000. Estudio del transporte en la Ría de Pontevedra (NOEspaña) mediante un modelo 3D: Resultados preliminares, In: Estudos de Biogeoquímica na zona costeira ibérica, edited by A. Da Costa, C. Vale and R. Prego, Servicio de Publicaciones da Universidade de Aveiro (in press).

Tett, P., and H. Wilson, 2000. From biogeochemical to ecological models of marine microplankton, J. Marine Syst., 25, 431–446.

Thornton, K.W., and A.S. Lessen, 1978. A temperature algorithm for modifying biological rates, Trans. Am. Fish. Soc., 107 (2), 284–287.

Trancoso, A., S. Saraiva, L. Fernandes, P. Pina, P. Leitão, and R. Neves, 2005. Modelling Macroalgae using a 3D hydrodynamic ecological model in a shallow, temperate estuary, Ecological Modelling.

UNESCO, 1981. Tenth Report on the joint panel on oceanographic tables and standards, Technical Papers in Marine Science, No. 36, 24 pp.

Pérez-Villar, V., 1998. Evaluation of the seasonal variations in the residual patterns in the Ría de Vigo (NW Spain) by means of a 3D baroclinic model, Estuarine Coastal Shelf Sci., 47, 661–670.

Valiela, I., 1995. Marine Ecological Processes, Springer-Verlag, New York, 686 pp.

Vila, X., L.J. Colomer, and Garcia-Gil, 1996. Modelling spectral irradiance in freshwater in relation to phytoplankton and solar radiation, Ecol. Modelling, 87, 56–68.

Villarreal, M.R., P. Montero, R. Prego, J.J. Taboada, P. Leitao, M. Gómez-Gesteira, M. de Castro, and V. Pérez-Villar, 2000. Water Circulation in the Ria de Pontevedra under estuarine conditions using a 3d hydrodynamical model, Est. Coast. Shelf Sci. (submitted).

PART 4. EFFECTS OF CBRN AGENTS ON AQUATIC ECOSYSTEM

9. THE ECOSYSTEM APPROACH APPLIED TO THE MANAGEMENT OF THE COASTAL SOCIO-ECOLOGICAL SYSTEMS

Angheluta Vadineanu
University of Bucharest, Romania

9.1. Introduction

The future long-term trend in science, education and management, emerged from the overall policy goal of building a knowledge-based society and sustainable development. To achieve this goal, a new, revolutionary, theoretical development is needed, which allows: (i) the identification and understanding of complex organization across space and time scales of the nature and society; and (ii) the ecosystem (holistic) approach and adaptive management of the environment, human societies and their co-development under the pressure of different driving forces. Among a wide range of drivers and pressures, special attention should be paid to potential social instability and unexpected human behavior (e.g., terrorist attack).

This chapter is dealing with the basic structural and functional characteristics of the complex, dynamic and hierarchical socio-ecologic units, established across spatial scales, as a result of very long-term (centuries and millennia) interactions and co-evolution among natural and human systems, in particular those from the critical transition zones, like coastal zones, which integrate also the lagoon ecosystems.

After a brief presentation of what we strongly believe is the most powerful conceptual framework to describe the ecosystem or holistic approach, adaptive management and sustainability, there is a presentation of the analytical framework for effectively approaching the management of the lagoon-ecosystems as part of coastal socio-ecological systems. Within this framework, the most critical steps for the ecosystem approach and adaptive management are outlined.

Approaching complexity of nature and society: The theoretical and conceptual background adopted and used in this regard comes from the broad and integrated theory of systems ecology (Odum, 1993, 1997; Vadineanu, 1998, 2001, 2005; Botnariuc, 1999), or pan-anarchy (Holling, 2001; Gunderson and Holling, 2002), which emerged by the end of the 1990s, as a result of a

I. E. Gonenc et al. (eds.), Assessment of the Fate and Effects of Toxic Agents on Water Resources, 199–224.
© 2007 *Springer.*

long-term multidisciplinary synthesis (e.g., biological ecology, environmental science, human ecology, ecological economics, conservation biology and ecology, mathematics).

As the science of Systems Ecology emerged, a wide range of complementary theories, concepts and tools were melted and incorporated, among which are information theory, systems theory, chaos and non-linear theory, network theory, cultural theory, social-learning theory, conflict theory, stability, ecological succession, mathematical modeling, and mathematical theory of complex and non-linear dynamic systems. Based on that, a comprehensive theoretical and operational framework has been defined and promoted, which allows the approach, identification, understanding and management of complex organization of nature/environment and human societies. The basic elements that define this theoretical and operational framework have been outlined recently in the book focused on modeling and sustainable use of the lagoon ecosystems (Vadineanu, 2005; Gonenc and Wolflin, 2005) and in other published books and articles dealing with some general aspects regarding the complexity of the environment (Holling, 2001; Gunderson and Holling, 2002; Vadineanu, 1998, 1999, 2001, 2004; Walker et al., 2002; Walker and Meyers, 2004; Fischer-Kowalski and Weisz, 1999). Here, we re-state some of those which we consider the most important for our purpose, and we add new elements that clarify and extend the overall view.

- The physical, biological, man-controlled and man-created environment from the upper lithosphere, hydrosphere and troposphere (Figure 1) has a hierarchical organization across space and time scales, in a wide range of life supporting systems or ecological systems (Figure 2).

These very diverse ecological systems, which are nested across space and time, consist of: (a) natural and seminatural or self[1] maintained ecological systems (mainly elementary or local ecosystems, local landscapes, and local and regional water scapes); (b) man-transformed and controlled ecological systems, which require in different extents a man-driven input of matter, information and commercial energy, or the so-called subsidized ecological systems (e.g., agro-ecosystems, agricultural landscapes, fish ponds and waterscapes for aquaculture); (c) man-created systems or systems that are highly dependent on the material, energy and information input from the first two categories (urban ecosystems, industrial ecosystems).

---

[1]The meaning is that they are not dependent on man driven input of matter, energy and information. However, there are many natural mechanisms which support material, energy and information exchange among them.

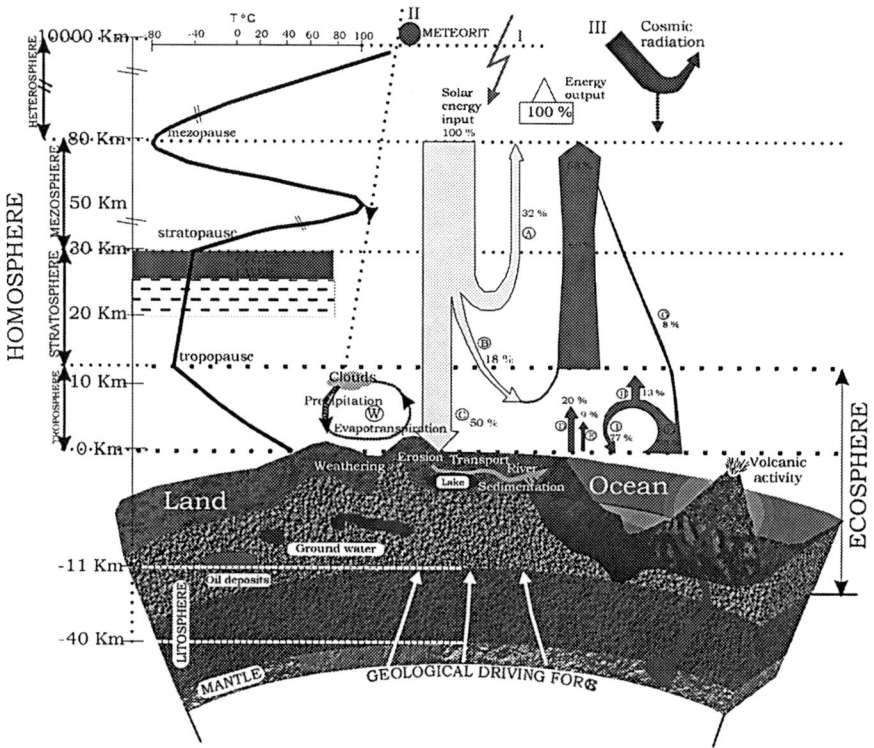

A. Reflected energy by lower atmospheric layers (32%);
B. Energy absorbed by the atmospheric layers (18%);
C. Incident energy at land and ocean surface (50%). Total radiated energy by the ecosphere (68%);
D. Latent heat (evapotranspiration)—20%;
E. Conduction and convection (kinetic energy—9%);
F. Radiated energy by land and ocean surface (98%);
G. Direct radiation to space from land and ocean surface (8%);
H. Radiation from troposphere to space (13%). Energy Output back to space (100%); consisting in reflected (32%) and radiated (68%) energy by ecosphere;
I. Contraradiation from troposphere back to land and ocean surface—77% (greenhouse loop).

Figure 1. Space distribution of the ecosphere within the Earth-Atmosphere System and the energy balance (*compiled from* Strahler and Strahler (1974) and Chiras (1991)).

According to the existing knowledge concerning the organization of life, one can discriminate among five hierarchical levels above biological individuals and four spatio-temporal levels within the ecological hierarchy. One must note that the three-dimensional space of the hierarchical organization integrates the upper lithosphere, the ocean basins, and the troposphere and

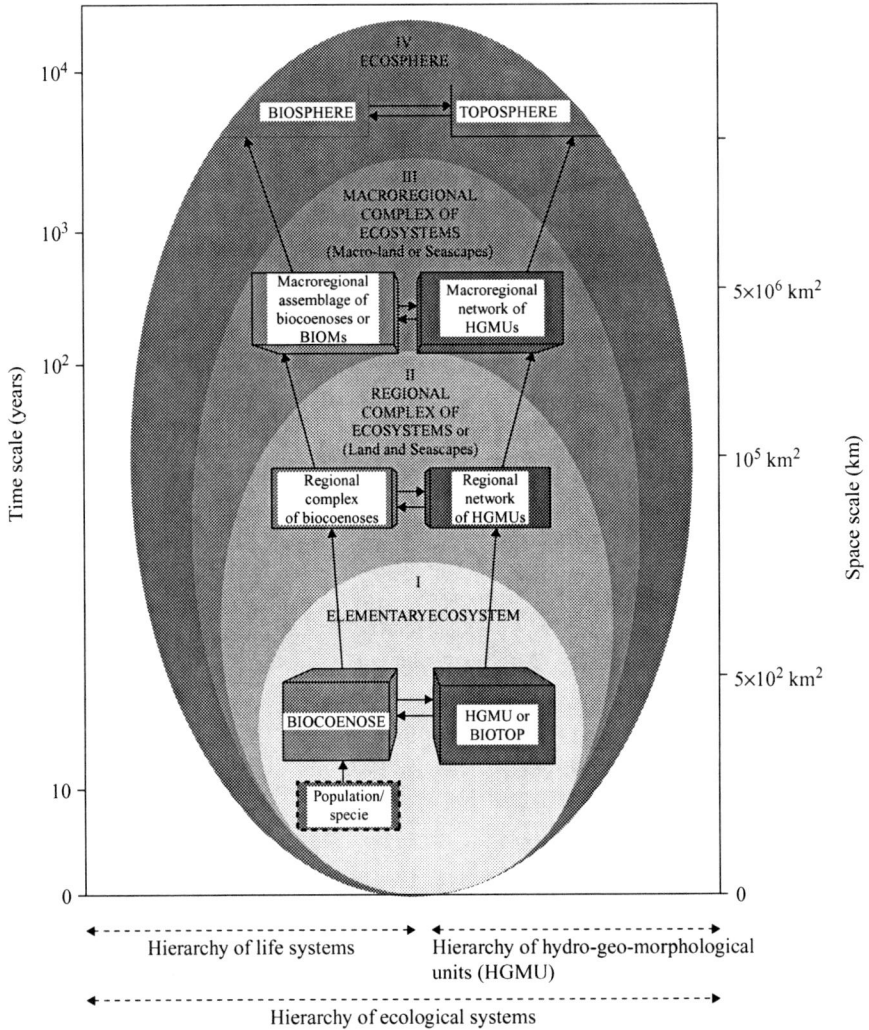

Figure 2. Hierarchical organization of natural, man-transformed, and man-created physical, chemical, and biological environments (after Vadineanu. A., Sustainable Development: Theory and Practice, Bucharest University Press, 1998. With permission)

that the time constants of ecological systems are measured in years, decades, centuries, or millennia.

• The Earth's nature consists of a mosaic of natural, seminatural, man-dominated and man-created ecological systems. Historically the interaction between man and nature led to the establishment at regional (hundreds and thousands of square kilometers), national (tens and hundreds of thousands of square kilometers) and global scales of mixed ecological complexes which integrate, in different proportion, self maintained, man-controlled and man-created ecosystems. In order to underline that the man or human populations, together with their created ecosystems, belong to nature, we have adapted the concept of "socio-environmental system" (Musters et al., 1998) into socio-ecological system (SEcS) for designating a mixed ecological complex (Vadineanu, 1998, 1999, 2001; Walker et al., 2002).

- The exponential increase of human population size is associated with a fast and extensive colonization of nature through:
  (a) transformation of natural ecosystems into man-controlled and subsidized or into man-created ecosystems; and
  (b) globalization of transport and communication networks allows us to consider that, at the beginning of the twenty-first century, the global nature system or ecosphere and its hierarchical organization have been transformed into a global socio-ecological system and a hierarchy of socio-ecological systems, nested within each other across space and time scales (Figure 3).

Within this hierarchy, the coastal socio-ecological systems established at local and regional scales occupy a particular position, being, on one side, the most populated, man-transformed and thus most vulnerable to disturbing factors and unexpected events, and on the other side, containing in the biophysical structure of their foundation the most productive natural and semi-natural ecosystems, such as lagoons, deltas, estuaries, continental shelves, mangroves and coral reefs.

Critical analysis and integration of the results of many research activities carried out in the last 4–5 decades on major biological, hydrological, biogeochimical and ecological or ecosystem and landscape processes prove that the costal zones and the large river systems, including their flood plains, delta or estuaries, are among the most efficient and important providers of services to the social and economic subsystems (Norberg, 1999; de Groot et al., 2002; Vadineanu et al., 2003, 2001; Vadineanu 2005; Pinay et al., 1990; Pinay and Decamp, 1988; Maltby et al., 1996; Costanza and Faber, 2002).

Recent estimates indicate that almost 70 percent of global human population lives within a coastal band of 100–150 km (Gonenc and Wolflin, 2005; Montgomery, 2006). Under these circumstances, the local, regional and national strategies and development programs should be targeted towards: (i) conserving their structural and functional integrity; (ii) restoring the degraded biophysical structure of natural capital; and (iii) balancing social and

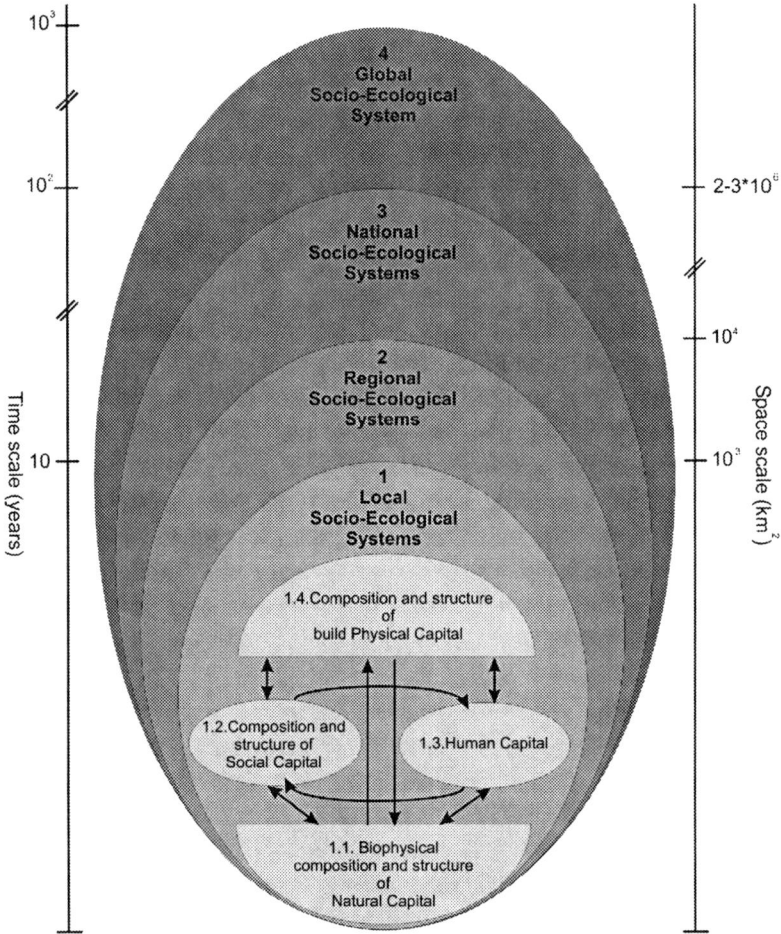

Figure 3. An alternative view upon the organization of nature/environment and human society across time and space scale, into socio-ecological systems

1. A local socio-ecological system and its capital structure, viewd as support for production, resilience and development
   1.1. Natural, seminaural and human manipulated and subsidized ecosystems which generate resources and services;
   1.2. Social organization, formal and informal institutions;
   1.3. Skilled human resources, scientific and traditional knowledge; practical expertise;
   1.4. Physical infrastructure of social and economic systems which supports social and industrial metabolism.
   Each type of socio-ecological system contains a similar capital structure.

economic structures and metabolism for sustainable use of the avaliable resources and services.

This requires well-equiped adaptive potential emboded in each of the four major compartments of the SEcS[s] (Figures 3 and 5). In order to meet these

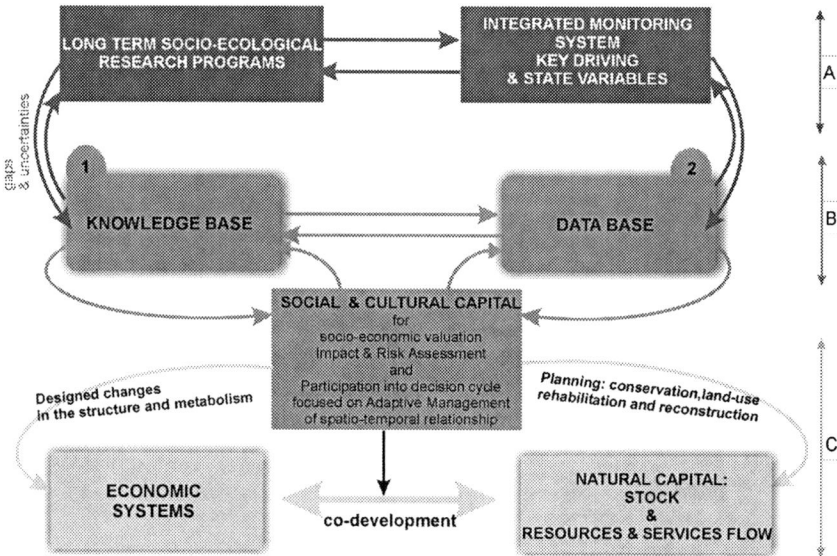

Figure 4. The flow diagram showing the relationships between: A) research and monitoring infrastructure; B) information system (core of Decision System) and C) adaptive management for co-development (sustainability) of Danube Delta and IBr$_s$ socio-ecological complexes (LDRS)

requirements, which can allow for ecosystem and adaptive management or co-development within or between SEcS^s, we have created and applied in the Lower Danube River Catchment (including the coastal delta and Razim—Sinoe lagoons) an overall conceptual framework (Figure 4) regarding the structure of a powerful and effective decision support infrastructure.

This framework recognizes the need for: (i) long-term socio-ecological research (LTSER) and integrated monitoring (IM) facilities and programs (Figure 4(A)); (ii) information regarding structural and functional changes in major drivers and pressures and reliable knowledge about the ecosystem processes and functions and the physical economy (Figure 4(B)); (iii) analytical frameworks and toolkits enabling scenario development and decision making (Figure 4(C)).

Indeed, a decision support infrastructure developed according to the proposed conceptual framework may allow policy and decision makers, managers and the public to make choices that center around finding the balance between the carrying capacity of the biophysical structure of NC and the many demands placed upon it by the different stakeholders and the entire social and economic subsystems that depend upon it.

Analytical approach and methods. It is now widely recognized among scientists, policy and decision makers, that a comprehensive analytical

Boundary conditions
of
Coastal Socio – Ecological Systems (SEcS)

STRUCTURAL ANALYSIS

Composition and structure of Natural Capital (NC) • natural, seminatural and man-subsidized terrestrial and aquatic ecosystems • Spatial distribution and connectivity	Composition and structure of build physical Capital (human settlements, water supply and sewage systems; infrastructure for solid waste collection and disposal; infrastructure for manufacturing activities; transport infrastructure)	Composition and structure of social capital (SC) (demography, stakeholders map, institutions)	Human Capital (Human resources, scientific and traditional knowledges, practical experience)

Building the Structural model which identify the coastal SEcS
Defining: a) the set of structural state variables (e.g. number of different types of habitats / ecosystems; number of human settlements; land cover; economic sectors or industrial units; primary, secondary and key stakeholders; institutions)
b) the internal and external natural and anthropic drivers and pressures

Functional analysis of Coastal SEcS

Land and water scape processes, functions, resources and services	Industrial and social metabolism	Knowledge based policy and decision cycle

Social valuation and estimation of total economic value (TEV) of NC	Physical economy of SEcS	Monetary flow

Adaptive Management of Coastal SEcS

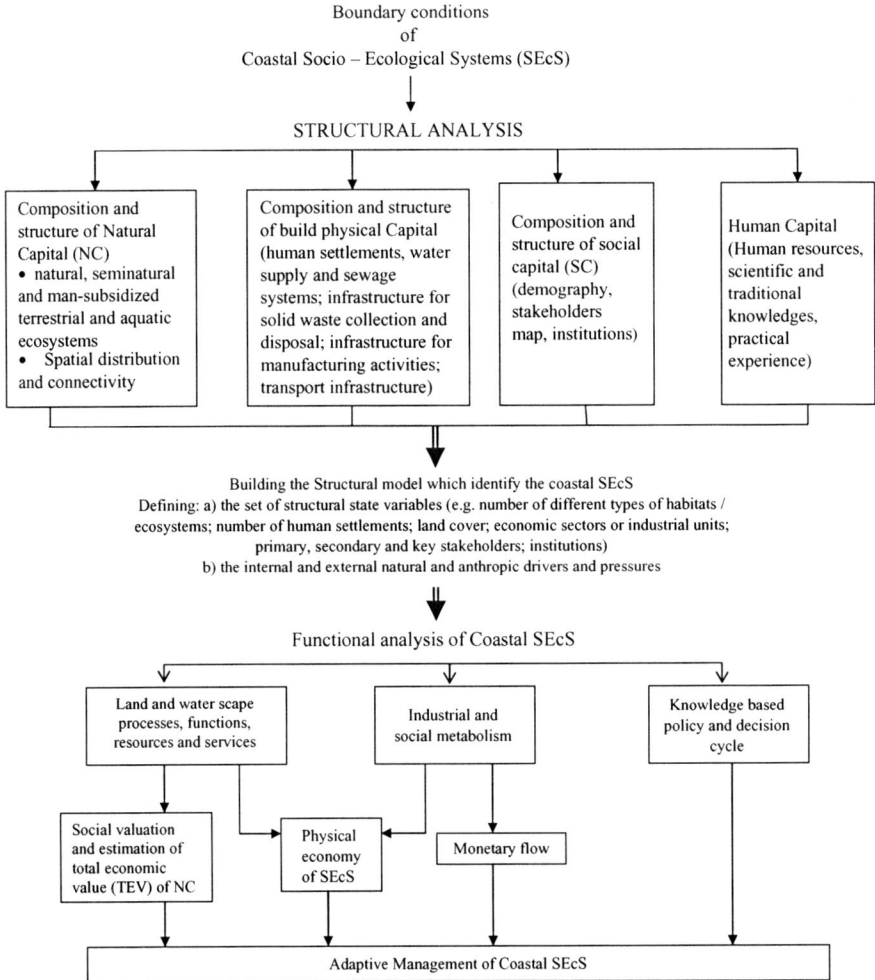

Figure 5. Flow diagram showing major steps for identification and description the dynamics of Coastal SEcS

framework and effective tools are needed to achieve the goal of sustainability (Figures 5 and 6). The most critical steps in the proposed analytical framework are dealing with: (i) defining boundary conditions and structural analysis for local and regional SEcS[s] (Figure 5) and for the major categories of elementary components (e.g. lagoon—Figure 6); (ii) building the structural homomorph models of coastal SEcS[s] and its constituent units (e.g., Lagoon TDM[s] network, stakeholders map) (Vadineanu, 2005; Pahl-Wostl, 1995); (iii) establishing the set of structural state variables and the key internal and external drivers and pressures; (iv) functional analysis for describing the physical economy of each category of component units (e.g., lagoon—Figure 6) and of the entire SEcS[s]

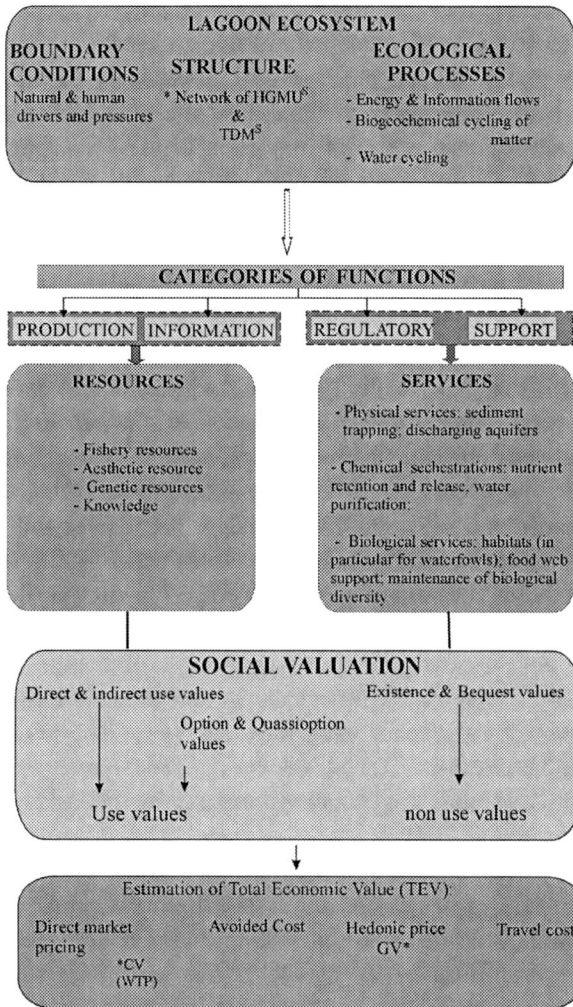

Figure 6. A proposal for analytical framework of lagoon functions, resources and services and their social and economic valuation
* HGMU—Hidrogemorphic units; TDM—Tropho-dynamic modules; CV—Contingent valuation; WTP—Willingness to pay; GV = Group Valuation

(Figure 5); (v) social analysis, conflict identification and resolution, and public participation; (vi) social valuation and estimation of TEV of NC (Figure 6).

9.2. Processes and Functions

Transfer of mass (e.g. biogeochemical cycles, water cycle), energy (flow through) and information are major ecosystem and landscape processes (functioning process) that account for the resilience of these systems and their capacity to provide resources and services. Experts in the field of ecology consider the function concept in terms of the capacity and workings of sets of coupled ecosystem processes. In that regard, the ecosystems and landscape functions explain some of the most important of their functioning features: resilience, productivity and carrying capacity.

Ecological economists and decision makers are taking into consideration the "functional values", which are defined in socio-economic terms, as relating to the actual provision of natural resources (goods) and services that satisfy needs and wants, both physiological and psychological (Burgess et al., 2001; Constanza et al., 1989; De Groot, 1992; O'Connor and Spash, 1997). Indeed, the performance of many human activities and the satisfaction of a wide range of physiological and psychological needs and wants, or in other terms the social and economic "metabolism", are totally dependent on the integrity and functioning capacity of a certain biophysical infrastructure, which provides resources and services. This concept recognizes the fundamental role of natural and semi-natural ecosystems in providing the foundation or "ecological footprint" (Wackernagel and Rees, 1996) to the socio-economic systems (Holling, 2001; Holling et al., 2002; Vadineanu, 2001).

By translating the ecosystem and landscape processes and functioning capacity into functional provisions to the socio-economic systems, the narrow concept of natural resources has been integrated in a much wider analytical framework that allows the adaptive management of socio-ecological complexes and finally operationalism of the concept of sustainability (Mohammadian, 2000).

In practice, reliable scientific data and information are required about the ecosystem (e.g., lagoons) and landscape funtioning processes in order to achieve conservation or to rehabilitate and restore the "good ecological status" of water bodies under the EU-WFD. In order to cope with the many difficulties in obtaining this information, different attempts have been made in the last decade to develop classification systems for ecosystem functions (Norberg, 1999; Maltby et al., 1996; de Groot, 1992, 2002). The classification proposed by De Groot in 1992 and improved in 2002 has been widely accepted and used in case studies, including the Millenium Ecosystem Assessment. It distinguishes four categories of ecosystem functions: (a) regulation functions; (b) carrier/support functions; (c) production functions; and (d) information functions (Figure 6).

Burgess et al. (2001) noticed that the rank order for De Groot's function categories is based on an econcentric perspective. From this perspective, the conservation of essential ecosystem and landscape processes, grouped as regulation and support functions, provide the preconditions for production and information functions. From a technocentric perspective, it might

be argued that the available living space (support/carrier functions) and resources (production functions) are most important in the short and medium
term. However, the actual holistic or ecosystemic approach is trying to avoid
ranking the importance of different categories of functions. That is because
it is well documented that the functions are inter-related, and their relative
importance changes with the adaptative transformations that are taking place
in the development cycle of ecological systems (Holling, 2001).

9.3. Functional Values and Socio-Economic Valuation

As has been already stated, due to the multifunctional characteristic of natural and seminatural ecosystems and landscapes, in particular of lagoons
and other coastal wetlands, the estimation of every function (service or resource flow) and of the linkages between them and consequently their socio-
economic valuation becomes a very difficult task. Addressing such difficulties, Barbier (1993, 1994) has proposed three aproaches to ecosystem change
valuation:

(a) impact analysis (assessing the damage from a particular impact);
(b) partial valuation (considering specific functions or categories of func-
    tions);
(c) total valuation (estimating the full present value of an ecosystem).

The experts in ecological economics consider that the key point in a comprehensive socio-ecomonic evaluation is to identify how society is affected by
the functions an ecosystem or landscape might perform, and by changes in
the ecosystem functioning (Burgess et al., 2001; Constanza et al., 1989, 1997;
De Groot, 2002). Thus the key to valuing a change in a ecosystem function
is establishing the link between that function and some service or resource
flow valued by people. The value of a change in an ecosystem function can be
derived from the change in the value of ecosystem service or resource flow it
provides (Figure 6).

The coastal land and waterscapes, together with their components and
diversity functioning processes (functions), are viewed as essential parts of
the "natural capital" (NC) stocks of any local and regional SEcS[s]. In the same
terms of ecological economics, the lagoon resources and services represent
the benefits that human populations derive from their sustained functioning.

Indeed, natural capital stocks and the associated flows of resources and services are combined with manufactured (man-made physical capital) and human capital goods and services in order to enhance human welfare. Although
the principles of neoclassical economics are still prevailing in the policy and
decision-making process, there is a growing new trend based on recognition

that natural capital and man-made capital are complements, rather than sub-
stitutes, in any socio-ecological complex. Thus, balancing spatio-temporal
relationships between them within hierarchical socio-ecological complexes is
the essence of sustainability.

In order to assess the lagoon functions, it is necessary to first establish
which of the potential functions are being performed and then to determine
the impact of the functions on human welfare. In that regard, a set of procedures named "Functional Assessment Procedures" (FAP[S]) have been developed, tested and validated during implementation (1995/2003) of four EU
research projects (Maltby et al., 1996; Burgess et al., 1998). These procedures are starting to be extensively used for wetland functions assessment.
Although FAP[S] distinguish only three categories of functions (hydrological,
biogeochemical and ecological or ecosystem maintenance), they currently
seem to be the only available tool for efficient identification and assessment
the functions of wetland ecosystems, including lagoon systems.

The socio-economic analysis and evaluation of coastal SEcS and lagoons
functions has to take into consideration the social context (institutional, political, socio-cultural) and the economic circumstances from the socio-ecological
complex in which they are integrated. De Groot (1992, 2002) has classified
values from a wide catchment perspective, where they are considered in terms
of the "full value" or "intrinsic value" of natural systems and the functions
they provide in terms of ecological, social and economic values. He argues
that the socio-economic value of any resource or service should be determined
by the extent to which it contributes to human welfare, which in the broad-
est possible sense includes material prosperity, environmental, physical and
mental health, employment and social contacts.

The aim of economic valuation methods and techniques is to provide
information on values that would normally be provided in the market (Burges
et al., 2001). This requires knowledge about the density flow of resources and
services from the river and stream systems, and about the values that society
places on them.

Each component of the Natural Capital Stock provides marketable resources and services (e.g., food, timber, drinking water, transportation, elec-
tricity etc.) and many non-marketable resources and services (e.g., biological diversity, clean water and air, aesthetic and cultural benefits) to a given
socio-economic system (Costanza et al., 1997; Wilson and Carpenter, 1999;
De Groot, 1994). Many attempts have been made in the last two decades
to provide economic valuation of the nonmarketable resources and services
produced by the natural, semi-natural and man-dominated components (e.g.,
ecosystems and landscapes) of the biophysical structure of NC, in order to
avoid or minimize the overestimation of the role of use values or the underes-
timation of the role of non-use values during the policy and decision-making

process (Freeman, 1993; Diamond and Hausman, 1994; Pate and Loomis, 1997; van der Straaten, 1998; Wilson and Carpenter, 1999; Campell and Luckert, 2002). As a result, many alternative methods and techniques were developed and applied to assign economic values as direct and indirect use values to the marketable resources and services as well as non-use values to the non-marketable resources and services, among which one can identify the optional, quasioptional, bequest and existence values (Figure 6).

Wilson and Carpenter (1999) have published the results of a critical analysis carried out on 30 research papers and study reports dealing with economic valuation of surface freshwater ecosystem resources and services. In all reviewed papers, the methods applied for economic valuation were developed according to the theory of neoclassical economics, which only recognizes human-centered values for the non-marketable resources and services. Based on the main conclusions of Wilson and Carpenter (1999) and many other authors working in the field of economic valuation of the ecosystem functions (Pimentel et al., 1997; Constanza et al., 1997; Turner et al., 2003; Holmes et al., 2004; Burgess et al., 1998), the following gaps and research needs should be considered in the near future:

(i)   Each valuation method targets a different function, or, in other terms its estimation potential, tends to be limited to some specific resources and services of the total range of a given component of the natural capital stock.
(ii) Economic valuation of the non-marketable resources and services tends to be specific to a particular method, type of ecological system and socio-economic circumstance.
(iii) The human-centred techniques of economic valuation grounded in the theory of neoclassical economics preserves a significant human bias.

Burgess et al. (1998) carried out a deep critical analysis of the role of Contingent Valuation (CV) methods (WTP and WTA) in environmental decision making. They have concluded that decisions about the environment should be based on social consensus about appropiate standards and acceptable choices, rather than on the individual WTP amounts elicited in CV surveys. Thus, group valuation (GV) is seen as a significant step to limit the human bias in economic valuation of natural capital.

9.4. Lower Danube Wetlands System Case study

Introduction: The Lower Danube Wetlands System (LDWS), comprising the river stretch of 840 km between the Black Sea and the Iron Gate II man-made reservoir together with the associated floodplains and inner and coastal deltas

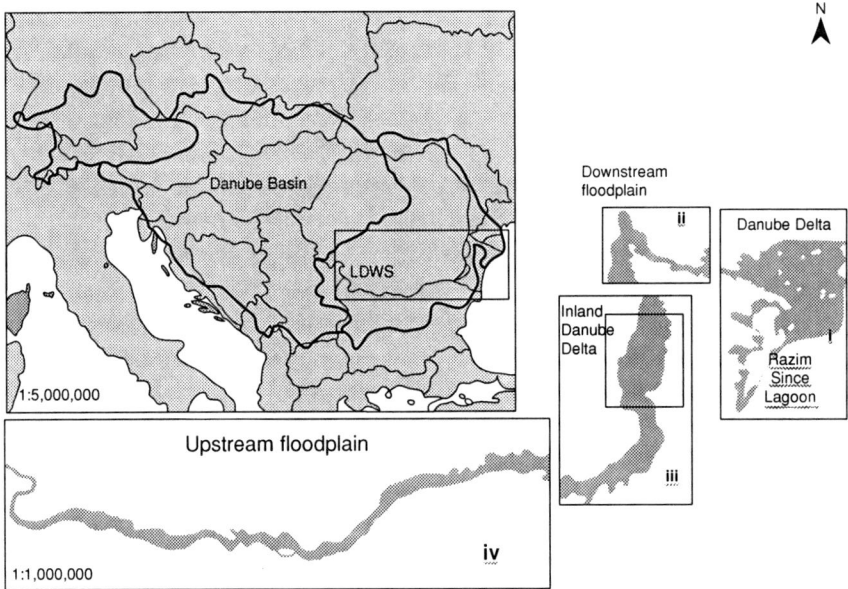

Figure 7. Danube River Catchment and Major units of the Lower Danube River System (LDRS)

functions as a key component of the second largest river system in Europe (Figure 7). It serves as a buffer system between the river catchment ($\sim$817,000 km^2) and the sea, bearing also the ecological footprint of the economies of the direct riverine countries: Romania, Bulgaria, Yugoslavia, Ukraine and Republic of Moldavia. In late 1950s the LDWS extended over a total surface of about 10,000 km^2 along the lower river stretch of 840 km (Figure 7). That consisted of four major wetland ecosystem units: (i) Coastal Danube Delta extended over 5193 km^2, from which 1015 km^2 are covered by the Razim-Sinoe Lagoon complex; (ii) floodplains (701 km^2) developed along river stretch of 92 km long, between coastal and inland Danube delta; (iii) Inland Danube delta which has developed along the river stretch between kilometers 170 and 365 and between Southern Romanian Plain and the Dobrogean Plateau, over a total surface of 2413 km^2 and (iv) the floodplains developed on the Romanian territory, along the Danube river stretch, between kilometers 365 and 840 or between Inland Delta and Iron Gate II man-made lake with a total surface of about 1500 km^2 (Vadineanu, 2001).

Ecological research and monitoring activities, which were carried out in the first six decades of the last century on LDWS, were sectoral (discipline based), reductionistic and inappropriately designed, in terms of space and time scales. The historical data and information derived from these have not helped to prevent the development and implementation of policies and management plans underpinned by basic principles of neo-classical economics and the reductionistic approach used in the fields of agriculture, civil engineering and water management. These principles suggested that wetlands should be considered as wastelands (Costanza, 1995; Arrow et al., 1995; Musters et al., 1998; Schot, 1999; Turner et al., 2000; Vadineanu, 1998, Vadineanu et al., 2003) and that the only "economic viable alternative" is their conversion into mono-functional (crop, fish or wood) and highly energy- and material-subsidized systems (Vadineanu et al., 1998; Vadineanu, 2001). Such conversion-based policy and management plans were "effectively" designed and implemented by a centralized, homogeneous and rigid social system that enabled a dictatorial command-control or linear decision-making process.

Since the 1970s, an extensive and long-term research programme was implemented on natural and semi-natural ecosystems, in particular those from two major areas of LDWS: (i) Coastal Danube delta and, (ii) Small Islands of Braila (a remnant of former inland Danube delta which preserves major structural and functional characteristics of former floodplain) (IBr) (Figure 7).

The ecosystem based research and monitoring carried out for more than four decades in these two characteristic areas of LDWS represents an extension of classical biological ecology research and monitoring activities carried out between 1900 and 1970 in coastal and inland Danube deltas. During the second phase of long-term ecological research, the activity was focused on the dynamics of composition structure and functioning processes of major categories of natural and semi-natural ecosystems (including biological diversity) under the influence of different driving factors.

The results produced in both phases of extensive and intensive ecological research activities allowed for identification and assessment of (i) the reference state (RS) of LDWS; (ii) the trends in the structural and functional changes under pressure of major anthropic driving forces; (iii) the estimation of impact upon the most relevant ecosystem functions and the respective flow of resources and services; and (iv) the goal and targets for future policies and management (Antipa, 1910; Botnariuc and Beldescu, 1961; Cristofor 1987; Cristofor et al., 1992, 1993; Vadineanu et al., 1987, 1992, 2000, 2001; Vadineanu and Vadineanu, 2004; Navodaru et al., 2002; Sarbu et al., 1997). An overview of the output of such synthesis that was developed according to the analytical DPSIR model (EEA/R_{20}/2001) is shown in Table 1. The analysis has been designed to allow for the identification of major internal drivers and pressures, as well as long-distance and cumulative effects, at local (IBr$_s$, Coastal Danube delta), regional (Lower Danube wetlands and catchment) and macro-regional (Danube River Catchment and N-W Black Sea) scale.

The information and knowledge derived from the integration and synthesis of a large bulk of ecological and biological data collected over more than six decades of research on LDWS and north-western Black Sea have

TABLE 1. Major drivers and pressures, all related to the policies and management activities applied at different space scales (from local to Danube river catchment), the Reference state (1950s) and the impacts on structure and functions of LDWS

Drivers	Pressures	Space scale	Reference state (1950s)	Impacts
• Land reclamation	• Conversion of floodplain into intensive agro-ecosystems	All	• Biophisical Structure: Almost 10,000 km², of natural and semi-natural wetlands (including Razim-Sinoe lagoon system)	• Erosion of biophysical structure
• Increase of crop production and livestock	• Over exploitation of natural resources (e.g. fish)	All	• Mezo-trophie	• Eutrophication associated with cyanobacteria species bloom and severe food web simplification
• Commercial energy production in hydropower plants	• Introduction of alien species (e.g. asian cyprinids, canadian poplar)	Local and regional	• Production function: (a) annual fish catches: 20–30 ktons; (b) >500 ktons of reed and reed mace; (c) up to 2×10^5 cubic meters of timber; (d) ~150 ktons of crops and animal products.	• Production function: dramatic reduction of fish productivity and catches (below 7 ktons per year) Regulation function: reduction of water storage capacity upstream coastal delta by more than 9 cubic kilometers and thus the flood duration and water retention time; reduction of nitrogen and phosphorous retention capacity by 40–45 ktons and respectively 2–3 ktons per year.

• Increase waterway transport capacity	• Intensification of auxiliary energy and material inputs into food production systems	All	• Regulation function: (a) flood detention capacity of >20 (up to 32) cubic kilometres; (b) nitrogen retention: 120–140 ktons per year; (c) phosphorous retention: 6–10 ktons per year.	• Support function: reduction with almost 78 per cent habitats for spawning, feeding and nesting fish and bird species. Increased number of threatened and endangered species.
• Urbanization and Industrialization	• Point and diffuse emission	Mainly at regional and river catchment	• Support function: (a) habitats for 1688 plant and 3735 animal species, in particular spawning, feeding or nesting habitats for migratory fish and bird species; species richness, including many ponto-caspian relicts	• Information: reduction of recreation potential upstream coastal delta.
	• Hydrotechnical works		• Information function: complexity and uniqueness of LDWS a huge recreation & knowledge potential	

TABLE 2. Legal and strategic framework for Lower Danube Wetlands System (LDWS) and North-West Black Sea

Conventions at the global level:

• The Convention on Biological Diversity.
• Convention on Wetlands of International Importance, Especially as Habitat for Waterfowl (Ramsar).
• Convention on Migratory Species (Bonn).
• World Heritage Convention.

Conventions and strategies at Pan-European and EU level:

• Berne Convention concerning the Conservation of the European Wildlife and natural Habitats.
• Pan-European strategy and action plan for biological and landscape diversity conservation.
• EU-Habitat Directive—92/43/CEE.
• Birds Directive—79/409/CEE.
• EU-WFD/2000/60/EC.

Convention at the regional level:

• Bucharest Convention for the protection of the Black Sea against pollution (Bucharest, 1992).
• Convention on the co-operation for the protection and sustainable use of the Danube (Sofia 1994).

Initiatives and programmes:

• Bucharest agreement on the establishment of Danube River Green Corridor/Bucharest/2000.
• Environmental Programme for the Danube River Basin (World Bank-GEF, EU-PHARE, EBRD).
• Action Plan for the Black Sea Rehabilitation (UNEP/UNDP/World Bank)/1998.

been used for adopting and development the legal and strategic framework (Table 2) applicable in this particular area. According to the sound scientific background regarding natural and seminatural ecosystems framework, the goal and objectives of policy and management were changed and focused on: (a) biodiversity conservation and sustainable use of their resources and services, and (b) protection against point and diffuse pollution, in particular against eutrophication. In that regard, the Coastal Danube Delta has been included in the network of Biosphere Reserves (1990) and network of world heritage and Ramsar Sites (1991), while the Small Islands of Braila have been designed as a natural park (2000), Ramsar Site (2001) and a special protected area for habitats and birds (EU—Habitat and Birds directives). In addition, at the end of 1990s, the legal representatives of the danubian countries agreed on the action plan for Black Sea rehabilitation, which includes bringing the total nitrogen and phosphorous discharges of the Danube River below 300 ktons per year and 16 ktons per year (a 40% reduction of highest discharges recorded at late 1980s) by 2010 (Vadineanu et al., 1998; EPDRB, 1995).

In spite of scientifically sound measures promoted in the new management plans, the achievements are still very far from what was expected. This is due to the lack of proper social and economic valuation of resources and services provided by biophysical components of Natural Capital from LDWS, the lack of communication to and among primary and most important stakeholders and their exclusion from decision making, and the diversion of resource and service flows out of local economies. It was recognized by the end of 1990s that for achieving the objectives of biodiversity conservation and sustainable use, ecological restoration and the goal of sustainability, there is a need not only for sound scientific certification based on disciplinary and interdisciplinary knowledge, but also for a wide and effective social certification assured by transdisciplinary knowledge, strong social capital, communication and public participation in the policy cycle.

Contribution from social science to LTSER (case study Danube Delta, IBr and LDWS): In the last years, the research staff of DSES—UniBuc has focused a significant part of the research effort on the development and implementation of a comprehensive analytical framework targeted for identification, assessment and valuation of the major functions and related flows of resources and services of the LDWS. In order to undertake this type of work, special attention was given to building the information and knowledge system (Figure 4(B)), by integrating reliable data and information related to the structure and functioning processes of LDWS, and to major strategies and policies applied at different spatial scales in the Danube River Catchment. The structural and functional changes that have occurred in the LDWS were identified by analyzing the state transition of LDWS between 1950s (reference state) and beginning of 1990s (Table 1) In particular, the analysis was focused on major functions: (i) production functions (e.g., fish catches; reed and reed mace; livestock, timber, crops, medicinal plants); (ii) regulation functions (nitrogen and phosphorous retention; flood detention); (iii) support functions (habitats, species richness and food web support); and (iv) information functions (tourism).

The crude estimation of total economic value of the remaining wetlands (1729 USD per ha per year) and the established polders was carried out using the direct market pricing, avoided cost (AC), replacement cost (RC); WTP and group valuation (GV) methods.

The outputs of such analysis clearly shows the economic performance of multifunctional and self–maintaining ecosystems of LDWS, compared with monofunctional and human-dependent agroecosystems established in the river floodplain after polderisation (Vadineanu et al. 2003; Vadineanu 2005).

It was stressed that the lack of this type of evaluation and the errors made in designing the structure and "metabolic" rate of the local and regional economic systems encouraged policy and decision makers from the past to perceive the floodplain of LDWS as "wastelands." The cost of extensive polderisation (>4000 km^2) of LDWS' floodplains reached at the end of 1980s a total amount of 4 billion USD, and since then we have estimated (by internalizing cumulative and long distance effects) an annual economic loss of about 500 million USD.

Bearing in mind the differences between the reference and current states of LDWS and the respective economic consequences as well as the long-term objectives of the new established policy in the region, which deals with: (a) biodiversity conservation; (b) 40% reduction of the potential nutrient discharges into Black Sea by 2010; and (c) sustainable development, an operational plan for holistic bio-economic management of these wetlands has been proposed. This plan is based on the reconstruction of 1500 km^2 of wetlands in the LDWS and the implementation of multifunctional farming in the Lower Danube Catchment and remaining polders.

The potential impacts of the implementation of such a plan in terms of the physical economy of LDWS and the economic benefits for local populations have also been estimated. Based on these estimates, expected benefits include an increase of the retention capacity of the LDWS by 22.5 ktones of TN and 1.3 ktones of TP per year and an increase of fish catches up to 12–14 ktones per year, as a result of the increase of nursery and production function for semi-migratory and migratory fish species. Increases of other renewable resources, such as timber, reed biomass, hay, medicinal herbs, are also anticipated.

The expected effects may allow also reintroduction of the traditional and extensive farming system of crop and vegetable production and animal raising adapted to the "natural hydrological" regime of the river. Biodiversity in the LDWS and North—Western Black Sea may also benefit significantly. The estimates of the impact of restoration in the LDWS upon local and regional economies, in particular on fishery and tourism sectors, indicate a potential annual net economic benefit that might exceed 300 millions USD (Vadineanu et al., 2003; Vadineanu unpublished).

The proposed operational plan provides strong support for the Inter–governmental Agreement among lower danubian countries (Bucharest, 2000), which involves designing and restoring the Green Corridor of Lower Danube River, and includes basic characteristics of a scenario centered on sustainability.

The Academy of Agricultural and Silvicultural Sciences was opposed to the "sustainability scenario" and promoted the "business as usual" scenario, which was intended to keep the status quo in the LDWS. In that regard, they proposed a project for "ecological restoration" of the established polders by consolidation of the system for flood defense, rehabilitation of the irrigation system, construction of a drainage system and strengthening the intensive crop production system, using the principles of neoclassical economics and sectoral approach. Although the economic valuation of major functions and flows of services of the natural and semi-natural ecosystems from Danube Delta and the Small Islands of Braila has brought clarification and additional strong economic arguments in favor of restoration the Green Corridor of Lower Danube River and extensive application of multifunctional farming on agricultural land from remaining polders and lower Danube catchment, some very strong constraints against it have been raised, which explains the confusion and delay in choosing a clear policy option.

The investigations carried out in recent years have identified major constraints that limit the effective transfer of good scientific knowledge into the decision cycle (Vadineanu and Palarie, 2004). This work has clearly shown that the final choice for one of the proposed scenarios and the action plan for successful implementation require further social research. In that regard, a complementary social research program has started to be financed and implemented in 2004 in four national LTSER pilot areas, among which are the IBr and the Coastal Danube delta. A consortium of thirteen partners is conducting intensive social research activities regarding: (i) the structure of social capital; (ii) people's perceptions, attitudes and behaviors related to biodiversity, wetlands restoration and sustainable use of ecosystem services; (iii) traditional knowledge and practical experience of primary stakeholders (local population); (iv) the education, information and communication systems; (v) the development of public awareness and raising level of involvement into decision-making process; and (vi) the diversity and origin of conflict of interests as well as the opportunities and mechanisms for conflict resolution and the re-designing of the structure and metabolism of local and regional socio-economic systems.

The analysis of partial results produced by the social research activities carried out in the LTSER platforms from LDWS has yielded several insights. First, it enabled us to identify, classify and map stakeholders who directly or indirectly are, or should be, involved in policy development and decision-making concerning future management of LDWS. Second, there is a need for improved bridges between two categories of scientific teams, each supporting one of the two scenarios regarding future development of local and regional socio-ecological complexes, Third, scientific stakeholders need to improve their communication of disciplinary information and knowledge to a large variety of primary and secondary stakeholders and work to create and communicate transdisciplinary knowledge. Fourth, a severe dichotomy exists between secondary stakeholders, represented by central and regional authorities, and even worse, between primary stakeholders, represented by the owners of newly established large (thousands of hectares) and very large (10,000–50,000 hectares) crop farms and by the owners of new large fishing companies on one side, and the social groups representing local communities on the other side. Lastly, the analysis has identified the sources and particular conditions that create conflicts between poor and marginalized local communities and very rich and powerful "colonists" representing or having very strong links with political and economic groups of interests.

The results are also showing that the most powerful and influential secondary and primary stakeholders who promote short term and sectoral policies and management plans are against extensive floodplain restoration and biodiversity conservation, while most primary stakeholders (poor and marginalized social groups representing local communities) are, or might be in favor of this scenario, if they are well informed about the potential benefits for their future wellbeing, and are deeply involved in all phases of policy cycle.

9.5. Conclusions

The proposed or selected theoretical and practical developments in the field of the ecosystem approach and adaptive management, briefly described and applied in this paper, allow for the formulation of two basic conclusions:

(i) The holistic conceptual and analytical frameworks help to address the complex organizational units of nature and human society, and to shift from a sectoral, reductionistic, linear and single equilibrium focused approach towards an ecosystem, non-linear and adaptive approach in both research and management of coastal SEcS[s]. In this context, any particular type of ecosystem (e.g., lagoon, estuary or farming system, urban ecosystem) should be investigated, understood and managed as a component of a local SEcS. The application of the proposed analytical framework for assessing and managing the risks and impacts on water resources related to major natural and human drivers and pressures should also be highly considered.

(ii) The effective application of the described analytical framework depends on:
   • Scientifically sound tools and methods and good scientific products, properly packed in flexible and accessible information and transdisciplinary knowledge systems;

- Effective integration of traditional and expert knowledge into information and knowledge systems;
- Adaptive management plans built on cost-effective technical solutions—including practical experience of local populations;
- Complex, adaptive and effective social capital in support of the decision cycle.

It is thus recognized and applied what Paavola and Adger (2005) stated recently about "the contribution of social capital to human welfare and well-being". In that regard, the social capital has a similar role as conventional factors of production and natural capital.

References

Antipa, Gr., 1910. Regiunea inundabila a Dunarii—Starea ei actuala si mijloacele de a pune in valoare (Danube floodplains: Actual status and means of valorification), Instit. de Arte Grafice Carol Gobi, Bucuresti, p. 317.

Arrow, K., B. Bolin, R. Constanza, P. Dasgupta, C. Folke, C.S. Holling, O. Jansson, S. Levin, K.G. Maler, C. Perrings, and D. Pimentel, 1995. Economic growth, carrying capacity and the environment, Ecol. Econ., 15 (2), 91–97.

Barbier, E.B., 1993. Sustainable use of wetlands. Valuing tropical wetland benefits: Economic methodologies and applications, Geogr. J., 159 (1), 22–32.

Barbier, E.B., 1994. Valuing environmental functions: Tropical wetlands, Land Econ., 70 (2), 155–173.

Botnariuc, N., 1999. Evolutia sistemelor biologice supraindividuale, Ed. Universitatii din Bucuresti, 205 pp.

Botnariuc, N., and D. Beldescu, 1961. Monography of the Crapina -Jijila complex of shallow lakes, Hydrobiologia (Bucharest), 11, 161–242.

Burgess, E.D., S. Cornell, K. Turner, and S. Georgiou, 2001. Framework for the socio-economic analysis of wetlands within EVALUWET, CSERGE, East Anglia University, Norwich, UK.

Burgess, J., J. Clark, and C.M. Harrison, 1998. Respondents' evaluation of a CV survey: A case study based on an economic valuation of the Wildlife Enhancement Scheme, Pevensey Levels, East Sussex. Area, 30 (1), 19–27.

Campell, J.B., and M. Luckert (eds), 2002. Uncovering the hidden harvest: Valuation Methods for Woodland and Forest Resources, Earthscan, London.

Costanza, R., 1995. Economic growth, carrying capacity and the environment, Ecol. Econ., 15 (2), 89–90.

Costanza, R., and S. Farber, 2002. Introduction to the special issue on the dynamics and value of ecosystem services: Integrating economic and ecological perspectives. Ecol. Econ., 41 (3), 367–374.

Costanza, R., C.S. Farber, and J. Maxwell, 1989. Valuation and management of wetland ecosystems, Ecol. Econ., 1, 335–361.

Costanza, R., R. d'Arge, R. de Groot, S. Farber, M. Grasso, B. Hannon, K. Limburg, S. Naeem, R. O'Niel, J. Parnelo, R.G. Raskin, P. Sulton, and M. van den Belt, 1997. The value of the world's ecosystem services and natural capital, Nature, 387, 253.

Cristofor, S., 1987. L'evolution de l'etat trophique des ecosystemes aquatiques caracteritiques du Delta du Danube. 6. Responses de la vegetation submerse en fonction de la reserve de nutrients et du regime hydrologique, Rev. Roum. Biol.-Biol Anim. (Bucharest), 32 (2), 129–138.

Cristofor, S., A. Vadineanu, and Gh. Ignat, 1992. Light penetration in the Danube Delta lakes in relationship with the main factors, Hidrobiologia, 20, 5–22.

Cristofor, S., A. Vadineanu, Gh. Ignat, and C. Ciubuc, 1993. Factors affecting light penetration in shallow lakes, Hydrobiologia, 275/276, 493–498.

Cristofor, S., A. Vadineanu, and Gh. Ignat, 1993. Importance of flood zones for nitrogen and phosphorus dynamics in the Danube Delta, Hydrobiologia, 251, 143–148.

Chiras, D.D., 1991. Environmental Science: Action for a Sustainable Future, Benjamin Cummings, Redwood City, California.

De Groot, R.S., 1992. Functions of Nature, Wolters-Noordhoff, Amsterdam.

De Groot, R.S., 1994. Environmental functions and the economic value of natural ecosystems, edited by M.A. Jansson, M. Hammer, C. Folke, C. Costanza, Investing in Natural Capital: The Ecological Economics Approach to Sustainability, Island Press, Washington, DC, pp. 151–168.

De Groot, R.S., M.A. Wilson, and R.M.J. Boumans, 2002. A typology for the classification, description and valuation of ecosystem functions, goods and services, Ecol. Econ., 41 (3), 393–408.

Diamond, P.A., and A.J. Hausman, 1994. Contingent valuation: Is some number better than no number? J. Econ. Persp., 8 (4), 45.

EPDRB, 1995. Strategic Action Plan for the Danube River Basin, Vienna, 1995–2005, p. 109.

Fischer-Kowalski, M., and H. Weisz, 1999. Society as hybrid between material and symbolic realms. Towards a theoretical framework of society—nature interrelation, Adv. Human Ecol., 8, 215–251.

Freeman, M., 1993. The Measurement of Environmental and Resource Values: Theory and Methods. RFF, Washington, DC.

Gonenc, E.I., and J.P. Wolflin (eds), 2005. Coastal Lagoons: Ecosystem Processes and Modeling for Sustainable Use and Development, CRC Press, Boca Raton, Florida.

Gunderson, H.L., and C.S. Holling (eds), 2002. Panarchy: Understanding Transformations in Human and Natural Systems, Island Press, Washington, DC.

Holling, C.S., 2001. Understanding the complexity of economic, ecological and social systems, Ecosystems, 4, 390–405.

Holling, C.S., H.L. Gunderson, and D.G. Peterson, 2002. Sustainability and panarchies, edited by H.L. Gunderson, C.S. Holling, Panarchy: Understanding Transformations in Human and Natural Systems, Island Press, Washington, pp. 25–62.

Holmes, P.T., C.J. Bergstrom, E. Huszar, S.B. Kask, and F. Orr, III, 2004. Contingent valuation, net marginal benefits, and the scale of riparian ecosystem restoration, Ecol. Econ., 49 (1), 19–30.

Maltby, E., V.D. Hogan, and J.R. McInnes, 1996. Functional analysis of European wetlands ecosystems. Phase I. (FAEWE) Ecosystems Research Report No. 18, Office for Official Publications of the EC, Luxemburg.

Mohammadian, M, 2000. Bioeconomics: Biological Economics Interdisciplinary Study of Biology Economics and Education, Edition Personal, Madrid.

Montgomery, W.C., 2006. Environmental Geology, McGraw-Hill Companies Inc., Boston.

Musters, C.J.M., H.J. DeGraaf, and W.J. Ter Keurs, 1998. Political and economic inequality and the environment, Ecol. Econ., 226 (3), 243–258.

Navodaru, I, D.A. Buije, and M. Staras, 2002. Effects of hydrology and water quality on the fish community in Danube delta lakes, Internal Rev. Hydrobiol., 87 (2–3), 329–348.

Norberg, J., 1999. Linking nature's services to ecosystems: some general ecological concepts, Ecol. Econ., 29 (2), 183.

O'Connor, M., and C. Spash (eds), 1997. Valuation and Environment: Principles and Practices, Edward Elgar, Aldershot, UK.

Odum, E.P., 1993. Ecology and Our Endangered Life-Support Systems, Sinauer Associates, Sunderland, Massachusetts.

Odum, E.P., 1997. Ecology: A Bridge Between Science and Society. Sinauer Associates Inc., Sunderland, Massachusetts, 330 pp.

Pahl-Wostl, C., 1995. The Organic Nature of Ecosystems. J. Wiley and Sons, New York.

Pate, J., and J. Loomis, 1997. The effect of distance on willingness to pay values: A case study of wetlands and Salmon in California, Ecol. Econ., 20, 199–207.

Paavola, J., and N.W. Adger, 2005. Institutional ecological economics, Ecol. Econ., 53 (3), 353–368.

Pimentel, D., C. Wilson, C. McCullum, R. Huang, P. Dwen, J. Flack, G. Tran, T. Saltman, and B. Cliff, 1997. Economic and environmental benefits of biodiversity, Bio-Science, 47, 747–757.

Pinay, G., and H. Décamp, 1988. The role of riparian woods in regulating nitrogen fluxes between the alluvial aquifer and surface water: A conceptual model. Regulated Rivers, 2, 507.

Pinay, G., H. Décamp, E. Chouvet, and E. Fustec, 1990. Functions of ecotones in fluvial systems, edited by R. Noiman, and H. Décamp, The Ecology and Management of Aquatic-Terrestrial Ecotones, Parthenon Press Publ., London, pp. 141–169.

Sarbu, A., S. Cristofor, A. Vadineanu, and C. Florescu, 1997. Changes in submerged macrophytes from Danube Delta, Acta Botanica Horti Bucurestiensis (1995/96), 85–91.

Schot, P.P., 1999. Wetlands, edited by B. Narth, L. Hens, P. Compton, D. Devuyst, Environment Management in Practice, Vol 3. Routledge, London, pp. 62–85.

Strahler, N.A., and H.A. Strahler, 1974. Introduction to Environmental Science, Hamilton Publishing, Santa Barbara.

Turner, R.K., I.J. Bateman, and W.N. Adger (eds), 2000. Economics of Coastal and Water Resources: Valuing Environmental Functions, Kluwer Acad. Publ., Dordrecht.

Turner, R.K., J. Paavola, P. Cooper, S. Faber, V. Jessamy, and S. Georgiou, 2003. Valuing nature: lessons learned and future research directions, Ecol. Econ., 46 (3), 493–510.

Vadineanu, A., 1993. Wetlands from western and central Europe: The Danube Delta, edited by P. Dugan, Wetlands in Danger. Reed Int. Books Ltd., London, pp. 161–183.

Vadineanu, A., 1998. Sustainable Development: Theory and Practice, Bucharest University Press.

Vadineanu, A., 1999. Balancing socio-economic development and biodiversity in Central and Eastern Europe, Eur. Nature: Mag. Interface Policy Sci., 3, 13–14.

Vadineanu, A., 2001. Decision making and decision support systems for balancing socio-economic and Natural Capital development, Observatorio Medioambiental, 4, 19–47.

Vadineanu, A., 2001. Sustainable development: theory and practice regarding the transition of socio-economic systems towards sustainability, Studies on Science and Culture, UNESCO (CEPES), Bucharest.

Vadineanu, A., 2004. Managementul dezvoltarii: O abordare ecosistemica, Editura Ars Docendi/Universitatea Bucuresti, Bucuresti, 394 pp.

Vadineanu, A., 2005. Identification of the lagoon ecosystems, edited by E.I. Gonenc, J.P. Wolflin, Coastal Lagoons: Ecosystem Processes and Modeling for Sustainable Use and Development, CRC Press, Boca Raton, Florida, pp. 7–37.

Vadineanu, A., and T. Palarie, 2004. Mobilising the European social research potential in support of biodiversity and ecosystem management, National Report for Romania, Contract EC/FP 6/GOCE-CT-2003-505429.

Vadineanu, A., S. Cristofor, and G. Ignat, 1987. L'évolution de l'état tropique des ecosystemes aquatique caracteristique du Delta du Danube: I, Le regime hydrologique, la transparence Secchi et la reserve de phosphore et d'azote, Rev. Roum. Biol.-Biol. Animal., 32 (2), 83–91.

Vadineanu, A., S. Cristofor, D. Nicolaescu, C. Dorobantu, and L. Gavrila, 1987. L'évolution de l'état trophique des ecosystèmes aquatique characteristique du Delta du Danube: III, La dynamique des fractions du carbone particulé, Rev. Roum. Biol.-Biol. Animal., 32 (2), 99–111.

Vadineanu, A., S. Cristofor, and G. Ignat, 1992. Phytoplankton and submerged macrophytes in the aquatic ecosystems of the Danube Delta, during 1980s, Hydrobiologia, 243/244, 141–146.

Vadineanu, A., I.C. Postolache, and I.V. Diaconu, 1998. Targets concerning socio-economic re-structuring emerged from the material accounting analysis on the national scale (Romania), edited by L. Hens, J.R. Borden, S. Suzuki, G. Caravello, Research in Human Ecology: An Interdisciplinary Overview, VUB Press, Bruxelles, pp. 289–314.

Vadineanu, A., S. Cristofor, G. Ignat, C. Ciubuc, G. Risnoveanu, F. Bodescu, and N. Botnariuc, 2000. Structural and functional changes within the benthic communities of Danube Delta lakes, Verh. Internal. Verein. Limnol., 27, 2571–2576.

Vadineanu, A., M.C. Adamescu, R.S. Vadineanu, S. Cristofor, and C. Negrei, 2003. Past and future management of Lower Danube Wetlands System: A bioeconomic appraisal, J. Interdisciplinary Econ., 14, 415–447.

Vadineanu, A., and St. R. Vadineanu, 2004. Our world, our Environment: How to protect our planet? edited by Ch. Susanne, Societal responsibilities in life sciences, Human Ecology Series, No. 12. Kamla-Raj Entreprises, Delhi, India, pp. 105–121.

Van der Straaten, J., 1998. The economic value of nature. Kathalike Universitate Braband.

Wackernagel, M., and R. Rees, 1996. Our Ecological Footprint: Reducing Human Impact on the Earth, New Society, Gabriola Island.

Walker, B., and A.J. Meyers, 2004. Thresholds in ecological and social-ecological systems: A developing database, Ecol. Soc., 9(2), 3 URL: http://www.ecologyandsociety.org//vol9/iss2/art3.

Walker, B., S. Carpenter, J. Anderies, N. Abel, G. Cumming, M. Janssen, L. Lebel, J. Norberg, D.G. Peterson, and R. Pritchard, 2002. Resilience management in social—ecological systems: A working hypothesis for a participatory approach, Conservation Ecol., 67 (1), 14; http://www.consecol.org/vol6/iss1/art14.

Wilson, A.M., and R.S. Carpenter, 1999. Economic valuation of freshwater ecosystems services in the US: 1971–1997, Ecol. Appl., 9 (3), 722.

10. DEVELOPING THE D–P–S–I–R FRAMEWORK OF INDICATORS FOR MANAGEMENT OF HUMAN IMPACT ON MARINE ECOSYSTEMS: BALTIC SEA EXAMPLE

Eugeniusz Andrulewicz
Department of Fisheries Oceanography and Marine Ecology, Sea Fisheries Institute in Gdynia, Poland

Abstract. The aim of this paper is to contribute to the developments of environmental indicators, which are an indispensable tool for ecosystem health assessment and ecosystem-based management. Most decision makers request scientific advice to be given in the form of indicators.

At present, indicators are under development for a number of European seas and coastal lagoons. The most advanced is the development of S (State) and P (Pressure) indicators (usually developed by "environmental experts"). Developing D (Driving force) and R (Response of society) indicators (usually undertaken by "socio-economic" experts) is less advanced. These developments should lead to the joint/combined approach to elaborate full D–P–S–I–R framework of indicators.

This paper describes basic ideas behind indicators and offers a set of illustrative examples of P–S–R indicators for the main environmental concerns in the Baltic Sea: eutrophication, overfishing, chemical contamination, oil pollution, invasion of non-indigenous species, sanitary conditions of coastal waters, dumping of dredge spoils, extraction of sand and gravel, offshore oil and gas production, contamination by artificial radionuclides, physical impact of large-scale constructions, and loss of biodiversity. Some illustrative examples of possible D-Driving force indicators are also presented.

A number of indicators have been taken from existing developments, but some of them are proposed according to the author's judgment. These indicators can well be used or adopted for marine and coastal management purposes.

10.1. Introduction

Humans have influenced marine ecosystems for many hundreds of years, but until recently this influence was easily absorbed. Unfortunately, rapid

225

I. E. Gonenc et al. (eds.), Assessment of the Fate and Effects of Toxic Agents on Water Resources, 225–243.
© 2007 *Springer.*

industrial, agricultural and population developments brought about significant ecosystem changes and a danger of irreversible damages.

Some countries, being aware of marine pollution and recognizing that relevant anti-pollution measures should be taken internationally, have signed respective conventions (e.g., Helsinki Convention, signed in 1974, Oslo Paris Convention signed in 1979). They started to monitor and assess effects of human impacts for preparation of anti-pollution measures. Monitoring results as well as scientific findings have been used to produce Quality Status Reports as well as various kinds of thematic assessments. These assessments, although usually presenting high scientific level, did not necessarily satisfy decision makers. Their overly scientific content made them time-consuming to read and difficult to comprehend.

To improve the link between science and decision makers, a number of other international commissions and agencies such as the European Environment Agency (EEA), Global International Water Assessments (GIWA), Helsinki Commission (HELCOM), Mediterranean Pollution (MED-POL), and Convention for the North Atlantic (OSPAR) have started to develop indicators to be able to produce indicator-based assessments. A conceptualized P–S–R (Pressure–State–Response) indicator framework originally adopted by the Organization for Economic Co-operation and Development (OECD 1993) was selected by the EU for the start-up towards development of a full D–P–S–I–R framework (EU COM 539, 2002).

This paper is based, to a large extent, on a review of HELCOM Periodical Assessments (HELCOM 1981, 1987, 1993, 1996, 2002) and on developments of Quality Status Reports (QSR) by various international commissions (Arctic Monitoring and Assessment Programme (AMAP), Black Sea (BUCHAREST) Convention, MED-POL, and OSPAR). The most recent developments of environmental indicators by EEA for the European seas and by GIWA for the Baltic Sea are also taken into account.

10.2. The Conceptual Frameworks of Environmental Quality Indicators and Indices

The term "indicator" has been given various definitions, but generally it refers to a measure of something. The OECD defined indicator as

A parameter, or a value derived from parameters, which points to/provides information about/describes the state of a phenomenon/environment/area with a significance extending beyond that directly associated with a parameter value (OECD 1993).

The relationship of indicators to the information on which they are based is illustrated in Hammond's information pyramid (Hammond 1995) (Figure 1).

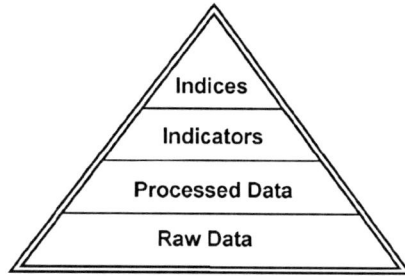

Figure 1. Information pyramid (Hammond, 1995)

Basically, "Raw Data" are elaborated and then published in the form of "Processed Data", i.e. scientific papers, reports, and assessments. In particular, peer-reviewed papers present high scientific quality, however, even such "Processed Data" are not very useful for laymen or decision makers. Therefore, there is a need to go further up in the integration in the information pyramid to the level of "Indicators."

The OECD developed a systematic framework for environmental indicators commonly referred to as "Pressure–State–Response" (OECD 1993) which is based on the following causality chain:

> Human activities exert pressures on the environment ("pressure") and change its quality and the quantity of natural resources ("state"). Society responds to these changes through environmental, general economic and sectoral policies (the "societal response"). The latter form a feedback loop to pressures through human activities (OECD 1993).

Indicators of pressure (P) can also be called indicators of stressors. The pollution load can be regarded as primary pressure indicator.

State (S) indicators are needed to properly assess the state of the marine ecosystem. This knowledge comes through research and monitoring. State indicators are measures of the state of environmental quality, such as concentrations of nutrients and contaminants.

Response (R) indicators should help the decision-making process, aid developing regulatory standards and identifying actions needed. Response indicators include government policies and regulatory efforts, as well as societal responses through individual and collective actions. These response activities usually result in changes in driving forces, thus closing the loop (Figure 2).

The recommended way to develop the full D–P–S–I–R framework is shown in Figure 2. It is clear from this scheme that only human activities (D—Driving forces) can be managed.

Indicator Selection: There might be hundreds of good indicators, however this large number has to be reduced to a logical/operational number according to

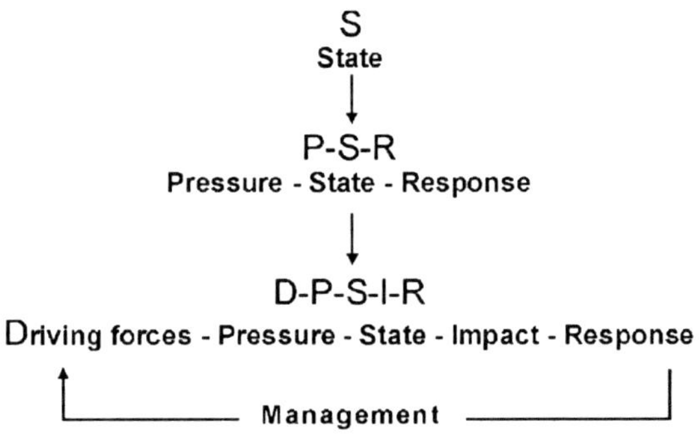

$$S$$
State
$$\downarrow$$
$$P\text{-}S\text{-}R$$
Pressure - State - Response
$$\downarrow$$
$$D\text{-}P\text{-}S\text{-}I\text{-}R$$
Driving forces - Pressure - State - Impact - Response

Management

Figure 2. Development and the use of D–P–S–I–R framework of indicators for management

established criteria. The U.S. EPA (US EPA 2002, 2004) applied the following selection criteria which imply that all operational indicators should be:

1. *Regionally Responsive*—The indicator must reflect changes in ecosystem conditions and respond to stressors (pressures) of concern across most resource classes and habitats within the monitored region.
2. *Unambiguously Interpretable*—Indicator must be related unambiguously to an assessment endpoint (relevant exposure/stressor/habitat variable) that forms part of the ecosystem's overall conceptual model of ecological structure and function.
3. *Simply Quantifiable*—Indicator can be quantified by synoptic monitoring or by cost-effective automated monitoring that can be adopted by all participants in the monitoring survey.
4. *Stable over the Sampling Period*—Indicator exhibits low measurement error and stability of regional cumulative frequency distribution during the sampling period (low temporal variation during the sampling period in regional statistics).
5. *Low Year-to-Year Variability*—Indicator must have sufficiently low natural year-to-year variation to detect ecologically significant changes within a reasonable time frame.
6. *Environmental Impact*—Sampling for the indicator should have minimal environmental impact.

Alternative selection criteria are possible, for example, ICES has decided that indicators should be: specific, measurable, achievable, realistic and time bound (ICES 2005). However, in all cases indicators should also be communicative, i.e., understandable by the layman and the educated public.

Development of "Indices": Already, under demand of various international organizations dealing with global assessments (e.g., Global International

Water Assessment, GIWA), as well as those dealing with large scale assessments (e.g., European Environmental Agency, EEA), have requested the development of a simple scoring system of indices that could help to compare the status of different seas and/or regions. Indices are developed in order to limit the amount of information that decision makers and other interested parties have to absorb. By their nature, indices are less informative than their indicator components, since the process of aggregation dampens the impact of the individual indicators. Indices represent the most integrated information with the least amount of detail. They are usually expressed as classes: no impact, slight impact, moderate impact, severe impact (UNEP, 2005). The EU WFD (2000) represents ecosystem quality as: high, good, moderate, poor, or bad. The U.S. EPA (2002, 2004) represents national coastal condition using the "traffic light approach"—good (green), fair (yellow), or poor (red). The OSPAR classification for eutrophication is even simpler: they recognize "problem area" and "no problem area." Indices are useful for comparing the quality of many water/marine systems, e.g., on a Pan-European scale, an American coast scale, and/or global scale, however, their development should inevitably be preceded by the development of indicators.

10.3. Examples of P–S–R Indicators for the Main Environmental Concerns

Generally recognized anthropogenic impacts on the marine environment are as follows: eutrophication, overfishing, chemical contamination, oil pollution, invasion of non-indigenous species, sanitary conditions of coastal waters, dumping of dredge spoils, extraction of sand and gravel, offshore oil and gas production, artificial radionuclides, physical impact of large technical construction, coastal degradation, and loss of biodiversity and habitat destruction. P–S–R indicators for each of these concerns are given below.

10.3.1. EUTROPHICATION

There are various definitions of eutrophication. ICES-ACMP (1992) defines eutrophication simply as "nutrient enrichment." According to EEA (1999):

> Eutrophication means the enrichment of water by nutrients, especially compounds of nitrogen and/or phosphorus causing an accelerated growth of algae and higher forms of plant life to produce an undesirable disturbance of the balance of organisms present in the water and to the quality of water concerned.

Eutrophication is primarily nutrient enrichment in the aquatic system. It also means a trophic status of the system in the chain: oligotrophic–eutrophic–

mesotrophic–polytrophic–hypertrophic. Eutrophication of marine systems is usually regarded in relation to anthropogenic activity, although it is a natural phenomenon, therefore, the most challenging issue is to distinguish between natural and anthropogenic eutrophication. Eutrophication is usually recognized as undesirable, although it is worth remembering that enhanced productivity also means more food in the system and therefore more fish. A number of HELCOM Recommendations regarding reduction of inputs of nutrients from wastewater treatment, agriculture and industrial emissions have been issued. The proposed P–S–R scheme is below.

Anthropogenic pressure (P) indicators	Environmental state (S) indicators	Government/society response (R) indicators
• Discharge magnitude of nutrients from point sources • Discharge magnitude of nutrients from diffuse sources • Amount of deposition of nitrogen from the atmosphere	• [Winter] concentrations of nitrogen and phosphorous in sea water • N/P ratio • Chlorophyll "a" concentration in surface waters • Secchi-depth • Depth range of macrophytes • Oxygen depletion (in historically oxygenated areas) • Number of phytoplankton indicator species (such as some blue–green algae) • Frequency of occurrence and presence of potentially toxic algal species • Increased biomass of fish above the halocline • Intensive and prolonged blue–green algae blooms	• Reduction of nutrient discharges from point sources • Construction of wastewater treatment facilities • Reduction of nutrient discharges from diffuse sources • Reduction in the use of fertilizers and detergents containing phosphorus • Buffer strips trapping nutrients to prevent eutrophication • Adoption of Good Agriculture Practices (sustainable agriculture)

10.3.2. OVERFISHING

Fish are almost exclusive living resources of market value in the Baltic Sea. The tradition of fishery management is well established, and it is usually controlled by national and international regulations, however, there are still problems related to realistic catch statistics, illegal catches and ineffective fishery control. Today, in many fishery areas, the main problem is overexploitation of available commercial fish resources.

For many years the exploitation objectives of fishery were targeted only to maintain commercial fish stocks. However, under the present policy of an *ecosystem-based fishery*, the focus is not only on commercial stocks but on the fishing pressure on marine ecosystem. This will involve a number of state and impact indicators.

The International Baltic Sea Fishery Commission (IBSFC 1998) has already adopted some indicators like sustainable stock biomass (SSB), fishing mortality, recruitment, landings, number of vessels, engine power, and number of full-time fishermen. However, they have not proposed fishing pressure indicators on the fished ecosystem, such as the amount and composition of non-target fish species, the number of birds and mammals caught, the amount of dumped fish offal and fish discards, and the number and quantity of key species removed from the environment. The proposed P–S–R scheme is below.

Anthropogenic pressure (P) indicators	Environmental state (S) indicators	Government/society response (R) indicators
• Number of fishing vessels per country • Landings of fish per country • Average engine power per country: total kilowatt of the fleet divided by the number of vessels • Number of full-time fishermen engaged in the area, by country • The amount and composition of non-target fish species • Number of birds and mammals caught • The amount of dumped fish offal and fish discards • Number and quantity of key species removed from the environment	• Spawning stock bio-mass (SSB) • Fishing mortality rate • Recruitment rate • The ratio between Yield and SSB • Number and size of bottom habitats affected by trawling • Changes in ecosystem structure due to removal of target, non-target and key species	• Regulation of landings (total allowable catches (TACs), per country) • Technical measures (fish gear, number and size of nets, etc) • Temporary closure of fishing (fishing grounds, time of fishing, etc) • Reduction in the number of licensed commercial fisherman

10.3.3. CONTAMINATION BY HARMFUL SUBSTANCES

Contamination and its adverse effects are caused by the presence of a substance or group of substances that are toxic, persistent, and liable to bioaccumulate. These include inorganic (heavy metals), organic (some biocides and industrial compounds, usually halogenated, and some polycyclic aromatic hydrocarbons) and organometallic compounds (organic compounds of mercury and tin).

Contamination—is used to describe the situation which exists where either the concentration of a natural substance (e.g. a metal) is clearly above normal, or the concentration of a purely man-made substance (e.g. DDT) is readily detectable, but where no judgment is passed as to existence of pollution (i.e. adverse effects) (ICES-ACMP 1991).

Anthropogenic pressure (P) indicators	Environmental state (S) indicators	Government/society response (R) indicators

• Magnitude of discharges from land-based/point sources
• Magnitude of discharges from land-based/diffuse sources
• Quantity of discharges of toxic compounds from sea-based sources
• Amount of atmospheric deposition of toxic compounds via atmosphere/fallout into the sea
	• Contamination (levels/concentrations in water sediments and biota)
	• Long-term trends in concentration levels
	• Bioaccumulation levels in organisms
	• Biomagnification rates in the food chain
	• Presence of ecotoxicological effects related to:
	  – Reproductive disturbances (e.g., imposex)
	  – Immunosupression (e.g., cortisol level)
	  – Carcinogenic effects (e.g., liver neoplasmia)
	  – Genotoxic effects (e.g., cytogenetic and DNA damage)
	  – Multiple stress factors (e.g., stress proteins)
		• Improvement and construction of wastewater treatment facilities
		• Ban on production of harmful substances (i.e., DDT)
		• Significant reduction in the use of harmful substances
		• Reduction of toxic emissions at sources

In some cases there is a need to identify specific chemical pollution problems in the marine environment. For example, if anti-fouling paints create environmental problem, specific indicators will be related to effects of organotin compounds: metallothioneins, intersex/imposex, and shell thickening. There can be a large number of indicators under ecotoxicological effects, e.g. related to histopathological changes, reproduction impairment, immunocompetence, genetic toxicity, carcinogenicity, endocrine-hormone disruption, etc.

10.3.4. OIL POLLUTION/MARINE TRANSPORT

After many years of studies on petroleum hydrocarbons at sea, there is still no consensus on the definition of oil pollution. Authors define these issues in different ways, which is why it is very often difficult to compare the results of their studies. For the purpose of this paper, the following definition is proposed:

> Oil (petroleum hydrocarbons) pollution includes pollution of the marine environment by crude oil, crude oil derivatives (except solvents) and polycyclic aromatic hydrocarbons (PAHs) derived by the combustion of fossil fuels *(Andrulewicz et al. 1996)*

This definition facilitates the interpretation of results of chemical analysis in which PAHs are usually determined. It also allows for a clear/univocal assessment of the level of marine environment contamination by compounds that are toxic, persistent and bioaccumulative or mutagenic and carcinogenic.

Anthropogenic pressure (P) indicators	Environmental state (S) indicators	Government/society response (R) indicators
• Frequency and amount of transported crude oil and oil derivatives (number and amount of discharges) • Number of accidents/collisions at sea/level at risk for spills • Amount of land based discharges (sewage out-falls and river run-off) • Amount of atmospheric deposition (from transport and combustion of fossil fuels)	• Levels of oil residues in sea water and sediments • Number of oil slicks on the sea surface • Concentrations of PAHs in water, sediments and some marine organisms • PAH-related effects in marine organisms (e.g. liver neoplasmia) • Number of oiled birds	• Regulations on transport (including ship requirements) • Reception facilities in ports • Inspections of marine transport activities (e.g., aerial surveillance) • Oil combating facilities in ports and oil combating vessels • Regulations on discharges (e.g., on offshore oil and gas industry)

In the case of oil/petroleum hydrocarbon pollution, the most effective course of action is to prevent oil spills. Preventive measures will not preclude entirely the events of oil spills, so an effective combat system is still needed. Additional information on the present level of petroleum hydrocarbons is needed in order to establish reference values for clean-up purposes, particularly in high-risk locations: ports, oil terminals, offshore oil rigs, and along shipping lanes.

10.3.5. INVASION OF NON-INDIGENOUS SPECIES

Invasive species, most probably entering the Baltic Sea with ballast waters, form a growing concern. However, at present their ecological significance is largely unknown.

Anthropogenic pressure (P) indicators	Environmental state (S) indicators	Government/society response (R) indicators
• Marine transport (amount and type of ballast waters) • Introduction programs • Natural migration	• Number and abundance of alien species • Nature of interaction with native species • Economical losses from interaction with native species and costs of combat measures	• Ballast water control and management technologies

10.3.6. SANITARY POLLUTION

Sanitary conditions are generally assessed by estimating the presence of pathogenic bacteria in bathing waters in a given coastal area. Existing classification systems are usually based on existing legal indicators, which are usually concentrations of coliform bacteria in bathing waters.

Anthropogenic pressure (P) indicators	Environmental state (S) indicators	Government/society response (R) indicators
• Amount of untreated sewage discharge • Run-off from polluted rivers and streams • Number of tourists • Lack of or insufficient sanitary facilities	• Fecal coliform index • Presence of other bacteria (e.g. *streptococci, salmonellas, vibrio*) • Presence of harmful parasites • Presence and amount of (decaying) algal mats	• Monitoring of bathing waters • Licensing beaches for bathing water quality • Closing beaches • Construction of sewage treatment facilities

10.3.7. DREDGE SPOILS DUMPING

Dumping of dredged spoils from port and navigation channels is allowed by international law; however it is regulated by national legislation acts as well as by international recommendations (e.g. HELCOM Rec. No. 13/1). There are national and international quality standards for dredged material; however, they are not related to selection of dumping sites or the fate of material after dumping. Therefore, ecological criteria regarding the proper disposal of dredged spoils have to be developed.

Anthropogenic pressure (P) indicators	Environmental state (S) indicators	Government/society response (R) indicators
• Amount of dumped dredged spoils • Quality of dredged spoils (e.g. concentration of harmful substances, amount of oxygen consuming substances)	• Fate of dredged material after dumping (dispersion/deposition rate) • Biological effects at dumping site (e.g. number of damaged organisms) • Recovery rate of dumping sites	• Obeying national and international regulations • Monitoring effects of dumping • Restricted amount of dumped material • Storage on land/landfilling

10.3.8. EXTRACTION OF SAND AND GRAVEL

Sand and gravel, apart from crude oil, are those currently exploited mineral resources in the Baltic Sea. Environmental policy regarding sand and gravel resource exploitation at sea, including guidelines and codes of practice, is well developed by ICES.

Anthropogenic pressure (P) indicators	Environmental state (S) indicators	Government/society response (R) indicators
• Number and size of extraction fields • Amount of extracted material per year	• Morphological changes on the bottom • Possible changes of the natural balance of currents • Effects on structure of bottom fauna and flora communities • Effects of suspended matter plumes on fish spawning grounds • Effects of suspended matter on marine vegetation	• Environmental impact assessments • Guidelines and /or code of practices for mineral exploitation • Licensing/ permission • Monitoring of effects/national or international surveillance activities

10.3.9. OFFSHORE OIL AND GAS PRODUCTION

Offshore oil and gas production in the Baltic Sea is at present limited to two production platforms (400 thou. tonnes/year), however this production might be increasing due to confirmed crude oil resources, particularly in the southern and eastern part of the Baltic. Discharges from offshore oil and gas production may include drilling mud, produced water, drill cuttings and incidental spills. The most important environmental problem may be related to produced water which would contain a composition of a variety of compounds: organic compounds, trace metals and radionuclides (Foyn, 1998).

Anthropogenic pressure (P) indicators	Environmental state (S) indicators	Government/society response (R) indicators
• Number and size of crude oil fields (resources) • Number of offshore oil and gas production platforms • Amount of discharges of produced water • Amount of discharges of and drilling mud	• Amount and quality of produced water • Amount and quality of drilling mud • Number and size of unintentional oil spills • Intensification of bunker transport and/or pipelines • Concentrations of contaminants in sediments and biota in the vicinity of oil rigs versus reference concentrations	• Regulations on discharges by offshore oil and gas industry • Oil combating facilities and assistance of combating vessels • Monitoring activities

Maximum allowable concentration of dispersed oil in 'produced' technical waters is 40 mg/l. Oil companies claim that they are able to reduce oil concentration to 20 mg/l. There is no regulation of all other compounds dissolved in technical water and therefore discharged to the sea. Discharge of "produced" water in the North Sea amounts to 4000 million m^3 (Foyn, 1998).

The discharge of drill cuttings and drilling muds can also create a problem, therefore HELCOM Recommendation 18/2 offshore activities, regulates the use of drilling muds as well as other operational details. In Norway, the use of oil-based drill mud is prohibited.

10.3.10. ARTIFICIAL RADIONUCLIDES

Radionuclides in marine ecosystems, similar to trace metals, are of both natural and anthropogenic origin. Long-lived artificial radionuclides (28–30 years) (^{137}Cs, ^{90}Sr) pose threats to human health; therefore they are the subject of monitoring/control. Short-lived artificial radionuclides (^{239}Pu, ^{240}Pu, ^{240}Am) are also harmful to human health and they usually are the subject of studies that follow contamination events.

Artificial radionuclides appear in the marine environment following nuclear bomb tests mostly from the atmospheric fallout; as well as operational releases from nuclear power plants. There have also been some incidental releases from nuclear power plants via the atmosphere, usually of local importance. The largest artificial radionuclide contamination incident was the Chernobyl disaster in 1986. Monitoring of radionuclides is necessary to determine levels of natural and artificial radionuclides in the ecosystem and to ensure that there is no growing trend of concentration of radionuclides in the marine environment. Apart from that, every country runs its own security service.

Anthropogenic pressure (P) indicators	Environmental state (S) indicators	Government/society response (R) indicators
• Enrichment of radioactive ores (radioactive wastes) • Operational discharges from the nuclear industry • Accidental emissions from nuclear power plants • Operational discharges from nuclear submarines • Radioactive ash as a by-product of coal combustion	• Contamination level of water with (^{137}Cs, ^{90}Sr) as an indicator of potential level of bioaccumulation • Contamination of fish and in some cases other biota (^{137}Cs, ^{90}Sr) • Contamination of sediments in deposition basins (^{137}Cs, ^{90}Sr) as an indicator of historical changes	• Restrictions on nuclear weapon production • Construction of safer nuclear power plants • Restrictions on atomic energy production • Safer storage of radioactive wastes

10.3.11. COASTAL DEGRADATION

There is growing anthropogenic pressure on the coastal zone, including a rapid increase in coastal population, growing coastal tourism, industrial development, coastal defense measures, and the drainage of coastal wetlands.

There are many different definitions of the coastal zone, but in every definition the coastal zone includes the marine/terrestrial interface (ecotone) with adjacent marine and terrestrial areas. The range of landward and seaward sites is arbitrarily defined, and therefore leads to important differences in definitions. This is not surprising, since these definitions are created for different purposes. In fact, there is no reason to give an overall/general definition of dimensions of the coastal zone, however in every case this has to be defined even for the purpose of common understanding.

All environmental indicators related to eutrophication, harmful substances, exploitation of fish and biological diversity are relevant to assess the environmental quality of the marine part of the coastal zone. Therefore indicators proposed here are related to terrestrial part of the coast. The key issue here is to ensure sustainable development, i.e., to continue existing and develop new economic activities and at the same time to preserve the natural state of the coast. This is usually an issue of integrated coastal zone management (ICZMs) plans.

Anthropogenic pressure (P) indicators	Environmental state (S) indicators	Government/society response (R) indicators
• Demographics-permanent population • Temporary population (tourism) • Marine-related land use (shipbuilding and repair, fisheries, marinas, etc) • Non-marine land use (housing, commercial use etc) • Coastal defense measures • Exploitation of mineral resources	• Percentage of the coastal zone without rural developments • Percentage of the coast with natural morphology • Natural (endogenous) plant and animal communities • Percentage of the coast with natural coastal dynamics (without "hard" protection-beach/sand bars, nourishments and other artificial protection measures) • Status of diversity of biotopes and species • Status of preservation of coastal wetlands and lagoons • Length of the coast covered by urbanization and industrial development	• Zoning (restrictions on use) of privately owned land • Designation of protected areas • Limits on anthropogenic activities • Restrictions/ licensing of specific uses (e.g. mineral extraction, mariculture facilities)

10.3.12. LOSS OF BIODIVERSITY AND HABITAT DESTRUCTION

Biodiversity has been defined as

> The variability among living organisms from all sources including terrestrial, marine and other aquatic ecosystems and the ecological complexes of which they are part; this includes diversity within species, between species and of ecosystems *(Rio Convention 1992)*.

According to this interpretation, biological diversity is defined at the levels of genes, species, ecosystems, and landscapes (Rio Convention, 1992). An Action Plan developed for the implementation of the Biodiversity Convention should contain baseline information about existing biological diversity, actions to control and restore biodiversity, and a monitoring program. The general objective is to preserve all types of marine diversity by applying the guidelines of the Rio Convention as well as national Action Plans that had to be developed accordingly.

Anthropogenic pressure (P) indicators	Environmental state (S) indicators	Government/society response (R) indicators
• Physical habitat destruction or fragmentation of habitats (e.g. marine aggregate extraction, large scale technical construction) • Discharges of nutrients • Discharge of toxic substances • Overexploitation of fish and benthic organisms • Destruction of bottom communities by heavy trawling • Human-induced transfer of non-native species	• Overall number of species • Overall number of biotopes • Overall number of landscape types • Genetic diversity (number of genotypes) • Presence of keystone species • Biological diversity indicators (e.g. Shanon-Wiener etc)	• Reduction of nutrient load • Reduction of harmful substances load • Legal protection of habitats • Legal protection of endangered species • Establishment of protected areas • Regulated catches of exploited species • Restoration of degraded habitats

10.4. Illustrative Examples of D-Type Indicators

D-type indicators should be developed for different economical sectors affecting the ecosystem health of coastal and lagoon systems: energy, agriculture, industry, transport, tourism and other relevant sectors. In order to determine the social/political/legal/economic reasons for the existing pressures, GIWA has proposed the application of "casual-chain analysis" methodology (UNEP, 2005). These developments may differ from one country another. However, a preliminary list of indicators can be produced and structured according to their importance. Examples of D-type indicators are given below:

Eutrophication

Fertilizer use per ha.
NO_x emission from stationary sources.
NO_x emission from mobile sources.
Supply or sale of mineral fertilizers.
Livestock density.
Manure tanks/reservoirs.

Contamination

Generation of industrial and municipal solid waste.
Household waste/garbage disposed per capita.

Generation of hazardous wastes.
Import/export of hazardous wastes.
Oil discharges into coastal waters.
Use of agricultural pesticides.
Number of cars per 100 inhabitants.

Overfishing

Number of full-time fishermen engaged in the Baltic Sea, per country.
Landings per country.
Number of fishing vessels per country.
Average engine power per country.
Marine fish consumption per capita per country.
Fishing subsidies and market failure.
Fishing gear modernization.
Privatization (followed by intensification of fishing effort) in former socialist
    countries.
Inappropriate assessment methods.
Inadequate fishery control.
Biased fishing statistics.

Biodiversity: The driving forces for "Loss of biodiversity and habitat destruction" are excessive eurtrophication and contamination by harmful substances, as well as fishing activities. However, some other human activities are also driving loss of biodiversity, such as

Population growth rate in coastal areas.
Inhabitation rate of the coast.
Change of coastal line usage.
Destruction/drainage of coastal wetlands.
Coastal constructions.
River damming.
Marine habitat change or alterations.

10.5. Summary and Conclusions

This paper presents an approach to classify, in the form of indicators, environmental problems of marine areas stemming from anthropogenic pressure. This approach is based on the example of the Baltic Sea. The list of indicators is by no means complete or in a final form, however these indicators illustrate the problem and an approach towards developing a complete, generally accepted set of indicators. This should help in understanding and assessing anthropogenic pressure on the ecosystem, preparation of holistic as well as regional assessments, and consequently help to implement ecosystem-based management.

Due to the complicated nature of ecosystems, indicators cannot perfectly represent the state of the environment or the complex interrelationships between the natural environment and anthropogenic activities. Indicator systems need to strike a balance between sophistication, as measured by the number of indicators and degree of functional representation, and simplicity and cost considerations. Very sophisticated systems including a large amount of data and complicated mathematical formulas may be a more accurate reflection of environmental conditions, but decision makers may not be able to make use of them.

It is clear that indicators contain less information than scientific papers or detailed scientific assessments. Therefore, the proper choice of indicators (informative and representative) is very demanding task. It will be a process lasting for some years before a "toolbox" of indicators is selected and generally accepted. It is worth noting that even a very good "toolbox of indicators" will not reduce expert participation in the assessment process and should not reduce monitoring effort. But it will produce "decision-maker friendly" environmental assessments and build better links to socio-economic driving forces and societal response, so that appropriate measures can be taken.

Indicators presented in this paper were selected to cover issues considered to be of major importance for the Baltic Sea area, and presented in a way to facilitate their use by decision makers as well as understanding by the general public. It is important that the usefulness of indicators to the ultimate users is kept in mind throughout the entire development process.

A step-by-step approach is proposed to achieve the full D–P–S–I–R framework of indicators beginning with State–Impact (S–I) indicators which should be developed in environmental groups of experts and Driving force–Pressure (D–P) indicators which should be developed in socio-economics groups of experts. These two developments should be merged by a combined group of experts to elaborate the full D–P–S–I–R framework, including society/government "Response" (R) indicators. D–P indicators should be taken as a separate process as "casual-chain analyses" in which socio-economic forces and different levels of pressure are analyzed in detail. Cause–effect relationships are barely understood and at present require separate approaches. There is also a practical reason to restrict D–P–S–I–R framework to P–S–R framework. This is to make the whole process of developing indicators simpler and easier.

The presented examples are by no means a complete identification of all related indicators. There is also a need to develop indicators of physical disturbances in the marine environment caused by coastal construction activities, wind-mill parks, cables, pipelines, and others. Challenging issues are the development of indicators for sustainable development, indicators of biological diversity, and indicators of coastal zone quality.

Acknowledgements

The author wish to thank Dr. Elzbieta Lysiak-Pastuszak, Institute of Meteorology and Water Management-Maritime Branch in Gdynia and Dr. Zbigniew Otremba, Maritime University in Gdynia, for reviewing this text, valuable discussions and support in technical preparations.

References

Andrulewicz, E., 1999. Development of Marine Environmental Quality Indicators—conceptual issues and practical examples, ICES-ACME Working Paper 1999/17.1.
Andrulewicz, E., G. Dahlman, B. Hägerhall, and G. Witt, 1996. Petroleum hydrocarbons, Baltic Sea Environ. Proc., 64B, 139–144.
Bergen Declaration, 2002. Ministerial Declaration of the Fifth International Conference on the Protection of the North Sea, Bergen, Norway, 20–21 March 2002.
EEA, 1999. Integration of information on Europe's marine environment, edited by T. Bokn, and H.R. Skjoldal, Technical Report No. 17.
EU COM 539, 2002. Towards a strategy to protect and conserve the marine environment, Communication from the Commission to the Council and the European Parliament, 65 pp.
EU WFD, 2000. EU Water Framework Directive, http://www.eucc-d.de/infos/ Water Framework Directive.pdf.
Foyn, L., 1998. Institute of Marine Research, Norway, Personal communication.
McGlade, J.M., 2002. The North Sea large marine ecosystem, edited by K. Shermann, H.R. Skjoldal, Large Marine Ecosystems of the North Atlantic. Changing States and Sustainability, Elsevier, pp. 339–412.
Hammond, A.L., 1995. Environmental Indicators, World Resource Institute, Washington, DC.
HELCOM, 1981. Assessments of the Effects of Pollution on the Natural Resources of the Baltic Sea, 1980 Baltic Sea Environment Proceedings No. 5B.
HELCOM, 1987. First Periodic Assessments of the State of the Marine Environment of the Baltic Sea, 1980–1985. Baltic Sea Environment Proceedings No. 17 B.
HELCOM, 1995. Second Periodic Assessments of the State of the Marine Environment of the Baltic Sea, 1986–1988. Baltic Sea Environment Proceedings No. 35 B.
HELCOM, 1996. Third Periodic Assessments of the State of the Marine Environment of the Baltic Sea, 1989–1993. Baltic Sea Environment Proceedings No. 64 B.
HELCOM, 2002. Fourth Periodic Assessments of the State of the Marine Environment of the Baltic Sea. Baltic Sea, 1994–1998. Baltic Sea Environment Proceedings No. 82 B.
IBSFC, 1998. Living Marine Resources Component, Baltic Sea GEF Project (mimeo).
ICES-ACMP, 1991. ICES Cooperative Research Report No. 177.
ICES-ACMP, 1992. ICES Cooperative Research Report No. 190.
Jansson, B.O., 1980. Natural Systems of the Baltic Sea, Vol. IX, No. 3–4, pp. 128–136.

OECD, 1993. OECD Core Set of Indicators for Environmental Performance Reviews Environ-
    mental Monographs No. 83. Synthesis Report OECD/GD (93)179.
Rio Convention, 1992. Convention on Biological Diversity, 5 June 1992.
Skjoldal, H.R., 1998. Overview Report on Ecological Quality (EcoQ) and Ecological Quality
    Objectives (EcoQOs), Institute of Marine Research, Bergen, Norway.
Sherman, K., 1994. Sustainability, biomass yields, and health of coastal ecosystems: an eco-
    logical perspective, Mar. Ecol. Prog. Ser., 112, 277–301.
Smeets, E., and R. Weterings, 1997. Environmental Indicators: Typology and Overview. TNO
    Center for Strategy, Technology and Policy, Apeldoorn, The Netherlands.
UNEP, 2005. Laane, A., E. Kraav, and G. Titova. Baltic Sea, GIWA Regional Assessment 17.
    University of Kalmar, Kalmar, Sweden.
US EPA, 2002. National Coastal Condition Report. EPA 620/R-01/005.
US EPA, 2004. National Coastal Condition Report II. EPA/620/R-03/002.
Wulff, F., and A. Niemi, 1992. Priorities for restoration of the Baltic Sea- a scientific perspective.
    AMBIO 2 (21): 193–195.

11. PHYTOPLANKTON ECOLOGICAL PROCESSES FOR ECOSYSTEM MODELING: SOME BASIC CONCEPTS

Javier Gilabert
Department of Chemical and Environmental Engineering Technical University of Cartagena, Alfonso XIII, 44, 30203-Cartagena, Spain

11.1. Introduction

Ecosystem modeling predicts the structure and functioning of ecosystems from initial conditions and knowledge of ecological processes. The compromise between the number of state variables and accuracy of predictions is the key issue for reliability of this technique. Although many ecosystem models are only focused on eutrophication and water quality, new technologies have increased the data acquisition rate for a larger number of state variables, which allows the complexity of simulations to increase. A comprehensive description of the main phytoplankton ecological processes included in most models relating hydrodynamics with eutrophication is developed in this section by considering the size of the cells as a variable from which many other state variables can be mirrored.

11.2. Phytoplankton Systematic and Functional Diversity

Planktonic organisms are those that by their lack of motility and weakness to swim against the currents exist in a drifting or floating state in the water. Plant-like forms are known as phytoplankton whereas animal-like as zooplankton. Depending on their trophic behavior they are classified into: autotrophs (photosynthetic organisms); heterotrophs (non-photosynthetic organisms) and mixotrophs (that being photosynthetic are also able to assimilate organic compounds). Whereas autotrophs are usually identified as phytoplankton and heterotrophs as zooplankton, some phytoplankton groups—e.g. dinoflagellates—can be identified as mixotrophs.

Phytoplankton is composed of a large diversity of single celled microorganisms comprising many different taxonomic groups. The most important are the Bacillariophyta (commonly known as diatoms) and Pyrrhophyta (dinoflagellates) with many other classes such as Cryptophyta (cryptomonads); Chrysophyta (golden-brown algae); Euglenophyta (euglenoids); Chlorophyta

I. E. Gonenc et al. (eds.), Assessment of the Fate and Effects of Toxic Agents on Water Resources, 245–258.

(green Algae) and Cyanophyta (blue–green algae or cyanobacteria) among others.

Diatoms are most abundant and diverse in cold, nutrient-rich (eutrophic) and well-mixed (turbulent) water. They are non-flagellated cells of about 5–500 μm, with both unicellular and colonial forms. External silica cells walls called frustules are made of two valves (epitheca and hypotheca) containing pores for entry of nutrients and gas exchange into the cell. They are able to prevent sinking by controlling their buoyancy using a variety of morphological (small size, high surface area to volume ratio—sometimes involving spines, wings or chain formation) and physiological strategies (e.g., accumulating gas and lipid storage). Their major pigments are Chlorophyll a, c, carotenes, and xanthophylls.

Dinoflagellates are the most dominant phytoplankton besides diatoms. Some ability to swim (up to 500 μm s^{-1}) is provided by two flagella: one longitudinal, propelling the cell forward, and another transversal, making the cell rotate by undulating motion. Size may range from 1 to 1000 μm with some larger species armored with cellulose plates. They can easily survive in oligotrophic (low nutrient) waters due to their large storage capacity for nitrogen (e.g. from 11 to more than 2,000 pgN cell^{-1} for cells ranging from 180 to 2.8 × 10^{5}μm^3) and phosphorous. Many species are adapted to low light conditions and may descend at night for nutrient uptake and ascend during the day for photosynthesis. They are extremely vulnerable to turbulence but can bloom in areas of restricted circulation. Chlorophyll a, c and β-carotene are their major pigments, with many others (e.g. peridinin, gyroxanthin) specific for some group of species. Dinoflagellates shows a complex life cycle involving both asexual and sexual reproduction. Adverse environmental conditions can initiate sexual reproduction and the formation of resting cysts. About 60 species from this group are reported to produce toxins with the ability to produce massive proliferations known as red tides.

Several other taxonomic groups worth mentioning due to their group-characteristic properties: Coccolithophorids—a mostly unicellular primarily marine group of less than 20 μm in diameter—are highly reflective due to their characteristic *calcareous* plates producing typical water discoloration when blooming; Cryptomonads, which contain a water-soluble biliprotein as an accessory pigment to the chlorophyll; Golden-brown algae, which appears brownish in color due to the dominance of β-carotene and some xanthophyll carotenoids in addition to Chlorophyll a; Euglenoids—a mostly freshwater group—that are a facultative heterotrophic group with nutrition supplemented by the uptake of ammonia and dissolved organic nitrogen; and Green algae, which appear green in color because chlorophyll is not masked by accessory pigments.

The blue–green algae (cyanobacteria), on the other hand, are prokary-otes (organisms without a cell nucleus containing their genetic material) that occur as unicellular or chain forming colonies commonly arranged in fila-ments called trichomes. Unlike other phytoplankton, which can only fix nitro-gen in form of nitrate, nitrite, and ammonia, some of these filament-forming cyanobacteria species can perform atmospheric gaseous N_2 fixation. They can be abundant in the temperate oligotrophic ocean and also in eutrophic fresh-water environments. Much smaller cyanobacteria of the genus *Synechococcus* (~ 1 µm in diameter, with phycocianin and phycoerithrin as major pigments) can be dominant in some environments, reaching high densities ($\sim 10^6$ cell ml^{-1}), as they are adapted to low light and capable of acquiring nutrients even at nanomolar concentrations. In the oligotrophic open sea the prochloro-phyte *Prochlorococcus* (0.4–0.8 µm diameter, containing divinyl derivatives of chlorophyll *a* and *b*, α-carotene, zeaxanthin, and a type of phycoerythrin) along with *Synechococcus* dominate the small size fraction of phytoplankton of the deep euphotic zone as they are able to absorb blue light efficiently at low light intensities.

Taxonomical complexity—with more than 5000 marine species described and from tens to hundreds of species (comprising 10^2–10^5 cells ml^{-1}) usually present in a small volume obviously imposes a limit to the specific description of the spatio-temporal phytoplankton community. To overcome this problem, a size-based classification scheme consisting of four categories is often used: pico- (0.2–2 µm diameter), nano- (2–20), micro- (20–200) and mesoplankton (>200). Although this classification is overly simplistic from a taxonomic point of view, it is successfully used to describe certain processes where taxonomic composition is not a crucial factor.

A universal descriptor of the phytoplankton biomass is the Chlorophyll *a* concentration, which is positively correlated to cell density and indirectly related to nutrient concentration. Chlorophyll concentration also reflects the degree of adaptation to the light field and the photosynthetic capacity. While it is obvious that introducing state variables for each single species is unre-alistic in practice, using bulk phytoplankton community state variables will undoubtedly produce only a crude representation of the reality. Based on the diatom's distinctive biology (e.g. carbon content per cell differs from diatoms to other phytoplankton groups and silica can limit diatoms growth as needed for making their valves) some ecosystem models achieve reasonable results by making a distinction between diatoms and non-diatoms phytoplankton.

Seasonal variability in environmental conditions—especially light in-tensity and nutrients concentration, but also turbulence, temperature and salinity—determines the seasonal dynamics of the phytoplankton commu-nity in any particular ecosystem. The large plasticity in the adaptation of individual species to the environmental conditions imposes a high degree of uncertainty on prediction at the seasonal time scale with current ecosystem models.

11.3. Environmental Factors Controlling Phytoplankton Biomass

Phytoplankton biomass increases as a result of two major metabolic processes: photosynthesis and nutrients uptake. Photosynthesis is the process by which absorbed light energy is used to oxidize water and reduce inorganic carbon dioxide to organic carbon compounds (primary production, which results in a increase in the number of cells).

$$CO_2 + 2H_2O \xrightarrow{8\,hv} (CH_2O) + H_2O + O_2.$$

Synthesis of cell matter requires, along with carbon, Nitrogen (N), Phosphorous (P) and other elements (e.g. silica for diatoms) acquired by active uptake from their water boundary layer. Available light, Nitrogen and Phosphorus are therefore major environmental factors controlling cell numbers. As both processes are highly enzymatic, biomass increase is also regulated by temperature.

Pigments are the molecules responsible for the absorption of solar energy to be transferred into the cell. They are organized in *photosystems* with reaction centers (chlorophyll *a* molecules) and *antenna* complexes (another pigments – mostly carotenoids and phycobiliproteins, able to transfer its sun's captured energy to the reaction centers). The molecular structure of the Chlorophyll is made up of a tetrapyrrole ring—with a magnesium ion in its center— and a phytol tail with several kinds of chlorophylls (a, b, c_1, c_2, c_3, and d), depending on the distribution of its chemical radicals. Carotenoids, on one hand, are made up of a chain of conjugated double bonds (C–C=C–C=C) able to absorb light with several kind of Carotenes (α and β), besides some oxidized forms called xanthophylls (e.g. Fucoxanthin; Diatoxanthin; Diadinoxanthin; Peridiniin; Zeaxanthin). Some Xanthophills are species-specific in several dinoflagellates. β-Carotene and Zeaxanthin play a photoprotective role to safeguard the photosynthetic apparatus against the excess of light. Phycobiliproteins (allophycocyanin, phycocyanin and, phycoerythrin), on the other hand, are made of a linear chain of four pyrrolic rings with the ability to absorb light in the blue–green region of the spectrum.

Photosynthesis takes place in cellular organelles called chloroplasts. It consists of both light and dark reactions. In the *light reactions* the electrons in the chlorophyll *a* molecules are excited and taken up by the *primary electron acceptors*. The products of the light reactions are used to fix carbon dioxide into organic carbon compounds in the *dark reaction*—a chain of biochemical steps known as the Calvin Cycle.

Light is extinguished in the water by absorption and scattering. Absorption is due to water itself, dissolved substances, and particles in the water (especially those containing pigments). Scattering is due to suspended materials (both live cells and non-living sestonic particles).

The relationship between Photosynthesis (P) and light intensity (E)—usually expressed with the so-called P versus E curves—can be fitted to several mathematical functions (e.g. rectangular hyperbola; quadratic; exponential; hyperbolic tangent). The photosynthesis rate at low light is proportional to the absorbed incident light. The slope of this part of the P versus E curve, which is approximately linear, is called photosynthetic efficiency and is usually designated by α. Photosynthetic efficiency is the product of the *chlorophyll-specific absorption coefficient* (a_{ph}^*) ($m^{-1}(mg\ Chl\ m^{-3})^{-1}$)—light absorbed by all active photosynthetic pigments normalized to chlorophyll a—and the quantum yield (Φ)—photons absorbed per reaction. As light increases, the photosynthetic rate reaches its maximum value (P_{max}) to fall off at higher light intensities when the excess of light can damage the photosynthetic apparatus (photoinhibition).

Phytoplankton cells adjust to both the intensity and spectral quality of light, thereby changing their P versus E curves (Figure 1). They also adjust their relative amounts of Carbon, Chlorophyll, Nitrogen and Phosphorus contents to maximize their growth under varying environmental conditions. Accordingly, cells synthesize less chlorophyll at high light and protect their photosynthetic apparatus by increasing their photoprotective pigment concentrations relative to chlorophyll a (Figure 2). In order to increase their

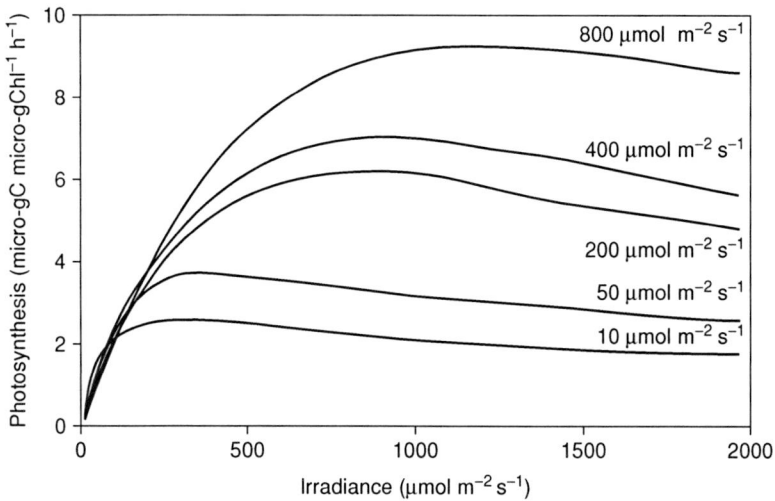

Figure 1. Variations of P versus E curves at different light intensities

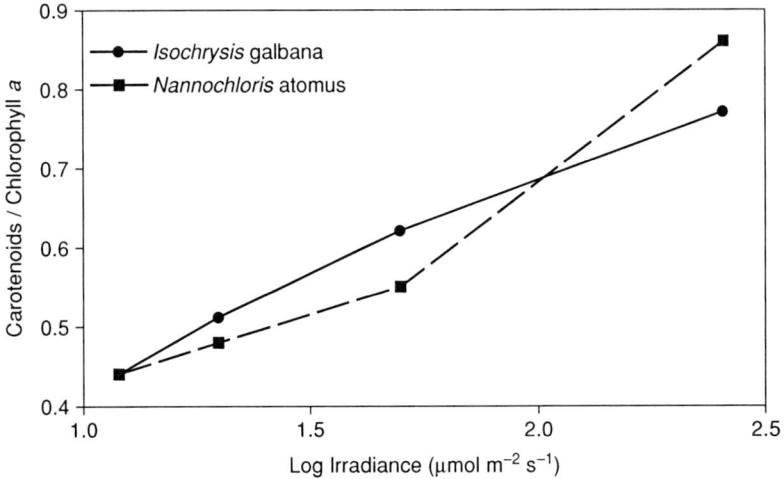

Figure 2. Variation of the Carotenoids/Chlorophyll *a* ratio with light intensity for two phyto-plankton species cultured in the laboratory

photosynthetic efficiency at low light, cells synthesize more chlorophyll per unit cell (greener cells), thus increasing the probability of capturing light photons. Increase in the chlorophyll content results in a lower carbon to chlorophyll cell ratio. Accumulation of chlorophyll molecules reduces their effectiveness in absorbing light due to self-shading (Figure 3)—the so-called chlorophyll package-effect. Photoacclimation produces a high variability in the Carbon to Chlorophyll *a* ratio (typically from 10 to 300) being also affected by temperature. Several formulations have been successfully applied into several ecosystem models to prediction of this ratio.

Phytoplankton acquire nutrients by active uptake from their water boundary layer (mainly N and P, but also Fe or Si, among others). Nitrogen is available in the water in its oxidized forms as nitrate (NO_3^-) and nitrite (NO_2^-) or reduced as ammonia (NH_4^+). Nitrate is first reduced to nitrite by the nitrate reductase (NR) enzyme. Nitrite is afterwards reduced to ammonia in the chloroplast by the nitrite reductase (NiR) enzyme. Most phytoplankton cells preferentially take up ammonia over nitrate, taking up the ammonia when the first is depleted. A relative preference index for ammonia versus nitrate can easily be included in the models. Phosphorous, on the other hand, is available only as phosphate (PO_4^-) in the water, being taken up by means of the alkaline phosphatase (AP) enzyme. Nutrients uptake is a highly temperature-dependent process as a consequence of the many enzymes involved.

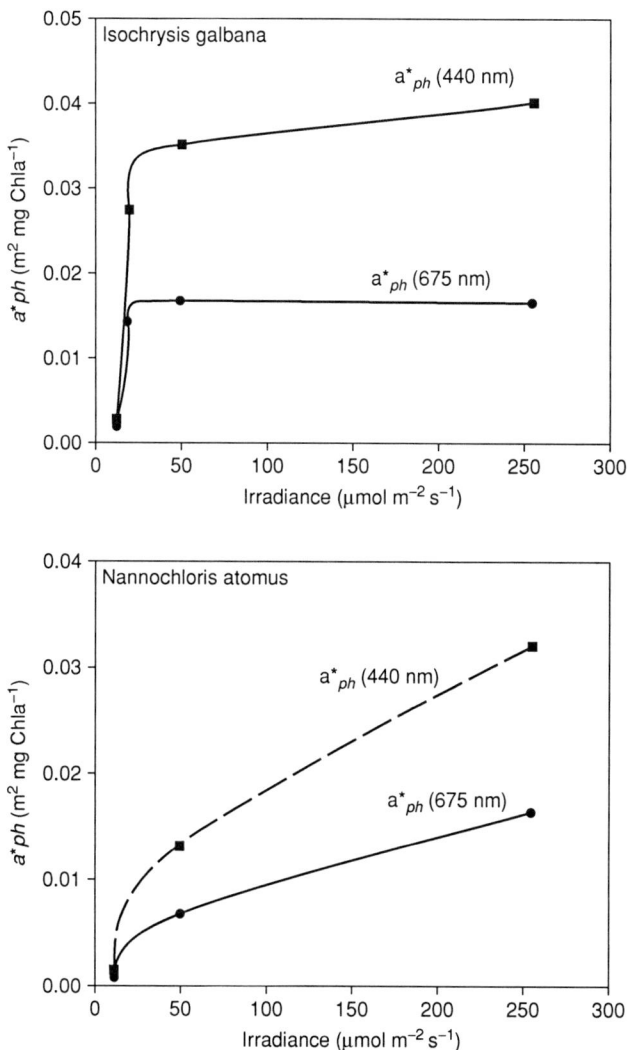

Figure 3. Variation of the *chlorophyll-specific absorption coefficient* (a_{ph}^*) with light at 440 nm and 675 nm for two phytoplankton species cultured in laboratory

The uptake rate (V) of any nutrient follows the well-known Michaelis–Menten enzymatic kinetics (rectangular hyperbola) (Figure 4) defined by the maximum uptake rate (V_{max}) and the half-saturation constant (K_s)—the substrate concentration at which $V = V_{max}/2$.

$$V = V_{max} \frac{S}{K_s + S}$$

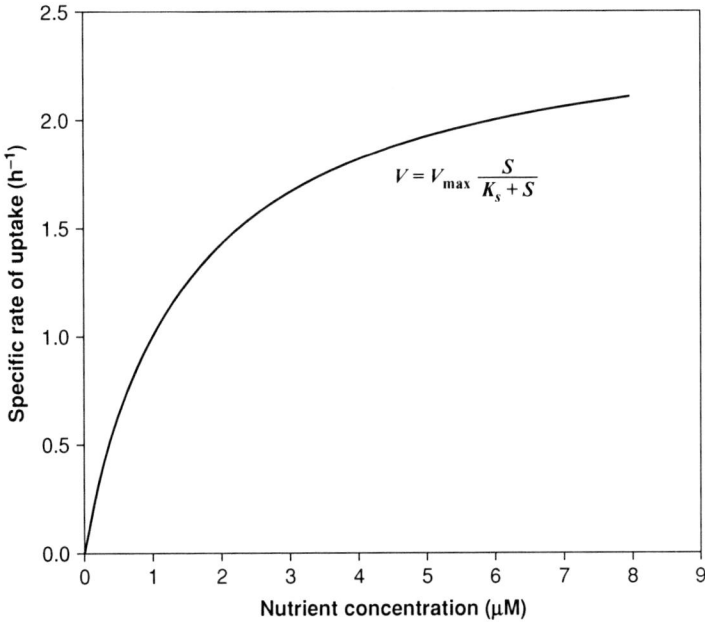

Figure 4. Specific rate of nutrient uptake in phytoplankton as a function of nutrient concentration

As phytoplankton growth follows nutrient uptake, it can be expresses by the Monod function following the Michaelis–Menten kinetics

$$\mu = \mu_{max}\frac{S}{K_s + S}$$

where μ is the specific growth rate and μ_{max} is the maximum specific growth rate.

Phytoplankton, however, can store nutrients in their internal pools to be used when they are temporarily scarce in the water—e.g. when cells move across a heterogeneous distribution of nutrients patches. Under this assumption, phytoplankton growth is better described by the Droop's cell quota model based on the internal concentration of limiting nutrient (Figure 5)

$$\mu' = \frac{\mu'_{max}(Q - Q_0)}{K_s + (Q - Q_0)}$$

where μ' is the nutrient limited specific growth rate; Q is the cell quota (amount of nutrient stored per cell); and Q_0 is the minimum cell quota for that nutrient (cell quota at $\mu = 0$). Cell quota can be experimentally estimated as the nutrient uptake to cell growth rate ratio.

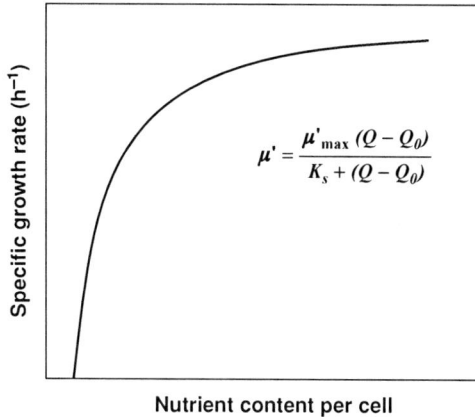

$$\mu' = \frac{\mu'_{max}\,(Q - Q_0)}{K_s + (Q - Q_0)}$$

Nutrient content per cell

Figure 5. Droop's cell quota model

As stated above, the increase of biomass is the end product of photo-synthesis lastly determined by the nutrient present in lowest concentration (limiting factor). Although nutrient uptake is also affected by light intensity showing a truncated hyperbolic function, the efficiency of nutrients uptake depends basically on two processes: the diffusive fluxes determining the rate at which externally-supplied nutrients can reach the cell surface layer; and the efficiency of the internal transport mechanisms. Phytoplankton acclimate not only to light intensity but also to nutrient levels. That is reflected, for in-stance, by finding lower K_s values in phytoplankton isolated from oligotrophic waters than those isolated from eutrophic environments. Under non-steady state conditions, cells adjust their metabolisms for carbohydrates, lipids, pig-ments and proteins cell quotas to buffer the internal machinery from exter-nal fluctuations while growing at rates as close to their μ_{max} as possible. Chemical composition of phytoplankton mirrors environmental conditions, e.g. by accumulating carbohydrates—molecules lacking nitrogen—when N is limited.

The equilibrium between dissolved nutrients in the water and the chem-ical composition of phytoplankton growing near their μ_{max} is, in average, described by the Redfield's ratio (C:N:P by atoms $= 106{:}16{:}1$). In many water bodies, however, the N:P ratio is continuously varying far from the Redfield ra-tio therefore requiring dynamic multi-nutrient mechanistic models to provide a better description of the phytoplankton growth rates.

Cell density can, moreover, be regulated top–down by zooplankton graz-ing. Most of the zooplankton organisms possess cilia or external mouthparts (e.g. copepods) that create turbulences to concentrate the phytoplankton to be ingested. The relationship between the predator ingestion rate and prey density can be represented by the Michaelis–Menten hyperbolic function for a broad range of organisms

$$I = I_{max} \frac{d}{K_d + d}$$

in which the ingestion rate (I) reaches its maximum value (I_{max}) at high prey densities (d), with K_d being the half saturation constant.

For some organisms, such as copepods, a minimum phytoplankton density (d_0) is required to induce filtration. Ingestion rate can be thus be described as a negative exponential function

$$I = I_m\left(1 - e^{-a(d-d_0)}\right)$$

where the constant a is the proportionality factor of the ingestion rate with phytoplankton concentration.

Zooplankton ingestion rates, as well as phytoplankton growth rates, are temperature sensitive processes. The effects of temperature on these rates can be represented as

$$\log(r_t) = \log(r_0) + \log Q_{10} \cdot \left(\frac{t - t_0}{10}\right)$$

where r_0 is the rate at temperature t_0 indicative for the metabolic process, r_t is the rate at temperature t, and Q_{10} is the Ahrrenius temperature coefficient, which has values typically varying from 1.5 to 4.

11.4. Size as an Alternative State Variable

Gain and loss terms for phytoplankton density are not often in the steady state. Seasonal climatic changes, mainly determined by seasonal fluctuations in solar irradiance and winds, produce changes in the water column structure, thus affecting the phytoplankton conditions for life. Species composition is readjusted to varying environmental conditions.

Seasonal sequence of physical properties concerning phytoplankton growth in many water masses can be summarized as follows: in winter, with low temperature, winds provide a well-mixed water column with high nutrient concentrations and low sedimentation rates. When temperature increases in spring, many phytoplankton—mainly diatoms—are stimulated to grow, producing what is known as the spring bloom. In the earlier summer, water stratifies with the formation of the thermocline (a strong gradient of temperature with depth), which acts as a barrier to diffusion of nutrients from deep water. Nutrients in the mixing (upper) layer are consumed by phytoplankton until they become limiting later in the summer. The lack of external supply

of nutrients from below the thermocline is somewhat compensated by rem-
ineralization (regenerated nutrients, mainly NH_4^+) from decomposition of the
spring bloom biomass with phytoplankton species composition also adjusting to this new source of nutrients. The sequence of species changes through
the year known as seasonal succession has been studied, since earlier studies on phytoplankton finding that factors controlling the seasonal cycle of
biomass in freshwater and oceans are quite similar to one another. In very
shallow waters where no stratification occurs, phytoplankton succession is
driven by temperature and turbulence together with some other advection
processes.

Although succession patterns differ from one geographical location to
another, depending on climate and trophic state of the water, major trends can
be summarized as follows: stage I) fast-growing, small-celled diatoms; stage
II) larger diatoms with lower growth rates; stage III) large dinoflagellates with
still lower growth rates.

The Margalef's classical "mandala" (Figure 6) summarizes this seasonal
sequence in a phase space describing the life-forms of phytoplankton as a two
dimensional nutrient-turbulence domain in which phytoplankton succession
follows a well defined path. From this point of view, production can be seen

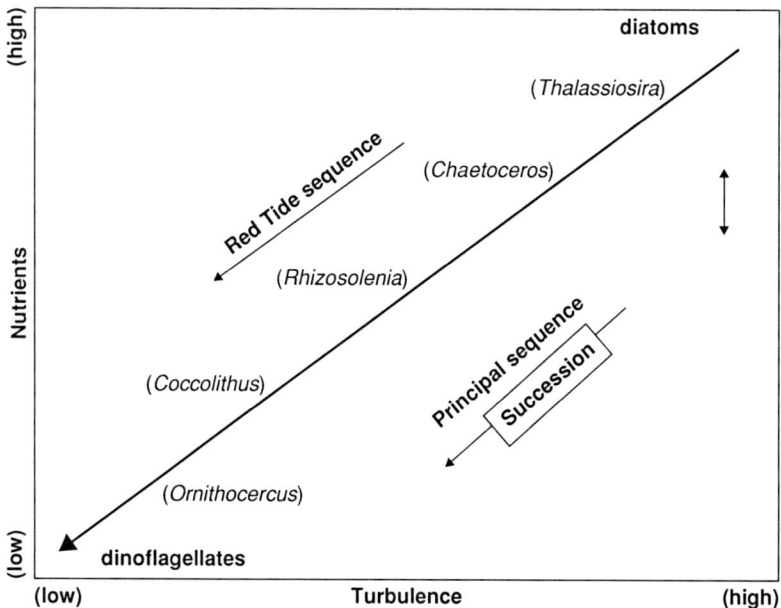

Figure 6. The so-called Margalef's "mandala" describing the seasonal sequence of changes in
phytoplankton succession related to nutrient concentration and turbulence

TABLE 1. Some phytoplankton size-dependent relationships

	Cell volume	
	Small	Large
Surface area to volume ratio	Large	Small
Half-saturation constant for NO_3^- uptake (K_s–NO_3^-)	Low	High
Maximum uptake rate (V_{max}) for NO_3^-	High	Low
Half-saturation constant for NH_4^+ uptake (K_s–NH_4^+)	Low	High
Maximum uptake rate (V_{max}) for NH_4^+	High	Low
Chlorophyll-specific absorption coefficient (a_{ph}^*)	High	Low
Chlorophyll concentration per cell volume unit	High	Low
Maximum growth rate (μ_{max})	High	Low
Maximum cell density	High	Low
Sinking velocity	Slow	Fast

as the result of interactions between advection and turbulence—as an external energy to the cells—and the covariance in distribution of factors of production

$$P = I + (A \times C)$$

where P is production (growth), I represents the advection inputs, A is turbulence, and C is the covariance.

According to this expression, the seasonal trend of changes in the phytoplankton community can be explained by two different, but not exclusive, kinds of mechanisms: physical processes (turbulence and diffusion of nutrients) and complex biological interactions between organisms. While making predictions of the exact number of species and cell density for each species is unrealistic in practice, describing the community composition in terms of phytoplankton cell size, growth rates, and sinking velocities seems an appropriate alternative to using only taxonomic composition.

Within this context, Table 1 summarizes some phytoplankton size-dependent physical and metabolic relationships.

To better understand the role of cell size as an expression of both physical and metabolic processes, temporal and spatial scales at which these processes take place should be taken into account. Fluctuations that operate at small spatial scales (μm to mm) and time scales less than the generation time (seconds to minutes) may have a larger impact on small phytoplankton cells than those acting at larger spatial and temporal scales. Small-scale fluctuations imply nutrient fluxes delivered by small and short pulses fluctuations (e.g., NH_4^+ excreted by zooplankton), resulting in a patchy environment sustained by turbulence. Fluctuations operating at larger scales (e.g., vertical diffusion of nutrients along the year) have a larger impact, inducing changes in larger cells. Processes taking place at small spatial and short time scales can effectively sustain nanoplankton (cells from 2 to 20 μm diameter) cell growth at higher rates than microplankton (cells from 20 to 200 μm diameter). Being small provides advantages for growing in environments with low nutrient fluxes and light regimes. The presence of a periodicity in the nutrient fluxes—external versus regenerated (based on ammonia) inputs—means that picoplankton can use the small scale flux while the larger, slow growing forms can use the longer term (days) scales that arise from atmospheric interactions. In late summer, and in persistently nutrient poor waters, this may be the reason explaining the coexistence of picoplankton (cells from 0.2 to 2 microm diameter) and large dinoflagellates.

The fate of small phytoplankton cells, unlike the larger microplankton that easily sink, is to be grazed by zooplankton. However, certain size fractions of phytoplankton are grazed more efficiently than others according to the size of zooplankton mouth parts. In oligotrophic waters—once the spring bloom has reduced the nutrients concentration in surface waters—the community metabolism switches from new production (based on NO_3^-) to regenerated sources of nutrients (based on NH_4^+), thus facilitating growth of smaller cells. A strong feedback between light, nutrients, grazing, sinking and cell-size is thus established. Accordingly, by mid-summer, when the surface water becomes oligotrophic due to the consumption of nutrients, the phytoplankton community would become dominated by both very small plankton (picoplankton 0.2–2 μm diameter) and by the larger plankton (mesoplankton 20–200 μm diameter) that escape zooplankton grazing.

The distribution of different sizes of organisms in different water types can efficiently be related to both the inherent constraints of predator/prey size ratios and external influence of the environmental fluctuations. Organisms of different size classes have the ability to integrate different environmental signals over space and time. Cell-size distributions are more predictable than species compositions and contain a high potential of information on the ecosystem structure and function. Cell size appears to be a promising alternative to specific composition in the development of a new generation of ecosystem models.

Further Readings

The material in this lecture was drawn from many sources of information, including electronic resources. Suggested further readings are indicated below:

Falkowski, P., and J.A. Raven, 1997. Aquatic Photosynthesis, Blackwell Science, Oxford.
Harris, G.P., 1986. Phytoplankton Ecology. Structure, Function and Fluctuation, Chapman and Hall, New York.

Kirk, J.T.O., 1994. Light and Photosynthesis in Aquatic Ecosystems, Cambridge University Press, Cambridge.

Lalli, C.M, and T.R. Parsons, 1993. Biological Oceanography: An Introduction, Pergamon Press, New York.

Parsons, T.R., M. Takahashi, and B. Hardgrave, 1995. Biological Oceanographic Processes, Butterworth Heinemann, Oxford.

Reynolds, C.S., 2006. Ecology of Phytoplankton, Cambridge University Press, Cambridge.

Valiela, I, 1995. Marine Ecological Processes, Springer, Amsterdam.

12. MODELING POSSIBLE IMPACTS OF TERRORIST ATTACKS IN COASTAL LAGOON ECOSYSTEMS WITH STELLA

Sofia Gamito
IMAR, Faculty of Marine and Environmental Sciences, University of Algarve, Campus de Gambelas, 8005-139 Faro, Portugal

Summary. In this chapter possible changes that would occur in the ecosystem of a lagoon in response to a terrorist attack will be identified and discussed. A very brief revision will be done on ecosystem organization, succession, effects of pollutants and biomagnification, and lagoon water renewal characteristics in order to draw possible scenarios under a terrorist attack. Simple Stella models will be used to simulate a lagoon ecosystem and demonstrate the possible consequences of an attack.

12.1. Ecosystem Organization

12.1.1. FOOD WEBS

Biological communities' organization can be schematized into trophic levels that represent ways in which energy is obtained from the environment, or "who eats whom". A web helps picture how a community is put together and how it works (Pimm et al., 1991). The first trophic level is usually assigned to primary producers (basal species), most often photosynthetic species. The primary consumers, the herbivores, constitute the second trophic level, and carnivores, together with parasites, pests or disease-causing organisms (a sub-class of predators) are assigned to the third or fourth trophic level. Decomposers and detritivores constitute the last but very important trophic level.

Within a food web, several paths or chains may be considered. For long food chains, fifth and even higher trophic levels are possible. A given species may occupy more than one trophic level, and a species can change its trophic position as it grows. However, the proportions of top, intermediate and basal species are nearly constant, regardless of web size (Briand and Cohen, 1984).

Depending on web size (number of S species in web) the number of possible interactions, or feeding relationships, can be very large (S^2). However, the connectance or the number of actual interactions in a food web divided by the number of possible interactions remains relatively constant (Martinez, 1991; Martinez, 1992). The result of this constant connectance is that each

259

I. E. Gonenc et al. (eds.), Assessment of the Fate and Effects of Toxic Agents on Water Resources, 259–277.
© 2007 *Springer.*

species is on average connected to more and more species as the species richness of the food web increases.

It seems that there is a limit on food chain length; few chains exceed eight or nine links or interactions, and the main chain length rarely exceeds five links (Krebs, 2001). Some hypotheses have been formulated to explain these attributes. The *energetic hypothesis* suggests that the length of food chains is limited by the inefficiency of energy transfer along the chain. The *dynamic stability hypothesis* explains that longer food chains are not stable, because fluctuations at lower levels are magnified at higher levels and top predators go extinct (Pimm, 1991).

For simplification, the species are sometimes grouped in trophic species, corresponding to groups of organisms that share the same predators and prey (Briand and Cohen, 1984). However, these organisms can have different growth and metabolic rates and different food preferences.

There is a general tendency to group organisms in large taxonomic groups, such as the phytoplankton group, or the insect group. As a consequence, omnivory can be overestimated. Lumping species together can increase the estimate of links that are classified as omnivorous (feeding in more than one trophic level). Furthermore, many organisms eat their way up the food chain as they grow in size. Also, the detritus that sustains some organisms originates in several trophic levels (Krebs, 2001).

Working with food webs is a tremendous task due to their complexity. The webs can easily attain hundreds or thousands of species. The possible interactions within the web can be millions. The number of chains within the web is also very large. Furthermore, the feedback mechanisms have also to be considered. For modeling purposes see Drossel and McKane (2003), Dune et al. (2005), Dune (2006) and Martinez (2006).

12.1.2. KEYSTONE SPECIES AND REPRODUCTIVE STRATEGIES

Within biological communities, certain species or groups of species may determine the ability of large numbers of other species to persist in the community. These keystone species affect the organization of the community to a far greater extent than one would predict, if considering only the number of individuals or the biomass of the keystone species (Power et al., 1996). Many top predators may be considered keystone species.

Dominant species constitute a large proportion of the biomass and affect many other species in proportion to their biomass. Some common species are plentiful in biomass, but have a relatively low impact on the community organization. Rare species have low impact in the community, proportional to their biomass (Primack, 2001).

Species have been classified, according to their reproductive strategies, as r and K species, although the species may not always maintain the same strategy, depending on the environmental conditions. An $r-$ K selection continuum is then considered, with an organism's position along it in a particular environment at a given instant of time (Pianka, 1970, 1994).

The r-strategy involves increased reproductive effort through early reproduction, small and numerous offspring with large dispersive capacity, short life span, and small body size of adults. The r-strategists have the ability to make use of temporary habitats. Many inhabit unstable or unpredictable environments where catastrophic mortality has occurred and they are able to exploit relatively uncompetitive situations.

In the other extreme, K-strategy species spend more energy on maintenance structures and adaptation, in a predictable environment, than on reproduction. Species with this kind of strategy have a larger body size and slower development, and are less abundant. They are specialists, strong competitors and efficient users of a particular environment, but poor colonizers. However, they often control the small opportunistic r-strategists (top–down control).

12.1.3. MATURE ECOSYSTEMS VS. DEVELOPMENTAL STAGES

Odum (1969) described several attributes to classify ecosystem maturity. In mature stages, biomass accumulates in large individuals and productivity is low, as opposed to developmental stages with low biomass and high production rates. The mature stages have weblike food chains that are predominantly detritic, while developmental stages have linear food chains. The species diversity and organization is high in mature stages. The species life cycle is complex and long, with narrow specialization. In developmental stages, few small opportunistic species dominates the community and diversity is low. The species have short life cycles and broad niche specializations. Recently, some software has been developed to quantify Odum's ecosystem attributes, such as Ecopath (Christensen et al., 2000).

Other holistic indexes and indicators have been recently developed, such as exergy, energy and ascendency (Jorgensen, 2002). However, their definition and determination is quite complex, requiring large amounts of data, which are usually not available.

12.2. Ecosystem Succession

After a large perturbation of the environment, there is a regression of the successional stages of a community. Depending on the degree of the perturbation, the community can recover or a new successional process can start. The end result can be different from the previous situation before perturbation.

The initial, or early successional species, often referred to as pioneer species, are usually characterized by high growth rates, smaller sizes, high degree of dispersal, and high rates of population growth (r-selected species). In contrast, the late successional species generally have lower rates of dispersal and colonization, slower growth rates, and are larger and long-lived (k-selected species) (Smith and Smith, 2001). In early stages of succession, biomass and biological diversity increase. In middle stages of succession, many species of different sizes may occur.

Several processes can occur during succession (Botkin and Keller, 2003):

Facilitation: Early successional species facilitate the ability of later successional species to become established. Knowing the role of facilitation can be useful in restoration of damaged areas. For example, plants that facilitate the presence of others should be planted first.

Interference: Sometimes certain early successional species prevent the entrance of other species for a period of time.

Chronic patchiness: Another possibility is that species do not interact and that succession does not take place. Earlier entering species neither help nor interfere with other species; instead, as in a desert, the physical environment dominates.

12.3. Pollutants and Bioamplification

Persistent pesticides and similar compounds accumulate in the tissues of one species and then are passed up the food web to other species where they become more concentrated. This process is called biomagnification or bioamplification.

The pollutants may affect wildlife in different ways. Their birth, death and growth rates may change, changing also their abundance (Botkin and Keller, 2003). The sensitive species decrease in abundance and the less sensitive may increase. Pollution, whether chronic or acute, usually tends to favor short-lived opportunistic species (Raffaelli and Hawkins, 1996).

The effect of pollutants may be similar to the effect of strong physical disturbance. In newly dumped dredge soil, immediately after disturbance (or close to the source of pollution) a few species of abundant, small, and productive polychaetes are found. These are followed by suspension-feeding or surface feeding mollusks either over time or space. The latter are replaced by large, slow-grazing species that live deeper in the sediment, feed on buried deposits, and oxidize the sediment by their activities. There is a close parallel to the gradient over space away from a grossly polluted site (Valiela, 1995). According to the Pearson-Rosenberg model, with increasing organic input there is an increase of abundance, biomass and species richness in a first step, and a progressive declining of species richness and biomass when eutrophication increases, while abundance (mainly of opportunistic species) continues rising (Pearson and Rosenberg, 1978).

12.4. Coastal Lagoons and Water Renewal Rate

Lagoon species assemblages are the result of the continuous or seasonal interactions between native lagoon inhabitants and sporadic, accidental, or periodic colonizers. The native species are physiologically adapted to stressed and changing environments, with reproductive capacity giving rise to new generations inside the lagoons. Individuals of colonizing species must compete with lagoon species for resources. Some of them will be able to survive and develop into adults but not reproduce inside the lagoon, and others can survive, reproduce, and become established but only under favorable conditions (Gamito et al., 2005).

Coastal lagoons offer a protected shallow habitat, which can be highly productive (Alongi, 1998). Well-structured communities controlled by k-strategists can develop and settle in leaky lagoons, that is, lagoons with wide entrance channels and tidal currents which guarantee a good water renewal. In these lagoons, biomass can accumulate in large organisms. In contrast, lagoons with a single narrow entrance, which may be closed for long periods, are characterized by persistent physical stress and are dominated by communities of small-sized r-strategists (Gamito, 2006; Gamito et al., 2005).

12.5. Terrorist Attack Consequences in The Lagoon Ecosystem

If a terrorist attack happens, a degradation of the ecosystem is expected. The worst case scenario will be the destruction of all living organisms. Expected ecological consequences are listed below.

Ecosystem organization:

- The top predators will die or move away.
- The dominant and the common species decrease in abundance, the more sensitive die.
- The opportunistic species will find space to develop rapidly (if the environmental is not too degraded).
- The food web will be shortened and simplified.

- Diversity will decrease.
- Most of the ecosystem attributes will change from a mature stage to a developmental stage.

Successional stages:

- The system returns to the first stages of succession.
- Pioneer/opportunistic species will increase in abundance.
- With time, if the system was not too degraded, other colonizers, stronger competitors, and long lived organisms will arrive and occupy the space, replacing the pioneer species.
- Top predators are expected to establish later and control the lower trophic levels.
- The successional stages can be shortened by artificially introducing the right pioneer species and/or by engineering works.
- The ecosystem may evolve to an ecosystem different from the previous one.

Pollution/Bioaccumulation (depending on the agent involved):

- Bioaccumulation can be expected, with concentration of the compound in the higher levels of the food chain.
- The compound can cause diseases and malformations in the next generations, decreasing growth rate and survival.
- The compound can be selective, affecting only some groups of organisms (possibly top predators), eventually related in a close evolutionary way to the target-species (man).
- If this last hypothesis is true, trophic food web organization changes are expected, with loss of the higher trophic levels and consequent loss of top–down control.
- The pollutant may cause a displacement of a well-organized community to a community dominated by few species of small opportunists, decreasing species richness and biomass.

Lagoon characteristics:
Depending on the water renewal rate in the lagoon:

- A high water renewal rate will increase the rate of recovery of the ecosystem by "diluting" the effects of the terrorist attack and by allowing the migration of potential new colonizers.
- If the water renewal is low, the recovery will be more difficult, but it can be improved by opening new or deeper connections with the sea.

12.6. Outlines for Modeling

After a terrorist attack, changes are expected in community organization. If there is a severe degradation of the environment, all species may die. In that case a recolonization process will take place, with small opportunistic species as the first colonizers. When the process starts depends on the agent involved. The final community will also depend on the agent involved, the type of lagoon (choked, leaky or restricted), and the lagoon's hydrological regime.

If the environmental degradation is not total, the most sensitive species, such as top predators, may die or move away and consequently the top–down control of the community is lost. Common or dominant species may also severely reduce their density. Depending on the severity of the attack, and the level of environmental degradation, the opportunistic species may develop in high densities. After some time, the environment may have recovered, and new colonizers will find conditions to develop and reproduce, following the natural stages of succession.

A model is always a simplification of nature. Complex models try to reproduce with more precision what happens in nature. But the more complex the model, the more difficult it is to know the proper rates and parameters needed to run it. Furthermore, little is known about trophic organization and food web control. One possible and simple approach is to model the effects of a terrorist attack on the different groups of species: the predators, the common/dominant species, and the opportunistic species using classical models of growth, competition and predation.

12.7. Ecological Modeling with Stella

STELLA software is a graphic, icon-based modeling software package from Isee Systems—The Visual Thinking Company (http://www.hps-inc.com), which can be used in the construction of relatively complex models.

The basic structure of the models includes stocks (which represent accumulation), inflow and outflow rates (into and out of a stock), and auxiliary variables (which help define the inflow and outflow rates).

A population growth model can easily be built, where the stock represents the population and the inflow and outflow rates represent the natality and the mortality processes (Figure 1).

There are three communicating layers that contain progressively more detailed information on the structure and functioning of the model. The lowest layer contains the equations, which are difference equations generated by the model structure in the middle level. The middle level represents the model

Population

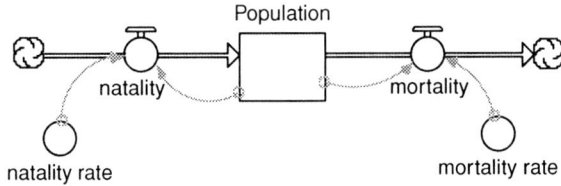

Figure 1. Conceptual diagram of a population

structure with icons, as previously shown. The graphic representation of these units is connected and manipulated on the screen to build the basic structure of the model (Costanza and Ruth, 1998). Once the basic structure of the model is completed, initial conditions, parameter values and functional relationships can be specified by simply clicking on the respective icons. The input of data can be done by graphical or mathematical functions. The highest layer is the "user interface". In the final stage users can easily access and operate the model from this level. With the use of slide-bars, a user can also immediately respond to the model output by choosing alternative parameter values as model runs to test alternative scenarios. The model output can be generated in tabular or graphical form.

12.8. Model Development

Model equations were based on Gamito (1998) and Gotelli (2001) and references there in.

12.8.1. EXPONENTIAL GROWTH MODEL

Returning to our very simple example, the growth of a population, if there is no migration the size of the population will vary according to the natality and mortality rates. If we maintain the natality and mortality rates constant, we have the exponential model:

$$dN/dt = nN - mN = (n - m)N = rN \qquad (1)$$

where N is the population density; n is the instantaneous natality rate; m is the instantaneous mortality rate; and r is the instantaneous rate of increase.

This model can than be represented as (Figure 2).

The population increases exponentially if $r > 0$, remains stable if $r = 0$, and follows a negative exponential decrease if $r < 0$.

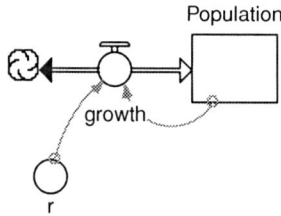

Figure 2. Conceptual diagram of the exponential growth model

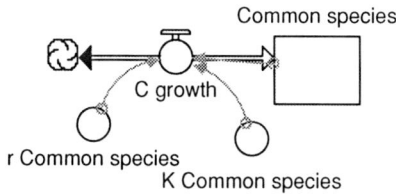

Figure 3. Conceptual diagram representing the common group of species growth

12.8.2. LOGISTIC GROWTH MODEL

Exponential growth cannot persist for long periods of time. Realistically, a population grows up to a certain limit, the *carrying capacity* (K):

$$dN/dt = rN \left(\frac{K - N}{K} \right). \tag{2}$$

In a coastal lagoon there is always a group of dominant or common species, within plankton, fish, benthic invertebrates, macroalgae, or birds. This group can be represented as a stock that can increase up to a certain limit, the carrying capacity of the lagoon for that group of species (Figure 3).

An opportunistic group of species can also be included. This group will probably have a higher growth rate and will grow up to a higher limit (Figure 4).

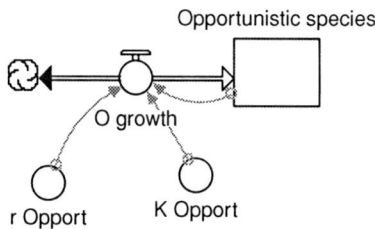

Figure 4. Conceptual diagram representing the opportunistic group of species growth

1: Common species   2: Opportunistic species

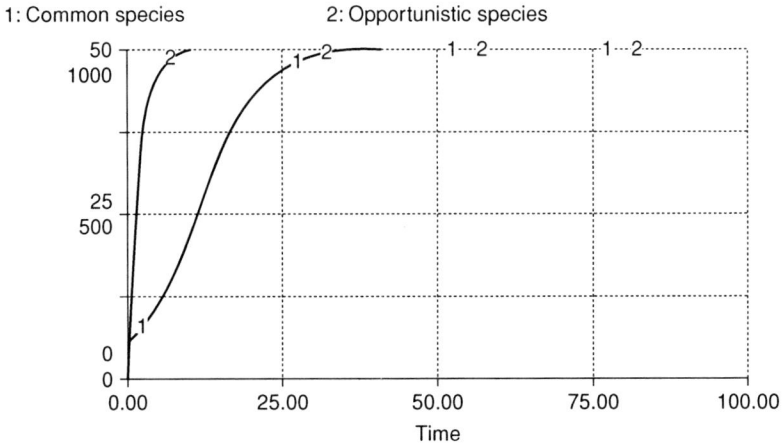

Figure 5. Simulation of growth of common and opportunistic groups

For the two groups of species a possible set of equations, initial values and growth rates were specified with Stella software:

Common_species(t) = Common_species($t - dt$) + (C_growth) * dt

Opportunistic_species(t) = Opportunistic_species($t - dt$) + (O_growth) * dt

O_growth = r_Opport*Opportunistic_species*(K_Opport-Opportunistic_species)/Opportunistic_species

C_growth = Common_species*r_Common_species*(1 −Common_species/K_Common_species)

INIT Common_species = 5

INIT Opportunistic_species = 5

K_Common_species = 50

K_Opport = 1000

r_Common_species = 0.2

r_Opport = 0.6.

The resulting simulation is represented in Figure 5.

12.8.3. COMPETITION MODEL

If these two groups of species are present together, some intergroup competition might occur. The opportunistic species are fast growing species with a lower competitive capacity. Their density decreases considerably in the presence of stronger competitors.

In the logistic model, population growth is reduced by intraspecific competition. In the Lotka–Volterra competition model, population growth is further depressed by interspecific competition (Gotelli, 2001) (Table 1).

TABLE 1. Growth equations of two populations where interspecific competition may occur. N_1 and N_2 correspond to the size of two populations, and α and β are competition coefficients. α is a measure of the effect of species 2 on the growth of species 1 and β is a measure of the effect of species 1 on the growth of species 2 (Based on Gotelli (2001)).

With no interspecific competition	With interspecific competition
$dN_1/dt = r_1 N_1 \left(\frac{K_1 - N_1}{K_1} \right)$	$dN_1/dt = r_1 N_1 \left(\frac{K_1 - N_1 - \alpha N_2}{K_1} \right)$
$dN_2/dt = r_2 N_2 \left(\frac{K_2 - N_2}{K_2} \right)$	$dN_2/dt = r_2 N_2 \left(\frac{K_2 - N_2 - \beta N_1}{K_2} \right)$

Depending on the competition coefficients, competition may occur (Gotelli, 2001). If $\alpha = 0$, there is no interspecific competitive effect (classical logistic model). If $\alpha < 1$, the intraspecific competition is more important then the interspecific competition. When $\alpha = 1$, then individuals of both species have an equal effect in depressing growth of species 1. When $\alpha = 4$, each individual of species 2 that is added to the environment depresses the growth of species 1 by the same amount as adding four individuals of species 1.

Different results are expected after the application of the Lotka–Volterra competition model (Table 2):

- One or the other group is extinct.
- Both groups of species can coexist in a stable equilibrium.
- Both groups of species can coexist in an unstable equilibrium.

Let us return to our lagoon and our groups of common and opportunistic species. We know that both groups coexist in the lagoons. In lagoons with good environmental conditions, the common species dominate and the opportunistic species are present in low densities. Since the two groups of

TABLE 2. Algebraic inequalities and competition outcome in the Lotka–Volterra equations (in Gotelli (2001))

Species 1 invades	Species 2 invades	Inequality	Outcome
Yes	No	$\frac{1}{\beta} < \frac{K_1}{K_2} > \alpha$	Species 1 wins
No	Yes	$\frac{1}{\beta} > \frac{K_1}{K_2} < \alpha$	Species 2 wins
Yes	Yes	$\frac{1}{\beta} > \frac{K_1}{K_2} > \alpha$	Stable coexistence
No	No	$\frac{1}{\beta} < \frac{K_1}{K_2} < \alpha$	Unstable equilibrium

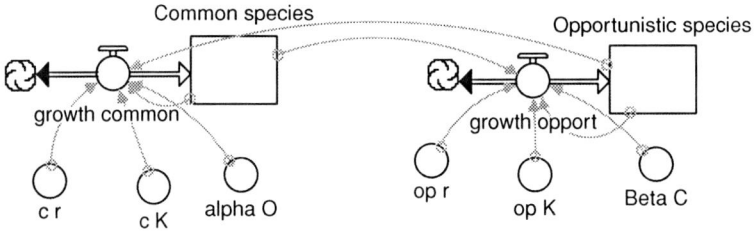

Figure 6. Conceptual diagram of a hypothetical competition model among the common and the opportunistic groups of species in a lagoon

species coexist, we have to obey the inequality for stable coexistence in Table 2.

Therefore, the competitive coefficient (alpha O in Stella code) must be small in relation to the common species group, but very large for the opportunistic group (Beta C in Stella code), meaning that this group of species has a low competitive ability. In a first run, a lagoon with good environmental conditions and dominance of the common species was considered (Figures 6 and 7 and model equations):

Competition model equations:
Common_species(t) = Common_species($t - dt$) + (growth_common) * dt
Opportunistic_species(t) = Opportunistic_species($t - dt$) + (growth_opport) * dt

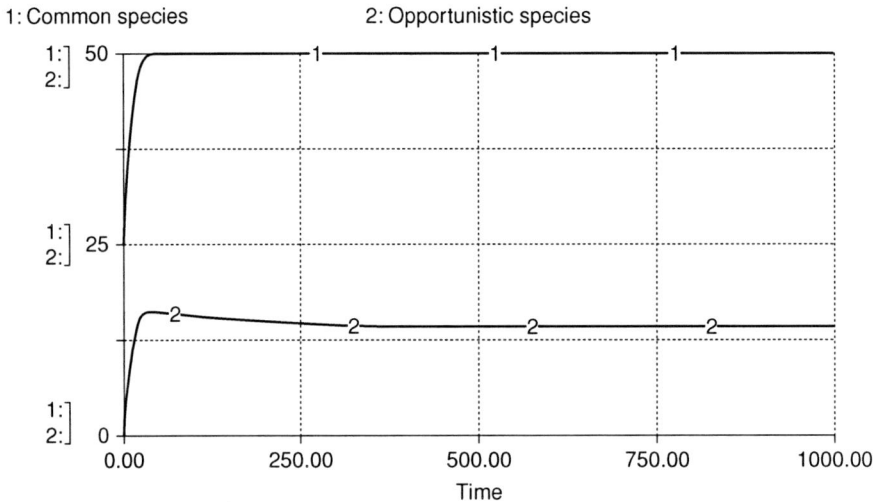

Figure 7. Simulation of two competing groups of species

growth_common = c_r*Common_species*((c_K-Common_species-alpha_O*
   Opportunistic_species)/c_K)
growth_opport = op_r*Opportunistic_species*((op_K-Opportunistic_species-
   Beta_C*Common_species)/op_K)
INIT Common_species = 25
INIT Opportunistic_species = 2
alpha_O = 0.005
Beta_C = 19.75
c_K = 50
c_r = 0.2
op_K = 1000
op_r = 0.6.

If a catastrophic event happens, the common group of species may decrease
and the opportunistic species increase. For example, changing the initial values of the common group to 10 and the opportunistic group to 1000, the
common species quickly depresses the opportunistic group and the equilibrium densities are attained at 49.9 and 13.9. The equilibrium point is also
attained when the initial values of the species groups are higher than their
carrying capacity. Whatever the initial values of both groups, the equilibrium
point is reached.

12.8.4. PREDATION MODEL

Both groups of species may be controlled by predators, which may act as
keystone species. There are two classical predator/prey models. One model
considers that in the absence of predators, prey populations will grow according to the logistic model. In this model, the prey population growth will be
further depressed due to predation (Gotelli, 2001):

$$dN_1/dt = r_1 N_1 \left(\frac{K_1 - N_1 - \alpha N_2}{K_1} \right) - \delta N_1 P. \qquad (3)$$

The predator group (P) will increase (or decrease) depending on prey density:

$$dP/dt = \gamma N_1 P - m P \qquad (4)$$

where δ and γ are the interaction coefficients; δ is the loss rate of prey individuals due to predation (predation rate in Stella code); γ is the increase
in predator's growth by feeding on prey individuals (conversion rate in Stella
code); m is the mortality rate of predator population.
   If we incorporate predation on both common and opportunistic groups of
species, our model becomes more complex (Figure 8).

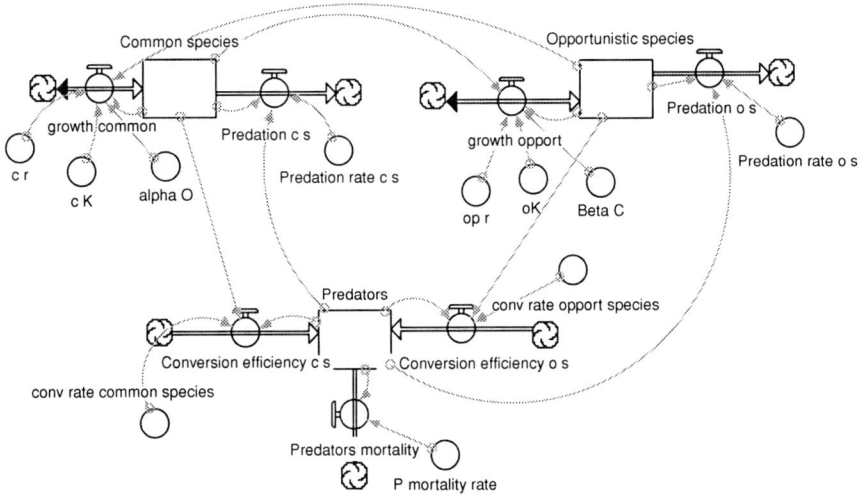

Figure 8. Conceptual diagram of a competition model with predation in a lagoon

A higher predation rate (interaction factor) on the opportunistic group was specified, as opportunistic species probably have fewer escape mechanisms than the common species group (Predation model equations, below). Keeping the same initial values and auxiliary constants of the previous model (Competition model equations), the predator group depresses growth of both the common and opportunistic groups of species (Figure 9):

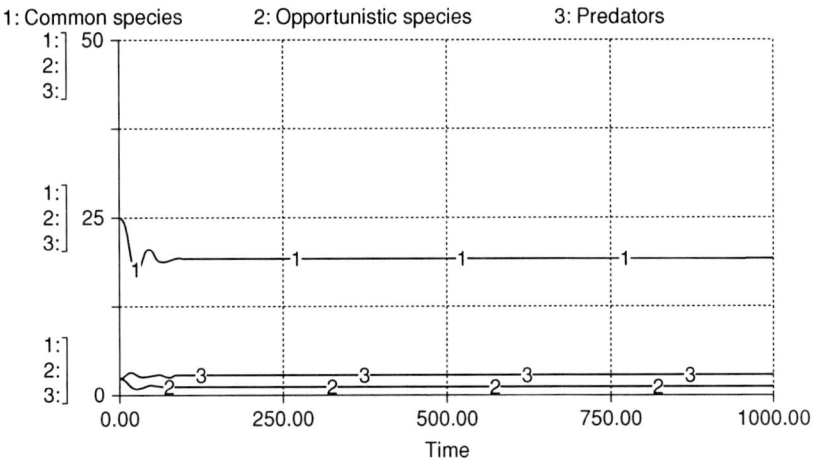

Figure 9. Simulation of two competing groups of species, the common and the opportunistic group, and the effect of a predation group on the previous groups

Predation model equations:

Common_species(t) = Common_species($t - \mathrm{d}t$) + (growth_common − Predation_c_s) * dt

Opportunistic_species(t) = Opportunistic_species($t - \mathrm{d}t$) + (growth_opport − Predation_o_s) * dt

Predators(t) = Predators($t - \mathrm{d}t$) + (Conversion_efficiency_c_s + Conversion_efficiency_o_s − Predators_mortality) * dt

growth_common = c_r*Common_species*((c_K − Common_species − alpha_O*Opportunistic_species)/c_K)

Predation_c_s = Common_species*Predators*Predation_rate_c_s

growth_opport = op_r*Opportunistic_species*((oK − Opportunistic_species -Beta_C*Common_species)/oK)

Predation_o_s = Opportunistic_species*Predators*Predation_rate_o_s

Conversion_efficiency_c_s = Common_species*Predators*conv_coef_common_species

Conversion_efficiency_o_s = Opportunistic_species*Predators*conv_coef_opport_species

Predators_mortality = P_mortality_rate*Predators

INIT Common_species = 25

INIT Opportunistic_species = 2

INIT Predators = 2

alpha_O = 0.005

Beta_C = 19.75

conv_rate_common_species = 0.01

conv_rate_opport_species = 0.01

c_K = 50

c_r = 0.2

oK = 1000

op_r = 0.6

Predation_rate_c_s = 0.05

Predation_rate_o_s = 0.151

P_mortality_rate = 0.2.

This model system could be applied to any large group of species, such as plankton, benthic invertebrates or fish. When in equilibrium, in each of these groups there are always common species, opportunistic species, and predators. The densities need to be adjusted to each of these groups, as well as the time—faster processes in plankton species, much slower processes in fish.

12.9. Simulation of Different Scenarios

In this section, three scenarios for the impacts of a terrorist attack on the lagoon system are illustrated using the STELLA model:

Scenario 1: A terrorist attack happens. All organisms die. After some time, some opportunistic species start to develop. If the lagoon is not isolated, for instance if it has good connections with the sea, some common species may migrate and find good conditions to develop. Later, predators might also enter the lagoon. After some time, the system reaches equilibrium again (Figure 10).

Scenario 2: A terrorist attack happens. The more sensitive species die or migrate to other systems. A reduction of the common species group is probable, and the extinction of the predator species group is also probable. The opportunistic group starts to develop, but then decreases rapidly when common species recover. If the lagoon is not isolated, some predators may migrate. After some time the system reaches equilibrium again (Figure 11).

Scenario 3: A terrorist attack happens. The environmental conditions degrade. The carrying capacity of the common and opportunistic groups de-

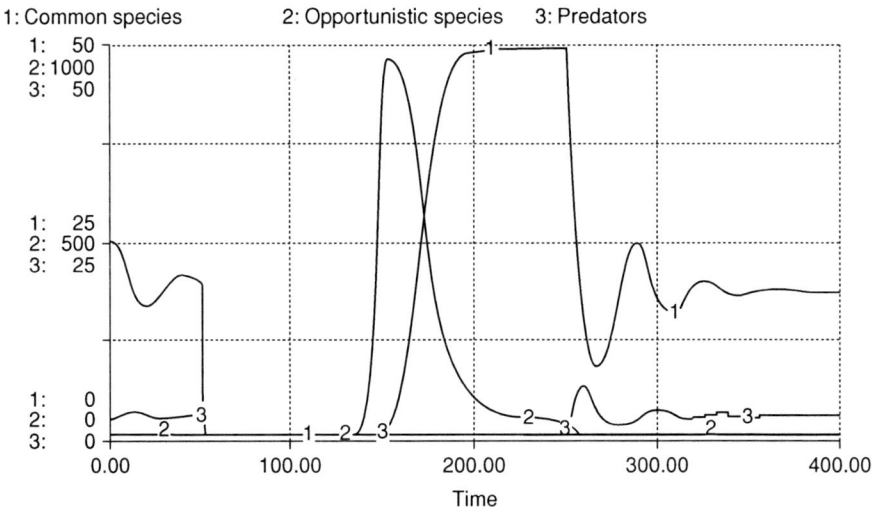

Figure 10. Simulation of the effect of a terrorist attack in the three groups of species (Scenario 1, see text for explanation)

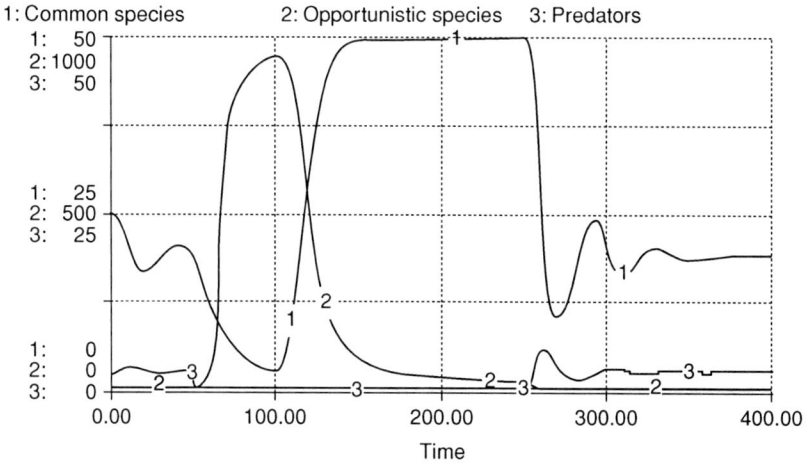

Figure 11. Simulation of the effect of a terrorist attack in the three groups of species (Scenario 2, see text for explanations)

creases by half by $t = 50$ to $t = 100$. The more sensitive species die or migrate to other systems. The opportunistic group starts to develop, slower than in Scenario 2, but decreases rapidly when common species recover. If the lagoon is not isolated, some predators may migrate. After some time the system reaches equilibrium again (Figure 12).

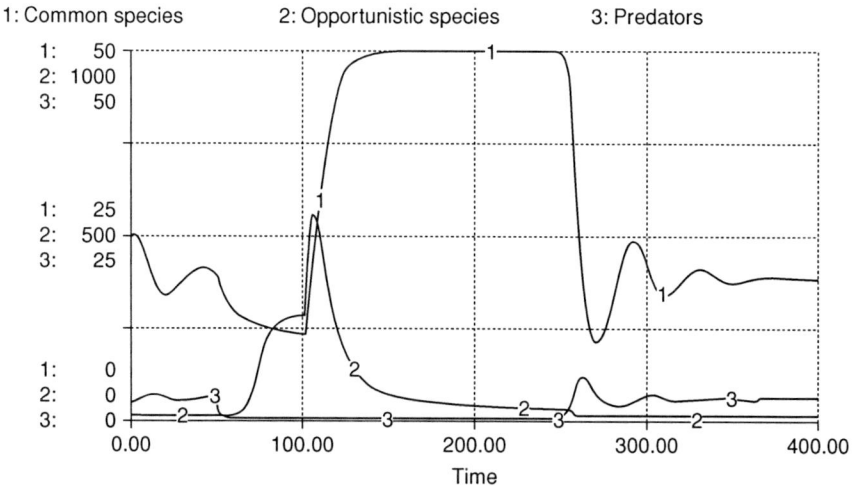

Figure 12. Simulation of the effect of a terrorist attack in the three groups of species (Scenario 3, see text for explanations)

12.10. Remarks

A model is a simplified representation of what we think that happens in the real world. However, we do not have data. We need experimentation to learn what might really happen after a terrorist attack. But we already know the possible succession of the main groups of species after a severe degradation of the environment.

We have used the classical growth/competition/predator-prey models to represent three groups of species interacting in a lagoon system. Some improvements could have been introduced such as including environmental variables that influence growth, such as temperature variation. Also, carrying capacity and growth rates of the different species groups may vary with environmental conditions.

Competition for nutrients or food and competition for space occur in communities. By adding predation and spatial patchiness, more realistic models of community organization can be constructed. The equilibrium model of community organization includes then competition, predation and spatial patchiness (Krebs, 2001). However, no spatial variation was considered in the model. Again, we have no information. Furthermore, models that include the distribution of organisms are exceptionally difficult to construct. When animals move—either because they are grazing, escaping from a predator, or migrating—it is very difficult to describe those movements correctly.

Neither diversity variation nor genetic variability loss were included in our simplified vision of the terrorist effects on lagoon ecology, nor were possible effects of bioaccumulation on the ecosystem.

However, in spite of the simplifications used in this model, it could show hypothetical changes at ecosystem level due to a terrorist attack.

Acknowledgements

Comments and English revision of the text by Dr. Brenda Rashleigh greatly improved the manuscript.

References

Alongi, D.M., 1998. Coastal Ecosystem Processes, CRC Press, Boca Raton.
Briand, F., and J.E. Cohen, 1984. Community food webs have scale-invariant structure, Nature, 307, 264–267.
Botkin, D., and E. Keller, 2003. Environmental Science. Earth as a Living Planet, 4th edn, John Wiley & Sons, New York.
Christensen, V., C.J. Walters, and P. Pauly, 2000. Ecopath with Ecosim A User's Guide, Fisheries Centre, University of British Columbia, Penang.
Costanza, R., and M. Ruth, 1998. Using dynamic modeling to scope environmental problems and built consensus, Environ. Manage., 22, 185–195.

Drossel, B., and A.J. McKane, 2003. Modelling food webs, edited by S. Bornholdt, and H.G. Schuster, Handbook of Graphs and Networks, Wiley-VCH, Berlin, pp. 218–247.

Dunne, J.A., 2006. The network structure of food webs, edited by M. Pascual, J.A. Dunne, Ecological Networks: Linking Structure to Dynamics in Food Webs, Oxford University Press, Oxford, pp. 27–86.

Dunne, J.A., U. Brose, R.J. Williams, and N.D. Martinez, 2005. Modeling food-web dynamics: Complexity-stability implications, edited by A. Belgrano, U.M. Scharler, J.A. Dunne, and R.E. Ulanowicz, Aquatic Food Webs; An Ecosystem Approach, Oxford University Press, Oxford, pp. 117–129.

Gamito, S., 1998. Growth models and their use in ecological modelling: An application to a fish population, Ecol. Model., 113, 83–94.

Gamito, S., 2006. Benthic ecology of semi-natural coastal lagoons, in the Ria Formosa (Southern Portugal), exposed to different water renewal regimes, Hydrobiologia, 555, 75–87.

Gamito, S., J. Gilabert, C. Marcos, and A. Pérez-Ruzafa, 2005. Effects of changing environmental conditions on lagoon ecology, edited by I. Gonenç, and J.P. Wolflin, Coastal Lagoons. Ecosystem Processes and Modeling for Sustainable Use and Development, CRC Press, Boca Raton, pp. 193–229.

Gotelli, N.J., 2001. A Primer of Ecology, 3rd edn, Sinauer Associates, Inc., Massachusetts.

Jorgensen, S.E., 2002. Integration of Ecosystem Theories: A Pattern, 3rd edn, Kluwer Academic Publishers, Boston.

Krebs, C.J., 2001. Ecology. The Experimental Analysis of Distribution and Abundance, 5th edn, Benjamin Cummings, San Francisco.

Martinez, N.D., 1991. Artifacts or attributes? Effects of resolution on the little rock lake food web, Ecol. Monogr. 61, 367–392.

Martinez, N.D., 1992. Constant connectance in community food webs, Am. Nat., 139, 1208–1218.

Martinez, N.D., 2006. Network evolution: exploring the change and adaptation of complex ecological systems over deep time, edited by M. Pascual, and J.A. Dunne, Ecological Networks: Linking Structure to Dynamics in Food Webs, Oxford University Press, Oxford, pp. 287–302.

Odum, E.P., 1969. The strategy of ecosystem development, Science, 164, 262–270.

Pearson, T.H., and R. Rosenberg, 1978. Macrobenthic succession in relation to organic enrichment and pollution of the marine environment, Oceanogr. Mar. Biol. Ann. Rev., 16, 229–311.

Pianka, E.R., 1970. On r and k selection, Am. Nat., 104, 592–597.

Pianka, E.R., 1994. Evolutionary Ecology, 5th edn, Harper Collins College Publishers, New York.

Pimm, S.L., 1991. The balance of nature? Ecological Issues in the Conservation of Species and Communities, The University of Chicago Press, Chicago.

Pimm, S.L., J.H. Lawton, and J.E. Cohen, 1991. Food web patterns and their consequences, Nature, 350, 669–674.

Power, M.E., D. Tilman, J.A. Estes, B.A. Menge, W.J. Bond, L.S. Mills, G. Daily, J.C. Castilla, J. Lubchenco, and R.T. Paine, 1996. Challenges in the quest for keystones, BioScience, 46, 609–620.

Primack, R.B., 2002. Essentials of Conservation Biology, 3rd edn, Sinauer Associates Inc, Sunderland.

Raffaelli, D., and S. Hawkins, 1996. Intertidal Ecology, Chapman & Hall, London.

Smith, R.L., and T.M. Smith, 2001. Ecology and Field Biology, 6th edn, Benjamin Cummings, San Francisco.

Valiela, I., 1995. Marine Ecological Processes, 2nd edn, Springer-Verlag, New York.

13. TROPHIC NETWORK MODELS AND PREDICTION OF TOXIC SUBSTANCES ACCUMULATION IN FOOD WEBS

Arturas Razinkovas
Klaipèda University, Lithuania

13.1. Introduction

The term food web or trophic network defines a set of interconnected food chains by which energy and materials circulate within an ecosystem. The classical food web could be divided into two broad categories: the grazing web, which typically begins with green plants, algae, or photosynthesizing plankton, and the detrital web, which begins with organic debris. In a grazing web, materials typically pass from plants to herbivores to flesh eaters. In a detrital web, materials pass from plant and animal matter to decomposers as fungi and bacteria, then to detritivores, and then to their predators. In water ecosystems, the classical food web is represented by the planktonic and benthic food webs, which are interconnected. Additionally, the "microbial loop" represents an alternative pathway of carbon flow that leads from bacteria to protozoa to metazoa, with dissolved organic matter (DOM) being utilized as substrate by the bacteria, which include nanoplankton (2–20 μm in size) and picoplankton (0.2–2 μm in size).

Food web models are generally built by reconstructing and quantifying the ecosystem trophic transfer network. This type of ecological model has been widely used for many applications, including fishery and stock management, eutrophication studies, and tracing bioaccumulation processes in water ecosystems. In many cases, once built, particular food web models could used for different applications such as for modeling both eutrophication and bioaccumulation (Koelmans et al., 2001), which often appear to be interrelated processes (Schaanning et al., 1996; Gourlay et al., 2005). The recent development of the ECOPATH project for dynamic simulation and the introduction of ECOSIM and ECOTRACE routines (Christensen and Walters, 2004) provides the opportunity to use a number of trophic models already developed for the modeling of persistent pollutants bioaccumulation in the trophic food webs.

I. E. Gonenc et al. (eds.), Assessment of the Fate and Effects of Toxic Agents on Water Resources, 279–289.
© 2007 *Springer.*

13.2. Bioaccumulation of Toxic Substances in Food Webs

There are two pathways of persistent pollutants in water ecosystems: physically and chemically facilitated pathways (hydraulic transport, chemical bind-
ing, settling and resuspension, absorption by suspended particles and sediment) and biologically induced pathways, or bioaccumulation. These two
pathways are interconnected. Persistent contaminants introduced into the
ecosystem can bind to particles and either remain suspended in the water column (particle-associated contaminants) or settle into the sediment (sediment-
associated contaminants). Benthic organisms are exposed to certain critical
pollutants also through their contact with the sediment. These pollutants may
then be introduced into the lower food chain, from which they can be transferred back up the aquatic food web. Demersal fish tend to accumulate persistent toxic substances in higher concentrations than pelagic species (Camanzo
et al., 1987; Rowan and Rasmussen, 1992; Pastor et al., 1996).

Bioaccumulation in living organisms is often explained by the two terms:
biomagnification and bioconcentration. Biomagnification is the bioaccumulation of a substance up the food chain by transfer of residues of the substance in
prey organisms. These processes result in an organism having higher concentrations of a substance than is present in its diet. Biomagnification can result in
higher concentrations of the substance than would be expected if water were
the only exposure mechanism. Accumulation of a substance only through
contact with water is known as bioconcentration (Figure 1). The Bioconcentration process could be defined as the net result of uptake and elimination;
thus, persistent contaminants are accumulated because uptake rates typically
exceed elimination rates (Shaw and Connell, 1986).

Trophic position frequently explains much of the observed variation in
contaminant levels among biota, due to the biomagnification (MacDonald
and Bewers, 1996). The process of biomagnification is evident in the well-
studied arctic cod (*Bormgadus saida*)—ringed seal (*Phoca hispida*)—polar
bear (*Ursus maritimus*) food chain (Muir et al., 1988). Concentrations increase by a factor of 6 between arctic cod and ringed seal and by a factor of
approximately 7 between ringed seal and polar bears (Muir et al., 1997). Several factors appear to influence the degree of contaminant bioaccumulation
in different species. Organism lipid content tends to be positively correlated
with bioaccumulation (Pastor et al., 1996). Attributes of different water bodies such as productivity, wave and current dynamics, and depth may also
influence bioaccumulation. Some authors found a significant negative correlation between biota and sediment accumulation factors (BSAFs) and lake
maximum depth (MacDonald et al., 1993). It is possible that resuspended sediments or benthic communities play a relatively larger role in the food webs
of shallow lakes, thereby facilitating contaminant transfer from sediments
(MacDonald et al., 1993). For instance, an approximately threefold increase

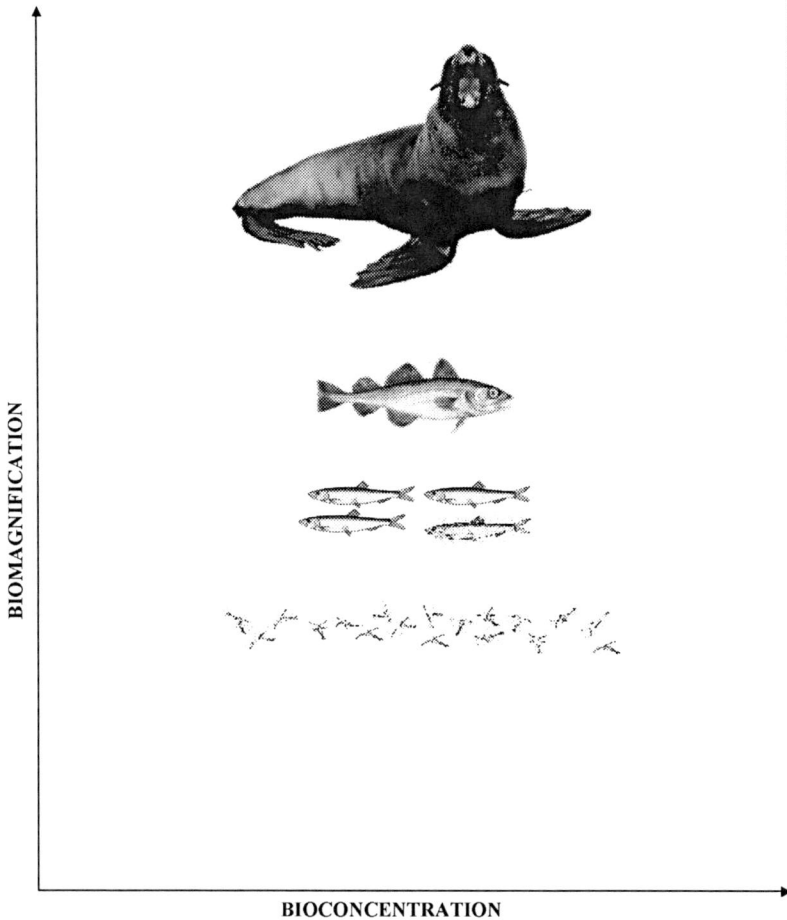

Figure 1. Bioconcentration is the accumulation of substance through contact with water; bio-magnification is the transfer of a substance up the food chain

in PCB concentrations occurs between zooplankton and prey fish (various species) in the Lake St. Clair food web, and a subsequent four- to fivefold increase between prey fish and predators (Haffner et al., 1994). Biomagnifica-tion, however, does not explain all species differences, as high concentrations were observed in carp (*Cyprinus carpio*), a direct sediment feeder. Evidently, there are multiple mechanisms involved in the bioaccumulation of pollutants by higher trophic level biota. These mechanisms can include direct uptake from water across gills or epidermis (bioconcentration), direct contact with sediment, and consumption of contaminated food (or sediment) (van der Oost et al., 1996). In certain species and in certain life cycle stages, a negative relationship between fish size and contaminant levels was observed (Sijm et al., 1992; Pastor et al., 1996). This phenomenon could be explained by the growth dilution effect. In the case of female fish, reproduction may also represent a significant elimination pathway (Sijm et al., 1992).

13.3. Bioaccumulation Web Models

Several types of models could be applied to predict bioaccumulation. The simplest ones represent only bioconcentration processes, using first order equations. These models predict the toxic substance concentrations in tissue and water by applying uptake and elimination rate constants. This type of model could be easily calibrated using laboratory experiments. Bioenergetic models add additional detail such as physiological attributes of fish (metabolic rates and growth), feeding rates, and dietary assimilation efficiencies (Lee, 1992). Norstrom et al. (1976) modeled PCB accumulation in yellow perch in the Ottawa River in this manner. Alternatives to these kinetic approaches are steady-state food web-type models, which assume that contaminant concentrations in biota are unchanging (Morrison et al., 1997). Morrison et al. (1997) developed this type of model for PCBs in Lake Erie, where relatively detailed information is available about predator-prey relationships. Another approach was taken by Carrer et al. (2000) to apply the static ECOPATH model in combination with an ecotoxicological model to model the bioaccumulation of dioxins. The ECOPATH software package has recently been expanded to include ECOSIM, a dynamic simulation package, and ECOTRACE, which predicts movement and accumulation of contaminants and tracers in food webs (Figure 2). As such, this set of models, which is described below, is a useful tool for modeling the prediction of toxic substances accumulation in food webs.

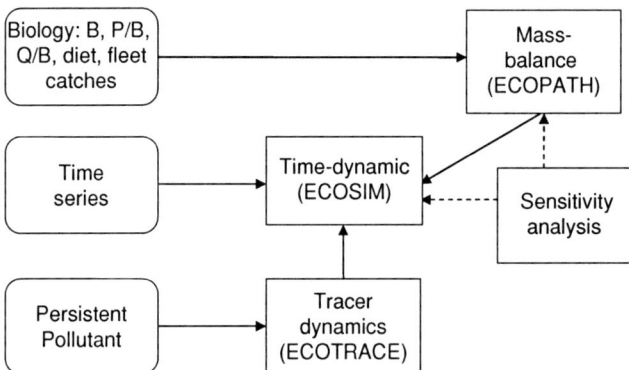

Figure 2. General scheme representing the links between ECOPATH, ECOSIM and ECOTRACE

13.4. ECOPATH

The core routine of ECOPATH is derived from the ECOPATH program of Polovina (1984a, 1984b) modified to render superfluous its original assumption of steady state. ECOPATH no longer assumes steady state but instead bases the parameterization on an assumption of mass balance over an arbitrary period, usually a year (but also see discussion about seasonal modeling). In its present implementation, ECOPATH is based on two master equations, one to describe the production term and one for the energy balance of each group. The first equation could be written as

Production = predation mortality + biomass accumulation + other mortality + catches + net migration; or in mathematical terms:

$$P_i - B_i M_i - \mathrm{BA}_i - P_i(1 - \mathrm{EE}_i) - Y_i - E_i = 0 \qquad (1)$$

where P_i is the total production rate of species i, M_i is the predation rate on group (i), B_i is the biomass of the group i, BA_i is the biomass accumulation rate for (i), while $P_i(1- \mathrm{EE}_i)$ is the 'other mortality' rate for group i. In this equation EE_i stands for the ecological efficiency, term describing the amount of energy in the biomass that is produced by one trophic level and is incorporated into the biomass produced by the next (higher) trophic level. Y_i is the total fishery catch rate of (i) and E_i is the net migration rate (balance between the emigration and immigration) for the group i.

This equation takes into the account most of the production processes widely known in hydrobiology, with the exception for the exuvia (in case of crustaceans) and generative production (gonads). Exuvia production is still out of the scope, but, still could be treated as the part of the general metabolism of crustaceans. Generative production in many cases is mostly consumed by other predatory groups, and, subsequently, could be included in either predation or other mortality.

It could be seen from that equation that biomass accumulation BA_i is the main term balancing the equation in case the group is not steady state, allowing the biomass to be increasing or decreasing.

Equation (1) could be rewritten as

$$B_i(P/B)_i - \mathrm{BA}_i - \sum_{j=1}^{n} B_j(Q/B_i)_j \mathrm{DC}_{ji}$$
$$-(P/B)_i B_i(1 - \mathrm{EE}_i) - Y_i - E_i = 0 \qquad (2)$$

where P/B_i is the production/biomass ratio, Q/B_i is the consumption/biomass ratio, and DC_{ji} is the fraction of prey (i) in the average diet of predator (j). That transformation gives an option to have as input to quantify the equation such parameters as P/B and Q/B, which could taken from the literature or approximated from taxonomically close groups. In that case most of the parameters needed to quantify the food web network are normally available as the output of routine sampling procedures or from the literature data. The biomasses of trophic compartments are the essential starting point for any food web model. Biomasses are obtained from relevant field studies. It is important to notice that even a scientific guess is better than no data. If biomass data are lacking for some trophic compartments, it is advisable to use biomasses from similar water ecosystems. The food matrix that should be given by quantifying the DC_{ji} terms could be taken from gut content studies or literature data. However, the food matrix in many cases is the most complicated type of data that should be feed to the ECOPATH model, especially regarding the fact that trophology of particular species could experience changes over the time so it is advisable to have when available the results of stable isotope trophology studies. The Main advantage of using natural stable isotope analysis is that the measurements are done within the natural environment, which integrates over time any variability of the abiotic and biotic environments, or due to ontogenetical, physiological or ethological changes of the consumer.

In a model, the energy input and output of all living groups must be balanced. These processes are described by the second ECOPATH equation:

Consumption = production + respiration + unassimilated food.

This equation originates from the productivity theory developed by Winberg (1956) who defined consumption as the sum of somatic and generative production, metabolic costs and waste products. However, as mentioned earlier, ECOPATH does not take into account such processes as exuvia and generative production. Standard input for ECOPATH requires production, consumption and the proportion of unassimilated food being entered for the model quantification; respiration is then calculated from the difference between consumption and the production and unassimilated food terms. However, if the respiration data are readily available there is a possibility to enter the respiration through the alternative input routine where the energy balance can be estimated using any given combination of the terms in the equation above.

13.5. ECOSIM

ECOSIM is the expansion of ECOPATH that provides a dynamic simulation capability at the ecosystem level, with key initial parameters inherited from the base ECOPATH model. ECOSIM uses the mass-balance results obtained in ECOPATH for parameter estimation. As the ECOSIM is the dynamic model it provides the variable speed splitting giving the different speed for the dynamics of 'fast' (picoplankton) and 'slow' groups (seals); It also includes biomass and size structure dynamics for key ecosystem groups, using a mix of differential and difference equations.

ECOSIM uses a system of differential equations that express biomass flux rates among pools as a function of time varying biomass and harvest rates, (for equations see Walters et al. (1997), and Christensen and Walters (2004)). Predator prey interactions are moderated by prey behavior to limit exposure to predation, such that biomass flux patterns can show either bottom-up or top down (trophic cascade) control.

As the ECOSIM is the dynamic model it normally requires time series data for the calibration by doing repeated simulations. For many of the groups to be incorporated in the model the time series data could be available from single species stock assessments. The time series fitting use either fishing effort or fishing mortality data as driving factors for the ECOSIM model runs. A statistical measure of goodness of fit to the time series data outlined above is generated each time ECOSIM is run. The model allows four types of analysis with the SS measure (Christensen et al., 2005):

1. Determine sensitivity of the goodness of fit measure to the critical ECOSIM parameters by changing each one slightly (1%) then rerunning the model to see how much the measure is changed (i.e., how sensitive the time series predictions 'supported' by data are to the changes);
2. Search for vulnerability estimates that give better 'fits' of ECOSIM to the time series data, with vulnerabilities 'blocked' by the user into sets that are expected to be similar;
3. Search for time series values of annual relative primary productivity that may represent historical productivity 'regime shifts' impacting biomasses throughout the ecosystem;
4. Estimate a probability distribution for the null hypothesis that all of the deviations between model and predicted abundances are due to chance alone, i.e. under the hypothesis that there are no real productivity anomalies.

13.6. ECOTRACE

ECOTRACE expands the ECOSIM models to predict movement and accumulation of contaminants and tracers in food webs. ECOSIM predicts temporal changes in flows of biomass among living and detritus pools using nonlinear functional relationships between flow rates and abundances of the interacting species. These flow rates (along with auxiliary information about factors such as isotope decay rate and physical exchange rates) can be used to predict changes in concentrations (and per-biomass burdens) of chemicals like organic contaminants and isotope tracers that 'flow' passively along with the biomass flows. The dynamic equations for such passive flow (and accumulation, e.g., 'bioamplification') are not the same as the biomass flow rate equations, and in fact are generally linear dynamical equations with time-varying rate coefficients that depend on the biomass flow rates; these linear equations are relatively easy to solve in parallel with the ECOSIM biomass dynamics equations.

In ECOSIM we allow parallel simulation of one tracer or contaminant type while the biomass dynamics equations are being solved. Tracer molecules are assumed to be either in the 'environment' or in the biota (in biomass and detritus pools) at any moment. Molecules are assumed to flow between pools at instantaneous rates (i.e., to be sampled along with biomass during biomass flows at rates) equal to the probabilities of being 'sampled' as part of the biomass flow: instantaneous rate = (flow)/(biomass in prey pool). We also allow for direct flows from the environment into pools, representing direct uptake or absorption of the tracer material, and for differential decomposition/decay/export rates by pool and from the environmental pool (Figure 3).

Users must specify additional parameters (using the ECOTRACE input form) in addition to those needed for ECOPATH/ECOSIM mass balance and biomass dynamics calculations (Figure 4): initial pool concentrations C_i, including environmental concentration C_0; direct uptake rate parameters u_i as rates per time per biomass per unit C_0; concentrations per biomass C_i in immigrating organisms; and metabolism/decay rates d_i.

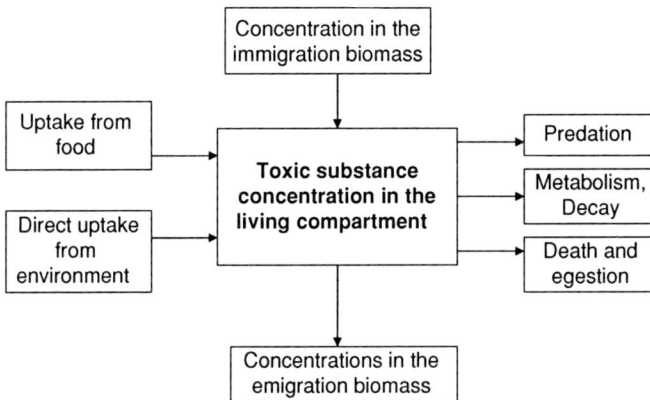

Figure 3. General scheme of processes modeled by ECOTRACE

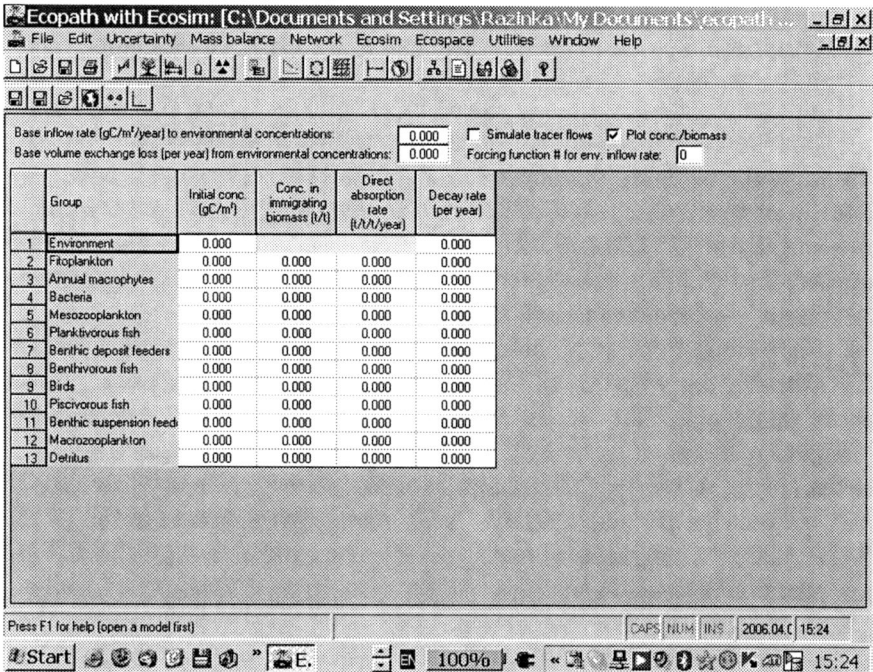

Figure 4. ECOTRACE input window

The rate equations for time changes in contaminant concentration in pool
i are represented as follows:

1. Uptake from food: $C_j \cdot GC_i \cdot Q_{ji}/B_j$, where C_j is the concentration in food
   object j, GC_i is the proportion of food assimilated by type i organisms,
   Q_{ji} is the biomass flow rate from j to i (estimated in ECOPATH as $B_i \cdot$
   $(Q/B)I \cdot DC_{ij}$), and B_j is the biomass of prey group j.
2. Direct uptake from environment: $u_i \cdot B_i \cdot C_o$, where u_i is the parameter
   representing uptake per biomass per time, per unit environmental concentration, B_i is the biomass of group i, C_o is the concentration in the
   environment.
3. Concentration in immigrating organisms: $C_i \cdot I_i$, where C_i is the tracer
   per unit biomass in immigrating biomass and I_i is the biomass of pool i
   immigrants per time.
4. Predation: $C_i \cdot Q_{ij}/B_i$, where C_i is the concentration in pool i, Q_{ij} is the
   consumption rate of type i organisms by predator type j, B_i is the group i
   biomass.
5. Detritus: $C_i \cdot MO_i + (1 - GC_i) \cdot S_j \cdot C_j \cdot Q_{ji}/B_j)$, where MO_i is the non-
   predation death rate of type i (per year), GC_i is the fraction of food intake
   assimilated, Q_{ji} is the intake rate of prey group j by the predator group i.

6. Emigration: $e_i \cdot C_i$, where e_i is the emigration rate.
7. Metabolism: $d_i \cdot C_i$, where d_i is the sum of the metabolism and the decay rate for the material while in pool i.

Unassimilated food and instantaneous non-predation death rates are estimated by the modeling routine, while the other parameters (GC_i, MO_i) are already entered or calculated in ECOPATH, and the Q_{ij} flows are calculated dynamically in ECOSIM. ECOTRACE users can also enter inflow, outflow (dispersal/advection exchange to outside systems), and the abiotic decay rates for the toxic substance concentration C_0 in the environment.

When the ECOTRACE simulation component is enabled (check box on ECOTRACE entry form), ECOSIM integrates the rate components listed above to generate time series of concentration C_i and concentration per biomass C_i/B_i for all ECOPATH/ECOSIM biomass pools. These results can be compared to data on bioaccumulation or tracer movement. Discrepancies between model and data may help identify weaknesses in the ECOPATH/ECOSIM trophic flow rate (Q_{ij}) estimates, and/or in the ECOTRACE rate parameters (u_i, d_i).

So far, there are very few examples of ECOTRACE applications. One of them is aimed to predict the bioaccumulation of mercury in the Faroe Islands marine ecosystem (Booth and Zeller, 2005). However, there are over one hundred calibrated ECOPATH/ECOSIM models that could be run with ECOTRACE. This combination of models provides a powerful system for representing the accumulation of toxic substances in the trophic networks of aquatic systems.

References

Booth, S., D. Zeller, 2005. Mercury, food webs, and marine mammals: implications of diet and climate change for human health, Environ. Health Perspect., 113, 521–526.

Camanzo, J., C.P. Rice, D.J. Jude, and R. Rossmann, 1987. Organic priority pollutants in nearshore fish fiorn 14 Lake Michigan tributaries and embayments, 1983, J. Great Lakes Res., 13 (3), 296–309.

Carrer, S., B. Halling-Sorensen, and G. Bendoricchio, 2000. Modelling the fate of dioxins in a trophic network by coupling an ecotoxicological and an Ecopath model, Ecol. Modelling, 126 (2–3), 201–223.

Christensen, V., and C.J. Walters, 2004. Ecopath with Ecosim: Methods, capabilities and limitations, Ecol. Modelling, 172, 109–139.

Christensen, Walters, and Pauly, 2005. Ecopath with Ecosim: A User's Guide, Univ. of British Columbia, Vancouver.

Gourlay, C., M.H. Tusseau-Vuillemin, J.M. Mouchel, and J. Garric, 2005. The ability of dissolved organic matter (DOM) to influence benzo[a]pyrene bioavailability increases with DOM, Ecotoxicol. Environ. Safety, 61, 74–82.

Haffner, G.D., M. Tomczak, and R. Lazar, 1994. Organic contaminant exposure in the Lake St. Clair food web, Hydrobiologia, 281, 19–27.

Koelmans, A.A., A. van der Heide, L.M. Knijff, and R.H. Aalderink, 2001. Integrated modelling of eutrophication and organic contaminant fate and effects in aquatic ecosystems: A review, Water Resources, 35, 3517–3536.

Lee, H., 1992. Chapter 12: Models, muddles, and mud: Predicting bioaccumulation of sediment associated pollutants, edited by G.A. Burton, Jr., Sediment Toxicity Assessment. Lewis Publishers, Boca Raton, pp. 267–393.

MacDonald, R.W., and J.M. Bewers, 1996. Contaminants in the arctic marine environment: priorities for protection, ICES J. Marine Sci., 53, 537–563.

MacDonald, C.R., C.D. Metcalfe, G.C. Balch, and T.L. Metcalfe, 1993. Distribution of PCB congeners in seven lake systems: Interactions between sediment and food-web transport, Environ. Toxicol. Chem., 12, 1991–2003.

Morrison, H.A., F.A.P.C. Gobas, R. Lazar, D.M. Whittle, and G.D. Hafier, 1997. Development and verification of a benthic/pelagic food web bioaccumulation model for PCB congeners in western Lake Erie, Environ. Sci. Technol., 31, 3267–3273.

Muir, D., B. Braune, B. DeMarch, R. Norstrom, R. Wagemann, M. Gamberg, K. Poole, R. Addison, D. Bright, M. Dodd, W. Dushenko, J. Earner, M. Evans, B. Elkin, S. Grundy, B. Hargrave, C. Hebert, R. Johnstone, K. Kidd, B. Koenig, L. Lockhart, J. Payne, J. Peddle, and K. Reimer, 1997. Chapter 3: Ecosystem Uptake and Effects, edited by F. Jensen, K. Adare, and R. Shearer, Canadian Arctic Contaminants Assessment Report, Indian and Northern Affairs Canada, Ottawa, No. 19, 1–294.

Muir, D.C.G., R.J. Norstrom, and M. Simon, 1988. Organochlorine contaminants in Arctic marine food chains: Accumulation of specific polychlorinated biphenyls and chlordane-related compounds, Environ. Sci.Technol., 22, 1071–1079.

Norstrom, R.J., A.E. McKinnon, A.S.W. DeFreitas, 1976. A bioenergetics-based model for pollutant accumulation by fish—Simulation of PCB and methylmercury residue levels in Ottawa River yellow perch (Perca flavescens), J. Fisheries Res. Board Canada, 33, 348–267.

Pastor, D., J. Boix, V. Fernandez, and J. Albaiges, 1996. Bioaccumulation of organochlori-nated contaminants in three estuarine fish species (Mullus barbatus, Mugil cephalus and Dicentrarcus labrax), Marine Pollut. Bull., 32 (3), 257–262.

Polovina, J.J., 1984a. Model of a coral reef ecosystems I. The ECOPATH model and its application to French Frigate Shoals, Coral Reefs, 3, 1–11.

Polovina, J.J., 1984b. An overview of the ECOPATH model, Fishbyte, 2, 5–7.

Rowan, D.J., and J.B. Rasmussen, 1992. Why don't Great Lakes fish reflect environmental concentrations of organic contaminants? An analysis of between-lake variability in the ecological partitioning of PCBs and DDT, J. Great Lakes Res., 18 (4), 724–741.

Schaanning, M.T., K. Hylland, D. Eriksen, T.D. Bergan, J.S. Gunnarson, and J. Skei, 1996. Interactions between eutrophication and contaminants. II. Mobilization and bioaccumulation of Hg and Cd from marine sediments, Marine Pollut. Bull., 33, 71–79.

Shaw, G.R., and D.W. Connelt, 1986. Chapter 6: Factors controlling bioaccumulation of PCBs, edited by J.S. Waid, PCBs and the Environment, CRC Press, Boca Raton, pp. 121–134.

Sijm, D.T.H.M., W. Selnen, and A. Opperhulzen, 1992. Life-cycle biomagnification study in fish, Environ. Sci.Technol., 26 (11), 2162–2174.

van der Oost, R., A. Opperhuizen, A.K. Satumalay, H. Heida, and N.P.E. Vermeulen, 1996. Biomonitoring aquatic pollution with feral eel (Anguilla anguilla) 1. Bioaccumulation: biota-sediment ratios of PCBs, OCPs, PCDDs and PCDFs, Aquatic Toxicol., 35, 21–46.

Walters, C., V. Christensen, and D. Pauly, 1997. Structuring dynamic models of exploited ecosystems from trophic mass-balance assessments Reviews in Fish Biology and Fisheries, 7, 139–172.

Winberg, G.G., 1956. Rate of metabolism and food requirements of fishes, Transl. Fish Res. Board Canada, 194, 1–253.

14. ASSESSMENT OF LAKE ECOSYSTEM RESPONSE TO TOXIC EVENTS WITH THE AQUATOX MODEL

Brenda Rashleigh
U.S. Envrionmental Protection Agency, 960 College Station Road, Athens, GA 30605, USA

14.1. Introduction

A terrorist attack involving a toxic chemical added to a water resource could have multiple effects on the aquatic ecosystem of that resource. This is particularly significant for systems such as lakes and reservoirs, where the residence time of water is long and there is more opportunity for organisms to be exposed to the chemical. A toxic chemical release in a lake ecosystem may cause bioaccumulation in fish tissues for a period of time after an attack. Bioaccumulation occurs due to direct uptake through the gill, as well as dietary uptake, and is important because the health of humans and wildlife may be affected through the consumption of contaminated fish (Mackay and Fraser, 2000). A toxic chemical may also affect fish biomass either through direct mortality, the reduction of growth and reproduction, or alteration of the food chain, where food items may also experience reduced biomass (Pastorok et al., 2001). The maintenance of fish biomass in a system can be important for a recreational, commercial, or subsistence fishery, as well as for the support of surrounding wildlife.

Ecosystem dynamics in lakes and reservoirs are quite complex due to interactions and feedbacks among habitat, feeding relationships, and other physiological processes. Because of the complexities of lake and reservoir ecosystems, these systems are often represented and studied with simulation models. The most comprehensive models that simulate these systems include the processes of nutrient and organic carbon cycling, the fate and transport of chemicals, and the accumulation and effects of contaminants in the food web (Koelmans et al., 2001). Koelmans et al. (2001) recently reviewed five existing integrated aquatic ecosystem models, including AQUATOX, GBMB, CATS-5, and IFEM, and rated the AQUATOX model as the most complete and versatile of these models. Although AQUATOX represents toxic chemicals that could be used in terrorist attacks, there are no published accounts of the AQUATOX model being used to assess the effects of such an attack. Here we use AQUATOX to model the effects of a toxic event on a lake ecosystem. Model applications such as this can be useful in contingency planning to

I. E. Gonenc et al. (eds.), Assessment of the Fate and Effects of Toxic Agents on Water Resources, 291–297.
© 2007 *Springer.*

assess ecosystem response to chemical releases under emergency conditions (Mackay, 1994).

14.2. The AQUATOX Model

AQUATOX is a process-based model for ecological risk assessment that can represent the effects of toxic chemicals and conventional pollutants on the aquatic ecosystem (Park et al. 2004). The model uses a daily timescale to simulate the physical environment (e.g., flow, light, and sediment) and the chemical environment (e.g., nutrients, oxygen, carbon dioxide, and pH). The dynamics of biotic components of detritus, algae, benthic invertebrates and fish can be simulated. The model is not spatially distributed but the system can stratify into two vertical layers based on temperature. AQUATOX represents dates as Month/Day/Year, which should be accounted for when using non-U.S. computers. The user is provided with default scenarios along with the download; alternatively the user can set up a new system. A detail accounting of setting up a new system can be found in the user's manual (Park et al., 2004). A Wizard is available to walk the user through this setup.

   AQUATOX Release 2.1 provides a chemical library with information on 55 chemicals; multiple chemicals can also be represented (Park et al., 2004). AQUATOX can simulate organic chemicals, but not metals or organometals, and it is most useful for organic chemicals with high octanol–water partition coefficients (e.g., $>10^4$) that can be expected to bioaccumulate. For a new chemical, data are required for molecular weight, octanol–water partition coefficient, Henry's Law constant, and transformation rates as appropriate. AQUATOX predicts the partitioning of a toxicant among water, sediment, and biota, and includes processes of microbial degradation, biotransformation, photolysis, hydrolysis, and volatilization.

   When a new study is set up, it is necessary for the user to either enter a toxicity record or map an existing toxicity record for each organism specified in the study. A species toxicity record includes the external concentration of toxicant at which 50% of population is killed (LC50, µg/l); the external toxicant concentrations that cause a 50% reduction in reproduction and in growth (EC50s, µg/l), which can be estimated from LC50s in the model; and an elimination rate constant (K_2, 1/d). Terrorist chemical attacks can be represented in the chemical loading screen as a one-day load bracketed by zeros for the days before and after the attack. The model gives daily output for both biomass and chemical concentration in the organisms. It would be best to track the percent difference in biomass. The expectation is that a toxic chemical would cause a reduction in biomass for a period of time; recovery can be assumed to occur when the percent difference in biomass is essentially zero. To track bioaccumulation, it is best to plot the concentration of toxicant in fish tissue in terms of ppb (wet weight). Recovery can be assumed to occur when the concentration of toxicant is very low (e.g., <0.1 ppb).

14.3. General AQUATOX Model Results: Factors Determining the Aquatic Ecosystem Response

The expectation is that fish biomass may decrease and bioaccumulation may occur for a period of time after an attack. Although this was generally true in AQUATOX, the exact nature and magnitude of the ecosystem response to a given chemical depends on characteristics of the species and the system and on the timing and amount of the chemical.

14.3.1. CHARACTERISTICS OF FISH SPECIES

The combination of diet, elimination rate, and lipid fraction controls bioaccumulation in the model. As expected, more bioaccumulation will occur for fish species that eat more contaminated food sources, such as benthic invertebrates that are exposed through the ingestion of contaminated sedimented detritus (Thomann et al., 1992). Fish species with lower elimination rates and higher lipid fractions will demonstrate more bioaccumulation. The biomass response for a fish species will be highly dependent on its LC50 and EC50 values. Higher values for LC50 and EC50s allow a fish species to survive, grow, and reproduce at higher toxicant levels. If biomass is drastically reduced as a result of the chemical, then the biomass loading rate specified by the user controls the recovery of sensitive species from reduced levels (e.g., fish species with higher loading rates, for example, through migration or drift from upstream waterbodies, show more rapid recovery back to initial biomass levels).

14.3.2. TIMING AND AMOUNT OF THE CHEMICAL

A toxic event simulated in a period or season of higher flow seems to have less of an effect on the system, due to the dilution effect. Toxic events of small to moderate amounts (compared to their toxicity) will generally show a small effect and then recovery. However, large amounts may cause a significant effect and it is possible that the system will never recover to its initial state. If complete mortality occurs, the system must rely on input of new individuals for recovery. Also, a large toxic sedimented biomass due to mortality causes buried detritus that can remain in the system indefinitely, according to the AQUATOX model.

14.4. Coralville Reservoir Example

Coralville Reservoir, located in Iowa, USA, was one of the original case studies for the AQUATOX model (U.S. EPA, 2000). The example is provided in the "Studies" folder that is automatically downloaded with the model (Release 2.1). Coralville Reservoir has a volume of 72,400,000 m^2, a mean depth of 2.5 m, mean epilimnion and hypolimnion temperatures of 16 °C and 13 °C, respectively, and it can stratify. The original case study was set up for the time period 10/15/1974 to 9/30/1977 with five species of fish, two of which (largemouth bass *Micropterus salmoides* and buffalofish *Ictiobus* spp.) were each represented with two age classes. Selected default model parameters are listed in Table 1. Biomass loading for all fish species was specified at 1 × 10^{-5} g/m^2. In the original case study, the pesticide dieldrin (CAS 60-57-1) was represented in the model as a time series, but for this example, dieldrin input is represented as a load of 10,000 g occurring on a single day. We considered two separate scenarios, an attack occurring on 3/2/1975, and one occurring on 8/2/1975.

Under the scenario of a terrorist attack occurring on 3/2/1975, none of the fish species showed a significant biomass response to dieldrin until 4/15/1975 (Figure 1, top left). Different fish species showed different responses to this event: bluegill (*Lepomis macrochirus*) biomass declined to nearly zero and showed no sign of recovery, while walleye (*Sander vitreus*) and largemouth bass adults declined and then slowly recovered to 100% and 68% of their background biomass, respectively. Buffalofish adults experienced an increase in biomass, which reflects their high LC50 values and lack of effects on growth and reproduction relative to the other fishes in the study. Shad (*Dorosoma* spp.) biomass declined sharply to nearly zero but recovered quickly, during a period of high water inflow. Shad biomass first recovered to a level much higher than in the background scenario, due to release from competition and predation, but matched the background biomass by the end of the simulation.

TABLE 1. Default AQUATOX parameters for fish and invertebrate species included in the Coralville Reservoir example simulation

Species	LC50 (µg/l, 96 h)	K_2 (1/d) (µg/l, 96 h)	EC50 growth	EC50 Repro (µg/l)	Lipid fraction
Bluegill	3.1	7.62×10^{-3}	0.31	0.16	0.045
Bass, walleye	3.5	8.29×10^{-3}	0.35	0.18	0.05
Shad	4.5	3.67×10^{-3}	0.46	0.23	0.05
Buffalofish, Tubifex	600	8.33×10^{-3}	0 (no effect)	0 (no effect)	0.12
Chironomid, Chaoborus	677 (48 h)	5.32×10^{-2}	68	34	0.05
Daphnia, Rotifer	250 (48 h)	7.35×10^{-2}	25	13	0.03

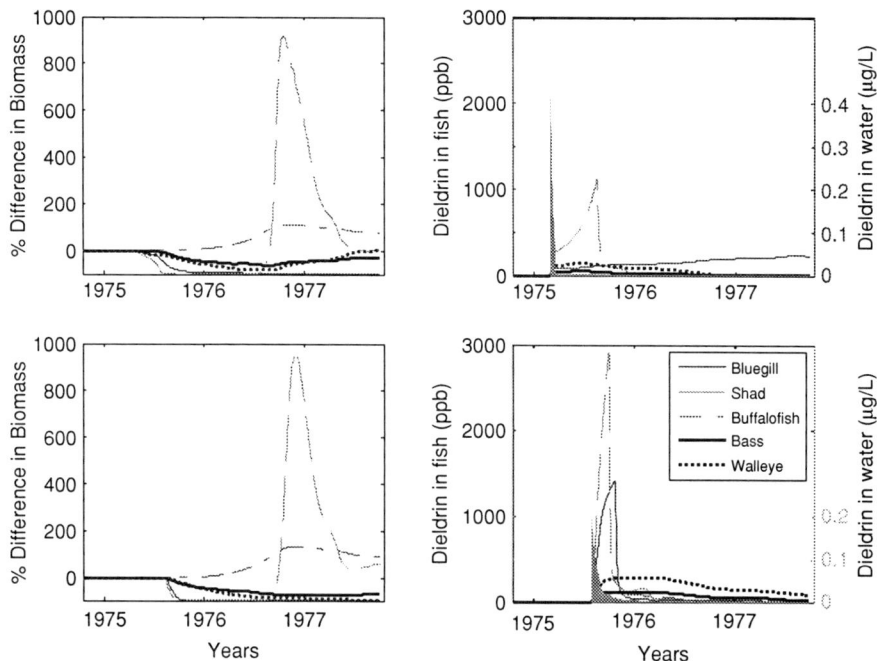

Figure 1. Coralville Lake ecosystem response to 10,000 g load of dieldrin: percent difference in fish species biomass between the background and toxicant scenarios and bioaccumulation (Y1) and concentration of dieldrin in water (Y2, shaded area) for a scenarios of a 3/2/1975 attack (top row) and an 8/2/1975 attack (bottom row)

Under the scenario of a 3/2/1975 attack, the concentration of dieldrin in water peaked quickly, reaching its highest concentration of 0.465 µg/l on the date of the spill (3/2/1975), but receded quickly. By 3/22/1975, the concentration of toxicant in water was <0.0005 µg/l (Figure 1, top right). The spill occurred before a high-flow event of 5.5×10^7 m³/d on 3/24/1975, compared to the mean flow of 3.5×10^6 m³/d, and this contributed to the rapid decline in water concentration. The greatest bioaccumulation occurred in shad, which is likely due to the observation that the elimination rate for shad is half as large as that of the other fishes in the study (Table 1). The dieldrin concentration in shad reached a maximum of 1148 ppb on 8/13/1975). The sharp decline in the toxicant concentration in shad resulted from the decline in shad biomass to the point where the population was largely sustained by the input of uncontaminated biomass. Toxicant concentration in walleye reached a peak of 162 ppb on 6/24/1975 and then slowly decreased. In all of the fishes other than bluegill, the toxicant concentration increased and then began to decrease. In bluegill, the concentration increased throughout the simulation, although this was an artifact of the very low bluegill biomass that persisted through the simulation.

If the terrorist attack is assumed to occur on 8/2/1975 (Figure 1, bottom row), there is a decline in the biomass of bluegill, bass, and walleye, with no indication of biomass recovery for these species in the simulation period. Because there was lower flow at this time, there was a longer residence time of the toxicant in water, which led to more rapid bioaccumulation at higher levels than for the 3/2/1975 example. The longer duration of bioaccumulation may also have resulted from a longer residence time of the toxicant in sedimented detritus. Under the 8/2/1975 attack scenario, the toxicant concentration in sedimented detritus did not drop below 5 ppm (dry weight) until September 1977, a period of 25 months, compared to 7 months in the 3/2/1975 attack scenario.

14.5. Summary

The EPA aquatic ecosystem model AQUATOX can be used to describe food web dynamics and characterize bioaccumulation under the condition of a toxic event, such as a terrorist attack to a water system. AQUATOX predicts the partitioning of a toxicant among water, sediment, and biota, and includes processes of microbial degradation, photolysis, hydrolysis, and volatilization. The user will most likely be interested in the effects and recovery of two endpoints: fish biomass and bioaccumulation of the chemical in fish tissue. In general, a toxic event will result in a decrease in biomass and an increase in toxicant concentration in biota for a period of time, followed by recovery of both measures to their initial states. However, as illustrated by an example for Coralville Reservoir in Iowa, USA, the response depends on the several factors: the octanol–water partition coefficient of the toxicant, which determines the rate of bioaccumulation; the characteristics of the species present, where less sensitive species may actually increase due to a release from competition; and the time of year, since a spill in a high-flow season will have less effect on biota. For valued water resources, a parameterized and calibrated AQUATOX model could be a useful tool in the response to a toxic event.

Acknowledgements

We are grateful for the assistance of Dick Park, Marjorie Wellman, Jon Clough, Earl Hayter, Bob Ambrose, Craig Barber and Ali Ertuck. This paper has been reviewed in accordance with the U.S. EPA's peer and administrative review policies and approved for publication.

References

Koelmans, A.A., A. van der Heide, L.M. Knijff, and R.H. Aalderink, 2001. Integrated modelling of eutrophication and organic contaminant fate and effects in aquatic ecosystems: A review, Water Resources, 35, 3517–3536.

MacKay, D., 1994. Fate models, edited by P. Calow, Handbook of Ecotoxicology (Vol. 2), Blackwell Scientific, London, pp. 348–367.

Mackay, D., and A. Fraser, 2000. Bioaccumulation of persistent organic chemicals: Mechanisms and models, Environ., Pollut., 110, 375–391.

U.S. EPA, 2000. AQUATOX for windows, Release 1. Vol. 3: Model Validation Reports, EPA-823-R-00-008, Office of Water, Washington, DC.

Park, R.A., and J.S. Clough, 2004. AQUATOX (Release 2) Vol. 2: Technical Documentation, EPA-823-R-04-002, Office of Water, Washington, DC.

Park, R.A., J.S. Clough, M.C. Wellman, 2004. AQUATOX (Release 2) Vol. 1: User's Manual, EPA-823-R-04-001, Office of Water, Washington, DC.

Pastorok, R.A., S. Bartell, S. Ferson, and L.R. Ginzburg, 2001. Ecological Modeling in Risk Assessment: Chemical Effects on Populations, Ecosystems, and Landscapes, Lewis Publishers, Boca Raton.

Thomann, R.V., J.P. Connolly, and T.F. Parkerton, 1992. An equilibrium model of organic chemical accumulation in aquatic food webs with sediment interaction, Environ. Toxicol. and Chem., 11, 615–629.

15. CHEMICAL WEAPONS DUMPED IN THE BALTIC SEA

Eugeniusz Andrulewicz
Department of Fisheries Oceanography and Marine Ecology, Sea Fisheries Institute in Gdynia, Poland

15.1. Introduction

Chemical weapons (CW) use toxic properties of chemical substances to kill, injure or incapacitate an enemy during warfare. Chemical weapons are classified by the United Nations (together with nuclear and bacteriological weapon) as weapons of mass destruction.

Chemical weapons include both chemical munitions (e.g. bombs, shells, grenades) and chemical warfare agents. Chemical warfare agents may be in liquid, gas or solid form. Liquid agents are set to be volatile (high vapor pressure) so they can be dispersed over a large region quickly. Solid (mostly plasticized) form is used rarely issued.

15.1.1. SHORT HISTORY OF USE OF CHEMICAL WEAPONS

Toxic properties of some natural agents were known and used as early as the stone age (e.g. toxic arc arrows). There were also attempts to use toxic compounds (mainly as smoke and fumes) against enemies in various ancient battles (e.g. by ancient Chinese and ancient Greeks). During the course of later centuries AC, toxic agents were sometimes used in wars and battles, but it is worth noting that battle commanders often considered these as "perfidious and odious" weapons and refused to use them.

The first full-scale deployment of a chemical warfare agent, chlorine gas, was during the World War I, in the battle of Ypres (15 July, 1915) by the Germans to attack French troops (Figures 1 and 2). The use of chlorine gas caused 5000 death and 15,000 wounded cases. Two years later in 1917, also near Ypres, mustard gas (2,2′-dichloro-diethyl sulphide) was used (see also: http://www.firstworldwar.com/battles/ypres3.htm). From that time mustard gas was mostly known as *Yperite*. During the course of the whole World War I, there were 85,000 causalities and 1,176,500 wounded by chemical warfare. (see also http://en.wikipedia.org/wiki/Chemical_warfare)

After Word War I, the possible use of chemical weapons caused deep fear in the minds of most people at that time. In 1925, sixteen of the world's major

I. E. Gonenc et al. (eds.), Assessment of the Fate and Effects of Toxic Agents on Water Resources, 299–319.

Figure 1. Chlorine gas released at Ypres. http://www.tau.ac.il/~pet/html/history2.html

nations signed the Geneva Protocol, pledging never use gas or bacteriological methods of warfare, however chemical agents were occasionally used to
subdue populations and suppress rebellion. In 1922–1927, combined Spanish
and French forces dropped mustard gas bombs in Morocco. In 1935, Italian
Fascists used mustard gas in Ethiopia, causing 15,000 casualties mostly from
mustard gas.

During the World War II, the German Nazis developed and manufactured
large quantities of old and new chemical agents, but chemical warfare was not
used on a large scale by either Germans or Allies. However, the German Nazis

Figure 2. Chlorine gas attack. www.home.zonnet.nl/rene.brouwer/usa/usa.htm. *See also*
www.eyewitnesstohistory.com/gas.htm   and   http://www.firstworldwar.com/photos/graphics/
cnp_gas_aeroplane_01.jpg

used the insecticide Zyklon B to kill a large number of victims in concentration camps. During the course of World War II, the Japanese used mustard gas against Chinese troops.

After World War II, enormous resources were spent by the USA to develop nerve agents (known as "V-Series" nerve gases). During the 1960s, the US explored the use of incapacitating agents and defoliant agents in Vietnam. Very little information was available about developments of Soviet chemical weapons, however during the Gorbatchev time it was published that highly toxic agents were developed in Soviet Union in large amounts (Surikov, 1996; Wikipedia, 2006).

After the World War II, in Iran-Iraq war, about 100,000 Iranian soldiers were victims of Iraq's chemical attack, mostly hit by mustard gas. About 20,000 Iranian soldiers were killed by nerve gas. In 1998 Iraqi Kurdish were exposed to multiple chemical agents, which killed about 5000 people.

All in all, about 70 different chemicals have been used or stockpiled as Chemical Weapons (CW) during the whole 20th century. At present, their production and stockpiling is outlawed by the Chemical Weapons Convention (CWC, 1993).

15.1.2. CLASSIFICATION AND PROPERTIES OF CHEMICAL WARFARE AGENTS

There are different classifications of chemical warfare agents; however, they are most often classified according to the health effects:

- Tear gases (lachrymators): Chloroacetophenone.
- Nose and throat irritations: Clark I, Clark II, Adamsite.
- Lung irritations: Phosgene, Diphosgene.
- Blister gases (vesicants): Sulphur Mustard, Nitrogen Mustard, Lewisite.
- Nerve gases: Tabun.
- Additives, such as monochlorobenzene (are made to the warfare agents in order to change their physico-chemical properties.

Chemical warfare agents are usually highly toxic chlorinated aromatic and/or aliphatic compounds. Some properties of selected agents are given in Table 1.

15.1.3. CHEMICAL WARFARE AGENTS IN THE MARINE ENVIRONMENT

The behavior of warfare agents in the marine environment depends on physical–chemical properties of the substances and external/environmental factors: temperature, salinity, pH value and turbulence in water. For degradation of chemical warfare agents dissolution in water is the first and the most

TABLE 1. Chemical structures and physical–chemical properties of the chemical warfare agents (HELCOM CHEMU, 1994)

Name	Synonyms	Structure	Melting point (°C)	Boiling point (°C)	Vapor pressure (mmHg) 20 °C	Density (g/cm³)	Aqueous solubility (g/l)
Tear agents							
Chloroacetofenone (2-chloro-1-fenylethanone)	CN, Mace	[structure]	54–56	244	0.013	1.32	1
Nose and throat irritants							
Clark I (diphenyl arsine chloride)	Sternite	[structure]	38–44	307–333	0.0016	1.422	2
Clark II (diphenyl arsine cyanide)	Sternite	[structure]	30–35	290–346	0.000047	1.45	2
Adamsite (10-chloro-5-hydrophenarsazine)	Phenarsazine chloride	[structure]	195	410	2×10^{-13}	1.65	0.002
Lung irritants							
Phosgene (carbon dichloride oxidate)	Carbonyl chloride, CG	[structure]	−128	7.6	1178	3.4	9

Name	Common name	Structure				
Diphosgene (trichloromethyl chloroformate)	Perstoff	$O=C(Cl)-O-C(Cl)(Cl)Cl$	−57	127	10.3	1.65	
Blister gases (vesicants)							
Mustard gas (2,2′-dichloro-diethylo sulfide)	Yperit Lost Senfgas	$S(CH_2CH_2-Cl)(CH_2CH_2-Cl)$	14	228	0.72	127	0.8
Viscous mustard gas		Different mixtures, e.g. 63% mustard gas + 37% Lewisite					
Nitrogen mustard gas tri-(2-chloroetylo) amina	Trichlormethin	$N(CH_3CH_2Cl)_3$	−4	235	0.011	1.24	0.16
Lewisite (2-chlorowinylo-dichloroarsyna)	L	$Cl_2AsCH=CHCl$	−18	190	0.35	1.89	0.5
Nerve gases							
Tabun (P-cyano-N, N-dimethyl phosphonamid acid ethyl ester)	Trilon 83	$(H_3C)(H_3C)N-P(=O)(CN)-O-CH_2CH_3$	−50	246	0,07	1.07	120

important step. As regards solubility, the reaction of chemical warfare agents with water depends on hydrolysis. This process leads to formation of new compounds with properties different from those of the chemical warfare agents. During this process they lose properties of warfare agents as they decompose to non-toxic and/or less toxic compounds. Table 2 presents selected examples of results (in simplified manner) of degradation processes in sea water.

15.2. Chemical Ammunition

15.2.1. CLASSIFICATION OF CHEMICAL AMMUNITION

Chemical ammunition produced during the World War II in Germany was in the form of:

- Aircraft bombs;
- Artillery shells;
- High-explosive bombs;
- Mines;
- Encasements;
- Smoke grenades.

Some warfare agents (e.g., Cyclon B) were kept in metal containers.

15.2.2. DUMPING THE CHEMICAL WEAPON IN THE WORLD OCEAN

Utilization of large quantities of CW on land is very expensive and in fact inoperable. Some smaller quantities of warfare agents can be utilized on military polygons with chemical treatment. Incineration of CW is also expensive and causes other problems, such as toxic fumes. Storage on land is unsafe and expensive. The best choice (the most safe and operable method) seems to be to dump CW at sea. This "best" choice has lead to problems for us and for future generations (NATO ASI Series, 1995).

There is sufficient evidence that almost all oceans and seas were used for dumping chemical and/or convention ammunitions. Some of the dumping sites are well known and documented. However, some of them are still hidden in archives of marine military operations. Figure 3 shows the main dumping areas (marked with dots) in the world ocean (Otremba, 2006)

15.2.3. CHEMICAL WEAPON DUMPED IN THE BALTIC SEA

After World War II, Allies found about 300 thousand tonnes of chemical weapon in Germany. Some part of this weapon was taken by Allies for storage, and the rest was dumped in Skagerrak and the Baltic Sea.

TABLE 15.2. Simplified formulas of chemical degradation (hydrolysis) of warfare agents in sea water

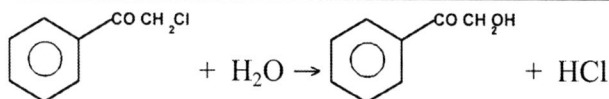

benzene ring—$CO\ CH_2Cl$ + $H_2O \rightarrow$ benzene ring—$CO\ CH_2OH$ + HCl

Chloroacetophenone (2-chloro-1-fenylethanon) after slow hydrolysis (dehalogenation) in sea water produces non toxic compounds. Hydrogen chloride reacts further to sodium chloride

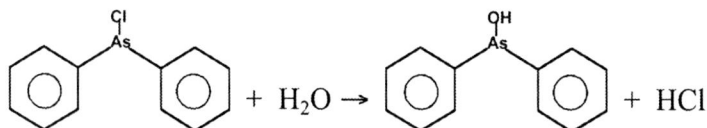

(diphenyl)—As—Cl + $H_2O \rightarrow$ (diphenyl)—As—OH + HCl

Clark I (diphenylochloroarsine) hydrolyses to a less toxic product, difenyloarsenios acid (which has no features of warfare agent) and hydrochloric acid. Toxic arsenic compounds are stable, however the will be diluted/dispersed in the marine environment

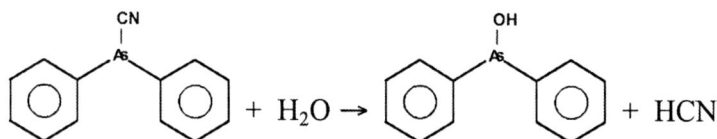

(diphenyl)—As—CN + $H_2O \rightarrow$ (diphenyl)—As—OH + HCN

Clark II (diphenylcyanoarsine) hydrolyses to diphenylarseniuos oxide and hydrogen cyanide. Toxic hydrogen cyanide is not stable in the marine environment and transforms further to formic acid and salts of formic acid

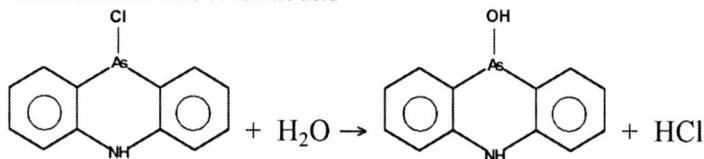

(acridine-type ring)—As—Cl, NH + $H_2O \rightarrow$ (acridine-type ring)—As—OH, NH + HCl

Adamsite (diphenyl-amino-chloro-arsine) slowly dissolves and hydrolyses in sea water. The final product will contain arsine

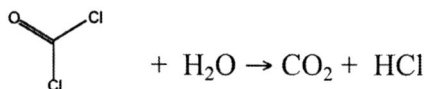

$O=$C(Cl)(Cl) + $H_2O \rightarrow CO_2 +$ HCl

Phosgene (carbonylchloride) transforms to non harmful substances

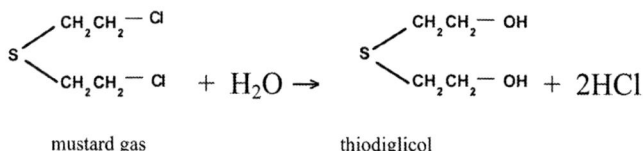

S(—CH_2CH_2—Cl)(—CH_2CH_2—Cl) + $H_2O \rightarrow$ S(—CH_2CH_2—OH)(—CH_2CH_2—OH) + 2HCl

mustard gas   thiodiglicol

Mustard gas (dichlorodiethyl sulfide) in water slowly hydrolyses and forms thiodiglicol and hydrochloric acid. Both final products are non-toxic

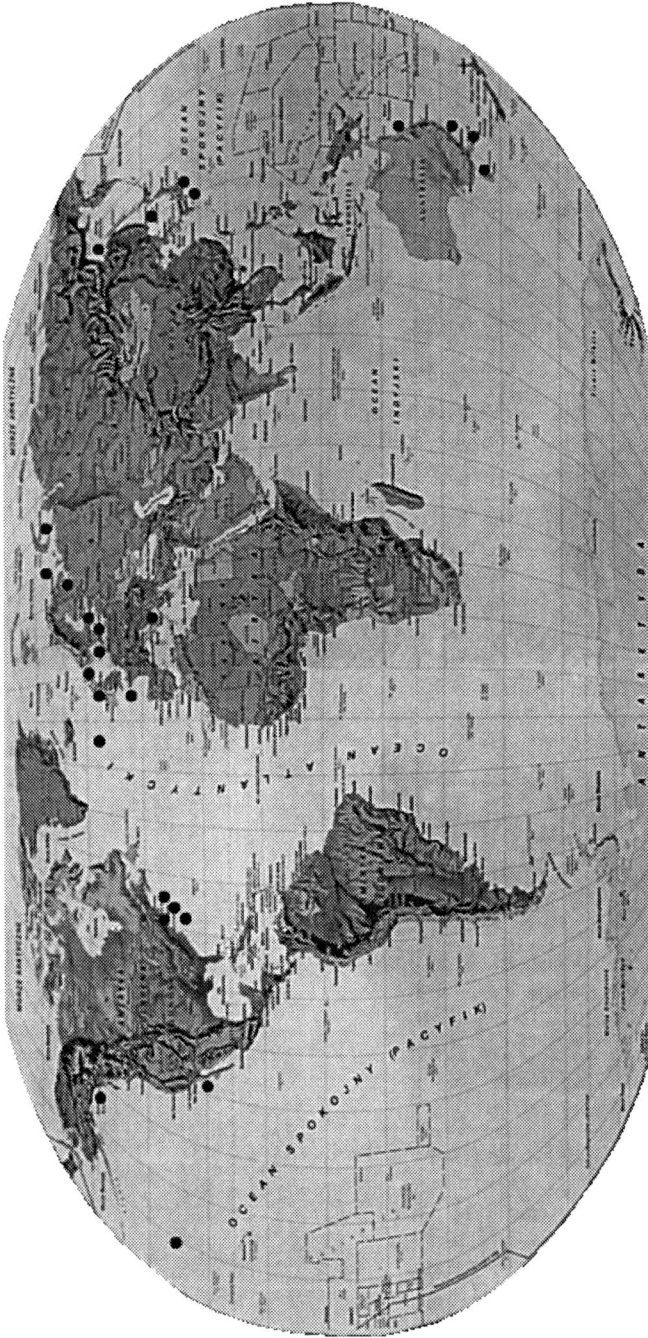

Figure 3. Dumpsites in global oceans (Otremba, 2006)

Figure 4. Dumping operation in the Beaufort Dyke (Irish Sea). The same dumping "technique" was used in the Baltic Sea http://www.manxman.co.im/cleague/archive/bombs.html

There were two different methods of dumping: throwing overboard (Figure 4), which was done in the Baltic, and sinking in old ships (Figure 5), which was done in the North Sea (Skagerrak).

Altogether about 65 thousand tonnes of chemical weapon were dumped in the Baltic Sea (Figure 6) in form of artillery shells, aircraft bombs (Figure 7), mines, smoke grenades, encasements, containers, drums. This is equivalent to about 13 thousand tonnes of toxic chemical agents: chloroacetophenone, clark I, clark II, yperite, phosgene, adamsite, lewisite, and tabun (HELCOM CHEMU, 1994). However, there were cases of throwing munitions overboard during the transport to dumpsites.

Figure 5. Sinking old military ship loaded with CW (US archives). This technique was used in the North Sea (Skagerrak)

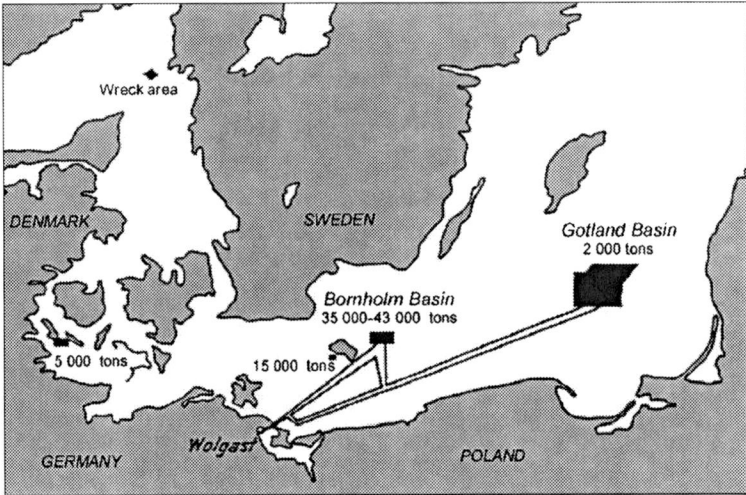

Figure 6. Known dumping areas in the Baltic Sea (HELCOM CHEMU, 1994)

15.2.4. IMMEDIATE EFFECTS AFTER DUMPING

Following the dumping operations of chemical ammunitions, there were numerous findings of CW on beaches of the Southern Baltic Sea and many cases of serious injuries (including injuries of children) in Sweden, Germany, Poland and Denmark (Andrulewicz, 1996; Glasby, 1997; Kantolahti, 1994).

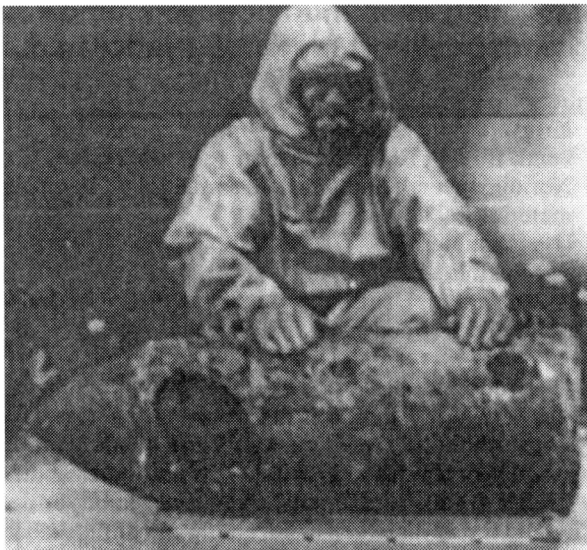

Figure 7. Handling of corroded chemical bomb (Fiskeri Arbogen, 2000)

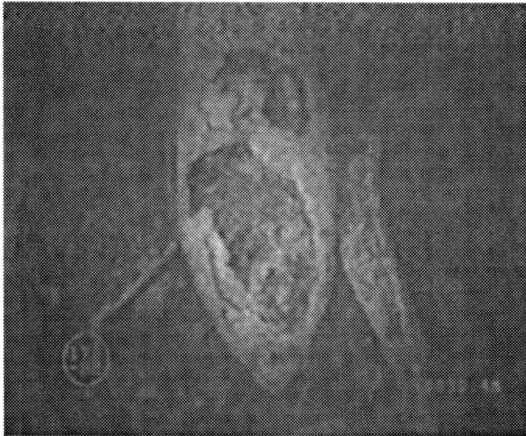

Figure 8. Picture of chemical weapons obtained by ROV during the Russian studies in the Bornholm Basin (Paka and Spiridonov, 2002)

More than fifty years after the dumping took place, such incidences on beaches are not recorded, however, they cannot be excluded (Korzeniewski, 1994).

Dumped CWs are probably partly buried in sediments, partly lying on sediment surface and therefore still visible (Figures 8 and 9).

Dumpsites in deep basins of the Baltic Sea are mostly under anoxic conditions, with the rate of sedimentation approximately 1 mm/year. Due to trawling operations some unknown part of chemical munitions has been redistributed on large parts of sea bottom. Over time, the brackish water of the Baltic Sea causes the shell castings/metal covers to corrode (Figures 8 and 9) and

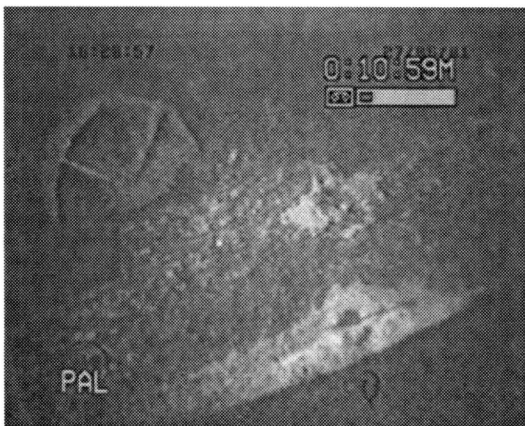

Figure 9. Picture of chemical weapons obtained by ROV during the Russian studies in the Bornholm Basin (Paka and Spiridonov, 2002)

310 EUGENIUSZ ANDRULEWICZ

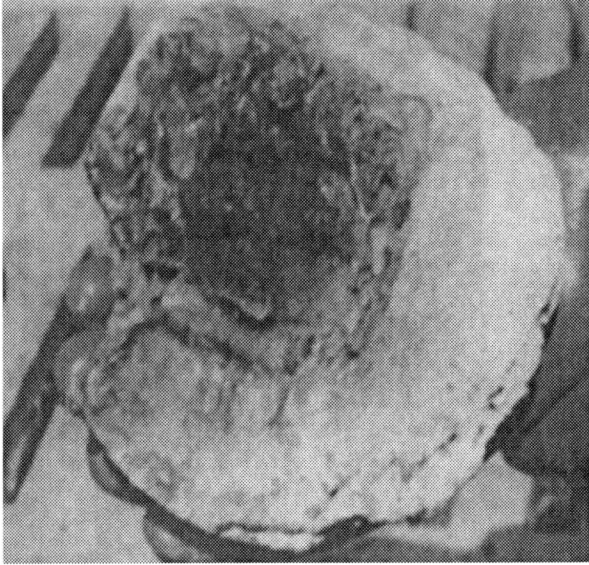

Figure 10. Lump of yperite caught by fisherman in the Baltic Sea (Fiskeri Arbogen, 2000)

release chemical agents to marine environment. Due to hydrolysis, chemical agents are transformed to non toxic or less toxic compounds. However, there are some suspicions that yperite may be transformed to a wax-like solidified form, sometimes resembling amber. The agent is presumably stable in this form at the sea bottom, yet active enough to cause severe contact burns (Figures 10 and 11).

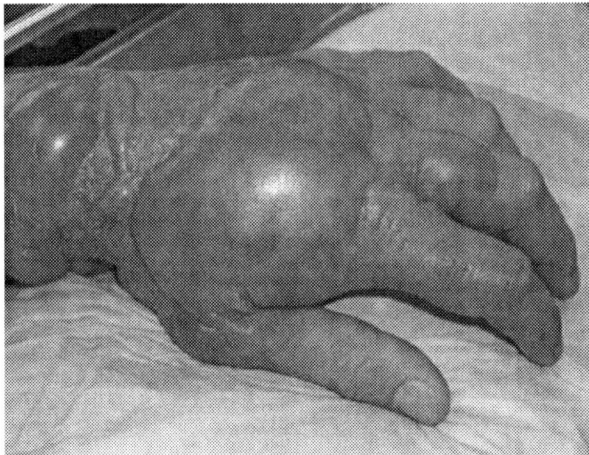

Figure 11. After contact with yperite (Fiskeri Arbogen, 2000)

15.3. Selected Cases of Contacts with Chemical Munitions in The Baltic Sea Area

15.3.1. INJURIES OF CHILDREN, 1955

In July 1955, children from an organized summer holiday were playing on the beach of the southern Baltic Sea. They found a barrel and rolled it along the beach. After 30 minutes, the first skin burning symptoms appeared. All in all, 102 children suffered skin burns and four suffered severe injuries to their eyes (Korzeniewski, 1994). The agent causing this effect was never identified.

15.3.2. INJURIES OF FISHERMEN, 1997

In January 1997, the Polish fishing cutter WLA-206 was trawling for cod and flatfish within fishing rectangle R-9 of the Polish Economic Zone, about 18 miles north of the Polish coast (approximate position 55°12′N; 18°38′E). Fish were collected into containers, but a substance resembling clay (estimated as 5–7 kg) was left on deck. This lump was dumped in a port rubbish container and finally brought to the city scrap yard. The next day, all of the fishermen experienced adverse skin reactions, a sort of burning sensation, skin lesions and reddening. Most of the doctors diagnosed these burns as caused by an unknown substance. After examination, this substance was identified by the navy experts as yperite.

A specialized chemical division of the Polish Navy decontaminated the fishing vessel, the area around the rubbish container and the road connecting the port and the disposal site. The toxic "clay-looking lump" was found and decontaminated on a military polygon; afterwards a sample was taken for chemical analyses. Four fishermen were hospitalized due to severe skin burns and released home after few weeks, another four were medically treated and released.

15.4. Research on Dumped Chemical Weapon in The Baltic Sea

The Working Group on Dumped Chemical Munitions (HELCOM CHEMU, 1994) issued several recommendations. Out of thirteen recommendations, four were related to the need of research on chemical munitions.

1. Regarding verification of existing official dumping sites: *"search for location of chemical munitions could be conducted on national basis."*
2. Related to investigations on the chemical processes and ecological effects of warfare agents: *"further investigation on these processes and effects,*

especially on poorly soluble compounds such as viscous mustard gas and arsenic compounds, should be undertaken."

3. Related to the state of chemical weapon—after more than fifty years on sea bottom: *"investigations on these issues should be carried out in selected parts of the dumping areas."*

4. Related to field investigations on selected dumping areas: *"investigations including water, sediment and biota should be conducted in selected dumping areas."*

Following HELCOM recommendations, two field studies were organized—by Germany—mainly on transport routes of CW (HELCOM, 1996a, 1996b) and by Russia mainly in the Bormnholm Basin (Paka, 2001; Paka and Spiridonov, 2002; Paka, 2004). There were also some laboratory studies conducted in Denmark, Poland, and Sweden.

15.4.1. RESULTS OF GERMAN STUDIES, 1994–1996

In 1996, Germany presented preliminary results of studies taken by the German Hydrographic Service (HELCOM, 1996b). They were based on magnetometric and hydroacoustic surveys (side scan sonar, high resolution sonar) within the German part of the Baltic transport routes from port Wolgast to a CWs dumping area located east of Bornholm. The following findings were reported for transport routes across 4000 nautical miles:

• 900 magnetic anomalies (7 very large anomalies-like ship wrecks; 50 big anomalies-like bombs on the sediment surface; 130 significant anomalies-like bombs covered by sediments).
• 30 other anomalies (most of the anomalies were found on transport routes and on the Odra Bank).
• 1300 contacts by side scan sonar (550 could be a natural objects, e.g. stone reefs).

After the above studies, there was proposed continuation of the research by a remotely operated vessel (ROV); however no results from these studies are available.

15.4.2. RESULTS OF RUSSIAN RESEARCH, 2001

Russian studies by the Institute of Oceanography in Kaliningrad on dumpsites were performed in 1994–1995 (HELCOM EC MON, 1996; HELCOM, 1996a) and in 1998–2001in the Gotland Deep, Bornholm Deep and Kattegat (Paka, 2001) (Figure 12). They were related to evaluating contamination of sediments in dumpsites (Figure 13) and collecting bottom documentation by ROV. These studies showed a considerable rate of corrosion of chemical

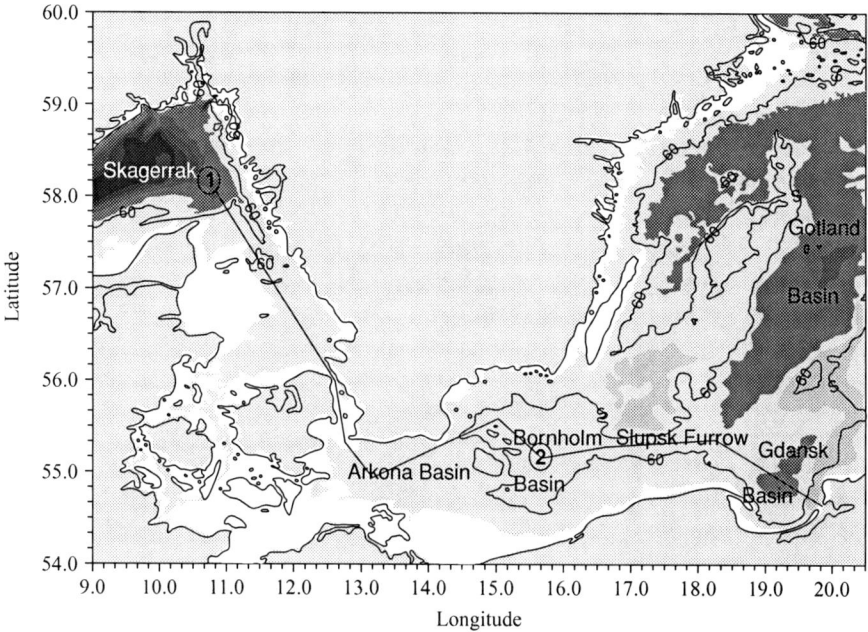

Figure 12. The Russian studies on dumpsites of the Baltic Sea (Paka, 2001)

weapons (Figures 8 and 9), as well as the contamination of sediment dump sites by arsenic.

15.4.3. RESULTS OF STUDIES IN OTHER COUNTRIES

Polish studies were performed following the catch of lump of yperitre by Polish fishermen in February 1997. These were laboratory analyses of chemical composition of yperite lump. Thin layer chromatography and gas chromatography

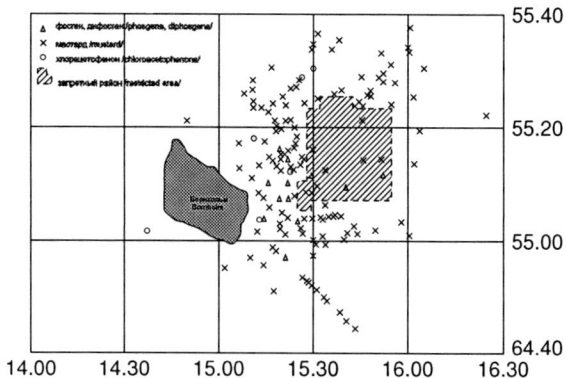

Figure 13. Sediment sampling sites in Bornhom Basin (Paka, 2001)

coupled with mass spectrometry and atomic absorption spectrometry were applied (Technical Military Academy in Warsaw). These techniques identified 20 toxic compounds of different toxicity in the lump of yperite (Witkiewicz, 1996).

Sweden reported having detected mustard gas at sea (Granbom, 1996). The highest concentration of mustard gas agent (119 ppt) was found in a sediment sample one kilometer away from a wreck containing chemical munitions in the Skagerrack. Denmark informed HELCOM about their methodology for determination of physio-chemical parameters of selected organoarsenic species (HELCOM EC 6/9/2, 1995).

15.4.4. NEW/ONGOING RESEARCH PROJECTS

In 2005–2006, the EU established a research project on "Modeling of Ecological Risk of Sea dumped Chemical Weapons (MERCW)" (Project manager: Dr. Vadim Paka, Shirshow Institute of Oceanography in Kaliningrad). Project objectives were:

- To develop hydro chemical, hydrographical and hydro biological investigations;
- To evaluate models of the release, migration and degradation of toxic chemical agents;
- To develop a regional ecological risk assessment model.

Recently, Russia has established a monitoring programme on Baltic dumpsites that will be carried out by AtlantNIRO and Shirshov Institute of Oceanology in Kaliningrad (Paka, 2006).

15.5. Lessons Learned from Unintentional Human Contact with Dumped Chemical Munitions

A brief summary/recapitulation of events recorded at a scoping session at the Marine Court after bringing a lump of warfare agents into the city of Władysławowo (Poland) (1997) is described below:

- A series of unforeseen events may lead to an accident, even if there is sufficient awareness about dumped chemical weapon at sea.

  The fishermen from Polish fishing cutter caught a lump of substance resembling clay and left in deck due serious engine problems, low air temperature and rough weather. Under "normal" circumstance, the lamp of clay looking substance would have been washed out at sea.

- It was never before described that a lump of dangerous material may look like ordinary, safe material.

    The physical and chemical properties of elasticized (non-gas) yperite remaining on the sea-bed may change to such a degree that it cannot be easily distinguished from other naturally occurring items, such as clay or amber.

- Adverse effects may happen even without having direct/physical contact with the warfare agent.

    The fisherman who took and disposed the lump of yperite was not as badly affected as were others who had not even see it. Those fishermen severely affected by yperite came into contact with a towel/cloth which was previously used to clean some traces of dirt.

- The effects of warfare agent will depend on outside air conditions, particularly air temperature.

    Negative effects were limited to physical contact with the dangerous substance due to the extremely cold air temperature—approximately -20 °C. At higher temperatures, the effects of contact with yperite could be much more serious—similar to that occurring from its use as a weapon.

- Chemical warfare agents in the Baltic Sea are not only found within the official dumpsites.

    Fish trawling did not take place within or near known dump sites or along transportation corridors. Present information regarding the location of dumped chemical munitions is inadequate to make firm estimates regarding where munitions are most likely to be found.

15.6. Designing Research on Dumped Chemical Weapons

Not many studies were performed on CWs until now (although some unpublished military research may exist). It is not because of lack of interest and/or low level of importance. These studies are simply expensive, difficult to perform, and require modern techniques that were not always available. This study can be dangerous and therefore it is not for just "ordinary scientists." Warfare agents or corroded ammunition must be handled by special military units.

Field investigations: Field survey for dumped CW will involve typical sea bottom mapping techniques:

- precise positioning system by digital global positioning system (DGPS);
- magnetometric techniques for detection of metal objects/magnetic anomalies (e.g. proton magnetometer);

- acoustic techniques: echo sounding (e.g. high resolution side scan sonar including automatic data processing) side-scanning, sub-bottom profiling and/or multibeam scanning;
- video techniques for field inspection and documentation, usually applying Remotely Operated Vessels (ROVs);
- bottom sediment sampling techniques for analysis of CW traces and/or their metabolites (e.g. arsenic compounds). It may be used standard sediment sampling (e.g. gravity corers). Sediment sampling may already require special safety conditions;
- analyses of fish and bottom macrofauna for CW traces.

The above-mentioned approach will allow the identification of metallic objects, and in addition, it will help to identify which objects are chemical munitions. In the case of identification of chemical munitions, decisions will have to be made as to whether or not to bring the warfare item to the surface.

Laboratory investigations: Lifting of warfare items from the bottom (e.g. for laboratory studies) can only be done when safety conditions are assured. Safety on board during sampling, handling and transport of samples to research laboratories will require military chemistry specialists. Research on actual items/samples recovered from the sea bottom will include the following analyses:

- rate of corrosion of metal walls;
- analyses of sediments for possible contamination by toxic warfare agents;
- products resulting from aging (hydrolyses and polymerization) of chemical munitions on sea-bed (by Gas Chromatography/Mass Spectrometry);
- sediment analysis for arsenic content (Atomic Absorption Spectrometry);
- toxicity tests.

The use of results: These studies will be useful for various reasons:

- Precise determination of the areas with dumped chemical munitions.
- Preparation of revised navigational charts for fishermen and others working on sea bottom.
- Determination of present status of CWs.
- Sea bottom mapping for growing demand of exploitation of sea bottom for minerals and different sub sea transmission lines.
- Comprehensive assessment of the potential threats posed by chemical munitions to the marine environment and human activities performed at sea (fishery, exploitation of mineral resources, cable laying and others).

- Recommendations on monitoring methods that will allow control of potential threats resulting from corrosion and chemical reactions of toxic warfare agents dumped at sea.
- Updating of guidelines for fishermen on how to recognize and deal with chemical munitions that have been accidentally brought on board with trawl catches.
- Preparing comprehensive plan defining which authorities and in what manner to deal with incidents where chemical munitions have been caught by fishermen, as well as how to avoid possible contamination of fish products with toxic warfare agents.

15.7. Conclusions and Comments

The following comments related to dumped warfare agents can be drafted:

- There is no realistic/practical possibility to utilize chemical weapon dumped in the Baltic Sea (and other seas) neither on sea bottom nor on land. Therefore present and future generations will have to leave with this "gift" from previous generations.
- There is an evidence that part of CW is corroded and hydrolyzed but some part will most probably stay there for hundreds of years (e.g. yperite in plasticized form).
- Massive ecological and human disaster due to dumped CW is unlikely. Local danger will appear when chemical weapon is lifted from the sea (accidentally or intentionally) during bottom trawling, touching it on the sea bottom or when stranded on the beach.
- There is a need for specific instructions about how to avoid contacts with chemical weapon and/chemical warfare agents, and/or or how to handle war items in case they are brought from the sea bottom. There is also needed to elaborate a rescue operation scheme in case of accidents with CW.
- There is a need for improvement of public awareness in case war items appear on the beach.
- There is a need for further research on dumpsites, present status of CWs and chemical transformation under marine conditions.
- The HELCOM "Report on Chemical Munitions Dumped in the Baltic Sea" was issued over ten years ago, so the report and recommendations should be revised and updated.
- The presence of CWs on the sea bottom is a serious obstruction for bottom constructions (oil and gas lines, high voltage power cables, windmill parks) and any kind of bottom trawling.

- Chemical weapons dumped at sea should be seriously considered by anti-terrorists measures. For many terrorist organizations, CWs might be considered an ideal choice for a mode of attack: CWs are cheap, relatively accessible, and easy to transport. Fortunately, the efficient use of CW on large scale is not easy.

There are some examples of use or attempted use of chemical agents by terrorists. The most well known is the case of March 20, 1995, when an apocalyptic group based in Japan released sarin into the Tokyo subway system, killing 12 and injuring over 5000 people.

Human medical treatment will depend on the type of CW used. The general rule is to escape towards fresh air and, if possible, to use oxygen as a breathing medium. In case of skin contacts with blister gases (yperite, lewisite), it is always necessary to wash out the injured area with plenty of water and, if available, a solution of chloramine.

References

Andrulewicz, E., 1996. War Gases and Ammunition in the Polish Area of the Baltic Sea, edited by A.V. Kaffka, Sea-Dumped Chemical Weapons: Aspects, Problems and Solutions, Kluwer Academic Publishers NATO ASI Series, Vol. 7, pp. 9–15.

BSH, 1993. Chemical Munitions in the Southern and Western Baltic Sea. Report by a Federal/Länder Government Working Group. The Federal Maritime and Hydrographic Agency—Bundesamt für Seeshiffart und Hydrographie (BSH), Hamburg.

CWC, 1993. Convention on the Prohibition on the Development, Production, Stockpiling, and Use of Chemical Weapons and on their Destruction (CWC).

Fiskeri, A., 2000. The Danish Fishery Guide. Arbog for den Danske Fiskerflade.

Glasby, G.P., 1997. Disposal of chemical weapon in the Baltic Sea. Sci. Total Environ., 206, 267–273.

Granbom, P.O., 1996. Investigation of a dumping area in the Skagerrak 1992, edited by A.V. Kaffka, Sea-dumped chemical weapons: Aspects, Problems and Solutions, Kluwer Academic Publishers, NATO ASI Series, Vol. 7, pp. 41–48.

HELCOM CHEMU, 1994. Report on Chemical Munitions Dumped in the Baltic Sea. Report to the 16th Meeting of Helsinki Commission from the ad hoc Working Group on Dumped Chemical Munitions, p. 43.

HELCOM, 1996a. Information on investigations of dumped chemical munition sites conducted by the Russian Federation in 1994–1995, EC MON 1/96 8/2.

HELCOM, 1996b. Results of magnetic anomaly—detection and hydroacoustic surveys on the German part of the transport routes from the Baltic port Wolgast to the chemical munitions dumping area east of Bornholm, EC 7/96, INF.16/Item 9.

HELCOM EC MON 1/96, 8/2, 1996. Information on investigations of dumped chemical munition sites conducted by the Russian Federation in 1994–1995.

Kantolahti, E., 1994. Chemical Munitions dumped in the Baltic Sea, in Proceedings of Symposium on NBC Defence'94, edited by Kari Nieminen, Finnish Scientific Committee of National Defense.

Korzeniewski, K., 1994. War Gases in the Southern Baltic Sea. Studia i Materiały Oceanolog-
iczne Nr. 67, Marine Chem., (10), 91–101.

NATO ASI Series, 1995. Sea-dumped chemical weapons: Aspects, Problems and Solutions,
edited by A.V. Kaffka, Kluwer Academic Publishers, NATO ASI Series, Vol. 7, Disarma-
ment Technologies.

Otremba, Z., 2006. Gdynia Maritime University, Personal communication.

Paka, V., 2001. An overview of the research of dumped chemical weapons made by r/v Professor
Shtokman in Gotland, Bornholm and Skagerrak dumpsites in 1997–2000.

Paka, V., and M. Spiridonov, 2002. Research of dumped chemical weapons made by R/V
'Professor Shtokman' in the Gotland, Bornholm & Skagerrak dump sites, edited by T. Mis-
sianaen, and J.P. Henriet, Chemical Munition Dump Sites in Coastal Environment, Federal
Office for Scientific, Technical and Cultural Affairs, Federal Ministry of Social Affairs,
Public Health and the Environment, Brussels, pp. 22–42.

Paka, V.T., 2004. Dumped chemical weapon: State of art, Russ. Chem. J., V.XLVIII (2), 99–109.

Paka, V., 2006. Shirshov Institute of Oceanography in Kaliningrad, Personal communication.

Surikov, B.T., 1996. How to Save the Baltic from Ecological Disaster, edited by A.V. Kaffka,
Sea-Dumped Chemical Weapons: Aspects, Problems and Solutions, Kluwer Academic
Publishers, NATO ASI Series, Vol. 7, pp. 67–70.

Wikipedia, 2006. The free encyclopedia, http://en.wikipedia.org/wiki/Chemical_weapon.

Witkiewicz, Z., 1996. Amunicja chemiczna zatopiona w Morzu Bałtyckim, Biuletyn WAT,
9, 115–120.

16. MONITORING AFTER ATTACK

NATALIA N. KAZANTSEVA
Federal State Unitary Enterprise "Keldysh Research Center", Moscow, Russian Federation

16.1. Introduction

The trend of monitoring natural water ecosystems is increasing the quantity of identified components and thus improving the methods of determination and identification. This is very important for long-term monitoring, however, another system must be used when dealing with ecosystems after a terrorist attack. In these situations, the aim is to verify surface water pollution and, if confirmed, identify the specific type of pollution, and all this should be done as quickly as possible.

Sources of surface water are more and more often being used to supply municipal and private water purification units. Large municipal water purification stations have laboratories that constantly control the quality of feed and potable water, but middle-sized units only control this from time to time, and small units do this very rarely. If the water source is heavily polluted, then existing technology cannot guarantee the production of suitable potable water. That is why it is necessary to detect pollution and to inform water purification producers and municipal water users and to introduce some changes into the water processing technology, to manage incidents that may compromise the integrity of potable water supplies.

Traditionally seawater was seen as the best defense against radiation and other damaging factors. It is well known that during the atomic bombing of Hiroshima part of the population managed to save themselves because they were by the seashore and they constantly poured sea water on themselves.

In many instances, it is not possible to get immediate accurate chemical analysis results from an attack. The reasons are:

1. *Limited facilities*. Monitoring of surface waters and other objects in the districts that may suffer from a terrorist attack, as is in the case in all other emergencies, is carried out by the means and tools of special state services. For example, in the Russian Federation, the Federal Service of Emergency Situations handles such emergencies. The Office of Hydrometeorology Service and Environmental Monitoring and Hydrometeorology Research

I. E. Gonenc et al. (eds.), Assessment of the Fate and Effects of Toxic Agents on Water Resources, 321–326.
© 2007 *Springer.*

Scientific Institute make a prognosis and an analysis in this sphere. Monitoring pollution of surface water sources in Russia is conducted constantly for more than 1200 water supplies including studies of radioactive pollution (i.e., gamma-radiation, percentage of strontium-90, tritium, and other radionuclids) and chemical agents. State sanitary-epidemiological control and productive ecological laboratories having complex analytical equipment, including mobile laboratories to fulfill this function. However, such monitoring systems are not available for all potable water supplies. Therefore, decision makers do not have time to wait for the analysis and need to make informed decisions based on incomplete information.

2. *Time limitations*. The current analytical methods do not allow for identification of all the necessary components during the short time frame of an attack.

16.2. Critical Points

The following are main components that should be considered as potential poisons (called "critical points"):

1. Warfare agents and industrial chemical poisons, which include:
   - blood agents (hydrogen cyanide),
   - nerve agents Tabun, Sarin, Soman and VX (listed as most toxic),
   - blistering agents Lewisite (arsenic fraction) and Sulfur Mustard,
   - hallucinogen BZ (3-quinuclidinyl benzilate) and LSD (lysergic acid diethylamide).

   The list should be extended to include:
   - *Inorganic chemicals* (metal salts, acids, and other substances). Most dangerous compounds include cyanide, arsenic, fluoracetate, antimony, cadmium and mercury;
   - *Organic chemicals, such as pesticides* (chlorinated dioxins and PCBs, organochlorines, organophosphates, phenolics, rodenticides, etc)—their effects on humans are nearly identical to that of nerve agents.

2. Pathogens and biological toxins.
   - *Pathogens* are live organisms, such as bacteria, viruses or protozoa (Anthrax, Brucellosis, Tularemia, C. Perfringers, Salmonella, Plague, Hepatitis A, Cryptosporidiosis). Many are stable in water and exhibit chlorine resistance (see Table 1)[1];

---

[1] Valcik, Jerry A., P.E. Biological warfare agents as potable water threats. Medical Issues Information Paper No. IP-31-017.

TABLE 1. Threat potential of pathogens and biological toxins

Agent	Type of the agent[a]	Produced as weapon	Stability in water	Infectious dose[b]	Chlorine tolerance[c]
Pathogens					
Anthrax	B	Yes	2 years spore	6,000	Spores resistant
Brucellosis	B	Yes	20–72 days	10,000	Unknown
C. Perfringers	B	Probable	Common in sewage	500,000	Resistant
Tularemia	B	Yes	<90 days	25	Inactivated, 1 ppm, 5 min
Cholera	B	Unknown	"Survives well"	1,000	"Easily killed"
Salmonella	B	Unknown	8 days	10,000	Inactivated
Plague	B	Possible	16 days	500	Unknown
Q Fever	R	Yes	Unknown	25	Unknown
Variola	V	Possible	Unknown	10	Unknown
Hepatitis A	V	Unknown	Unknown	30	Inactivated, 0.4 ppm, 30 min
Cryptosporidiosis	P	Unknown	Stable days or more	130	Oocysts resistant
Biological toxins					
Botulinum toxin		Yes	Stable	0.07 mg	Inactivated, 6 ppm, 20 min
T-2 mycotoxin		Probable	Stable	None given	Resistant
Aflatoxin		Yes	Probably stable	2 mg	Probably tolerant
Ricin		Yes	Unknown	None given	Resistant at 10 ppm
Staph enterotoxins		Probable	Probably stable	0.004 mg	Unknown
Microcystins		Possible	Probably stable	1 mg	Resistant at 100 ppm
Anatoxin A		Unknown	Inactivated in days	None given	Unknown
Tetrodotoxin		Possible	Unknown	1 mg	Inactivated, 50 ppm
Saxitoxin		Possible	Stable	0.3 mg	Resistant at 10 ppm

[a] B–bacteria, R–Rickettsia, V–virus, and P–protozoan.

[b] Infectious dose based on number of organisms or spores for bacteria, number of oocysts for Cryptosporiosis and viral units for virus.

[c] Chlorine resistance at free available chlorine (FAC) concentration of 2.0 parts per million (ppm). Most water systems maintain FAC concentrations of less than 1.5 ppm. Resistance at these levels is unknown. Many overseas bases may have no FAC as the host nation may not chlorinate its water.

- *Biological toxins*—chemicals derived from organisms, primarily bacteria and fungi. Many are environmentally stable and water-soluble. Effective doses are extremely small (see Table 1).
3. Radioactivity (Stroncium-90, Tritium, β-radiation).

16.3. Chemical and Biological Monitoring and Detection

The aim of chemical and biological monitoring is to determine as quickly as possible and using the simplest analysis techniques if there is water source pollution as a result of a terrorist act and identify what kind of poisonous agent was used. Water sampling and analysis should be conducted in accordance with International Organization of Standardization (ISO) of water quality and state requirements. Safety measures are required during sampling procedures.

Water sampling should be conducted immediately as close to the location of the terrorist attack as possible and in other nearby locations depending on water flow and wind direction. A minimum of 3 samples should be taken from different water depths. The frequency of water sampling depends on the results of the initial analysis and can vary from minutes to days, (depending on the agents) and continuing potential for impact to the water supply. Some of the water samples after corresponding conservation should be sent to specialized laboratories for complete analysis, however, a quick analysis should be carried out in the field and/or at the closest nearby laboratory.

> It should be noted that any significant change in observed value from the average statistics in corresponding seasons or daytimes in the sample must be carefully studied.

There is currently no real-time detection capability for biological weapon pathogens in water, and only a limited capacity for chemical detection. Rapid detection and quantification of chemical and/or bacteriological agents within a water source is difficult at best. It could take days if not weeks to positively identify and quantify the agent. In order to verify pollution, the following decisive points should be determined:

- Visual observation. Change of color and odor, presence of oil spots, etc.
- Fish and algae. Special attention needs to be paid to fish and algae since they are very sensitive to water quality. Although the method of biological identification demands a substantial amount of time, the effects of strong pollution can manifest itself quickly.
- Radioactivity. The simplest field method is to use the Geiger counter.
- Hydrogen ion concentration (pH-value) is a universal value.
- Total dissolved salt content can be assessed by specific electric conductivity.

- Toxicity of water in general can be determined by concentration of active oxygen. (Oxygen consumption by active silt microorganism decreases in the presence of toxic substances.)
- Permanganate index is a measurement for water pollution with organic and oxidizable non-organic compounds. Critical increase of this measurement can serve as an indicator of presence of pesticides, halogens, etc.
- For non-organic compounds, there is a set of standard test assortments and test strips for heavy metals (i.e., cyanide, arsenic, chlorine, iodine, bromine) along with traditional atomic absorption.
- Organic compounds have very small critical values. If extraction is possible using the appropriate solvents, gas chromatography or mass spectrometric determination, then concentrations of organic substances, including pesticides can be detected. Immunoassay—an analytical technique for determining both inorganic and organic compound concentrations in field conditions is very useful (method EPA, series 4000)[2]. Immunoassay test products use an antibody molecule to detect and quantify a substance in a test sample. The performance and accuracy of test products will vary depending on the manufacturer.
- Microbiological monitoring is based on testing indicator organisms such as total faecal coliform bacteria, Clostridia and Clostridia perfringers, and Enterococci. Unfortunately, microbiological analyses for biological weapon pathogens, from the time of collection to reporting results can vary from a day to weeks, depending on transportation logistics and analytical capability, and, if the pathogen is in spore or cyst form, it could be extremely difficult to identify and enumerate.

A special analytical division typically operates a coliform bacteria, *E. coli* as indicator organisms of fecal contamination. The use of membrane filtration techniques and combinations of different chromogenic substrates can result in a 24 h analysis for *E. coli*; however, this method is not very accurate for strongly polluted waters. Nevertheless, it can be used for express-analysis and for getting preliminary results. Unfortunately, a coliform analysis does not detect the most dangerous pathogens, as mentioned above.

A new method has been recently reported for testing *E. coli* using magnetic beads coated with antibodies.[3] This option is potentially less accurate because the method is to be done in less than an hour of impact without an enrichment step.

---

[2] Method 4000. Immunoassay. http://www.epa.gov/epaoswer/hazwaste/test/4_series.htm.
[3] Deininger, Rolf, A., Lee Ji Young, and Arvil Ancheta. 2002. Rapid determination of pathogenic bacteria in surface waters. School of Public Health The University of Michigan. http://www.deq.state.mi.us/documents/deq-ogl-mglpf-deininger.doc.

The following indicators are essential for a speedy preliminary analysis of a situation:

Physical–chemical and microbiological parameters	Parameters value	Method of analysis and inspection
1. Color, odor		Visual
2. Radioactivity	<2.2 Curie/l	Geiger counter
3. PH	6–9	Electrometer, *test-strips*
4. Conductivity (TDS)		Conductometer
5. Dissolved oxygen, % saturation O_2	80–120	Electrometric method (oxygen meter)
6. Permanganate index		Volumetric titration
7. Heavy metal ions, ppb, such as: As, Cd, Cr, Pb, Hg, etc		Atomic absorption possibly preceded by extraction, *assortments and tests-strips, immunoassays*
8. Organic compounds, ppb, such as pesticides, etc		Gas chromatography/mass spectrometry determination preceded by extraction, *immunoassays*
9. Microbiological indicators: *E. coli*/100 ml	<500	Membrane filtration and culture on an appropriate medium, subculturing and identification of the suspect colonies

The results of the tests can be used as a tool to gain a rough model of the ecological situation, aid in the prediction of future ecosystem behavior and needs for decision making.

Next actions

- Immediately informed if pollution is detected.
- An appropriate technological regime for potable water production must be identified (increasing of coagulant, flocculants, reagents contained active chlorine, prolongation of treatment, ultraviolet ray exposition, etc). Useful techniques include using multistage water treatment systems (personal, for house) and boiling drinking water.

The actions undertaken in due time (including change of water treatment regime on the water purification units) can prevent or minimize human illness and mortality.

**PART 5. DECISION MAKING IN RAPID ASSESSMENT
AND DIAGNOSIS OF CBRN EFFECTS ON
COASTAL LAGOONS**

.

17. CASE STUDY: DIELDRIN ATTACK IN DALYAN LAGOON

Ali Erturk[1], Robert B. Ambrose, Jr., P.E.[2], and Brenda Rashleigh[2]
[1]*Istanbul Technical University, Faculty of Civil Engineering, Department of Environmental Engineering, 34469 Maslak,Istanbul, Turkey*
[2]*U.S. Environmental Protection Agency, 960 College Station Road, Athens, GA 30605, USA*

17.1. Background

During the first two weeks of December of 2005, NATO sponsored an Advanced Study Institute (ASI) In Istanbul, Turkey. Part of this ASI involved a case study of a terrorist attack, where a chemical was assumed to be dumped into Sulunger Lake in Turkey. This chapter documents the response developed by the ASI participants to this scenario, in terms of hydrodynamic transport, ecosystem effects, and decision making.

17.1.1. ATTACK SCENARIO

At midnight (1 March 2006 at 0:00 a.m.), 10 tonnes of 50% pure dieldrin powder was spilled into the Eastern arm of Sulungur Lake (Figure 1) in a terrorist attack. The entire load of dieldrin mixed into the water within 8 h. Dalyan Lagoon and its vicinity are well-known tourist sites, and aquaculture in the lagoon is an economically important activity.

An Emergency Response Team (ERT) was established to develop an efficient action plan that minimizes damage to human and ecosystem health. Hydrodynamics and eco-toxicology experts were involved in order to provide information about the expected transport and transformation of the contaminant in this system, and the potential toxic effects on the ecosystem. The objectives of this exercise were to prepare, run, and interpret operational toxicant models of the Dalyan Lagoon System, predict the effects of the terrorist attack, and make appropriate short and long-term recommendations to protect human and ecological health.

17.1.2. DESCRIPTION OF DALYAN LAGOON

The Koycegiz Lake-Dalyan Lagoon System is situated in the southwest Mediterranean coast of Turkey. The watershed of interest, shown in Figure 2, covers approximately 1200 km². It consists of two subsystems, Koycegiz Lake

I. E. Gonenc et al. (eds.), Assessment of the Fate and Effects of Toxic Agents on Water Resources, 329–386.
© 2007 *Springer.*

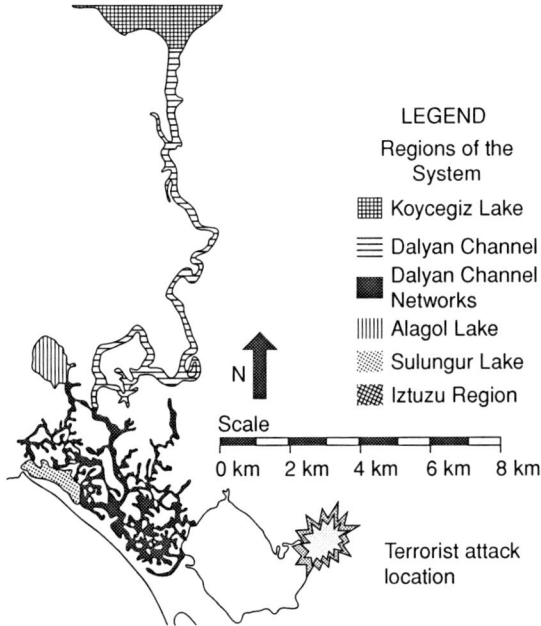

Figure 1. Map of the lagoon and location of the toxicant spill

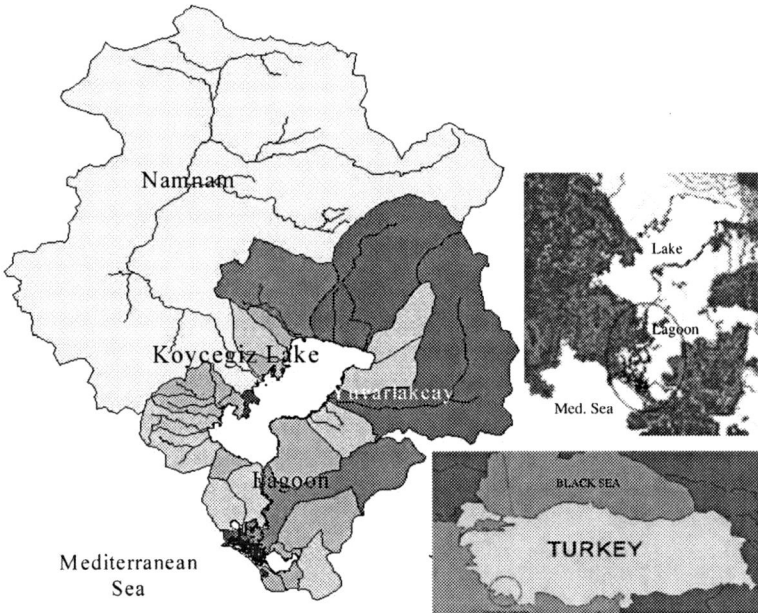

Figure 2. Location of Koycegiz-Dalyan Watershed and the sub-basins

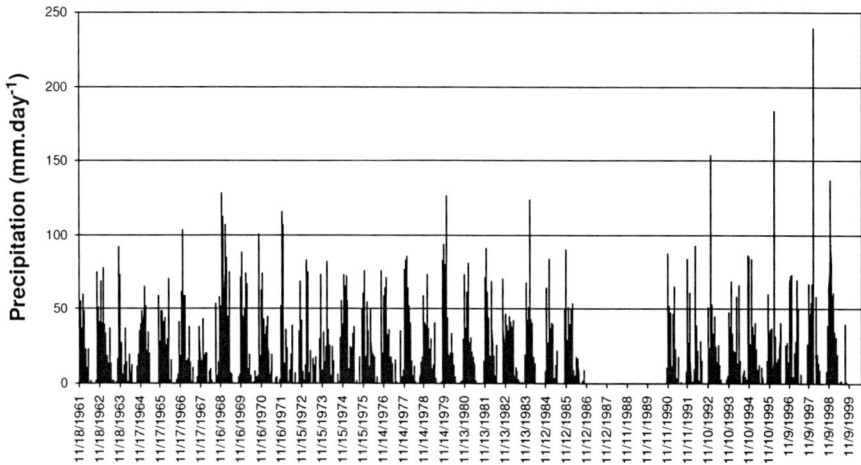

Figure 3. Daily precipitation

with a drainage area of 1070 km^2 in the northern part and Dalyan Lagoon with a drainage area of 130 km^2 in the southern part.

The channels of the Dalyan Lagoon connect Koycegiz Lake to the Mediterranean Sea. The system, especially the Dalyan Lagoon, is of complex structure. The watershed of Koycegiz-Dalyan Lagoon is a mountainous region with quite complex and variable land-use distribution. There are two main streams (one perennial and one ephemeral) and eight smaller ephemeral streams feeding the receiving water system.

When modelling a complex system with spatial heterogeneity, it is important to identify similar water bodies/regions. Regions with different geomorphology and hydrodynamic properties are expected to have different transport properties such as advection/dispersion, retention time or mixing and turbulent effects. These differences should be represented in the model for more realistic simulations.

Time series related to meteorological and hydrological data are shown in Figures 3–5.

Data analyses for the Dalyan Lagoon and its channels indicate a two layered (surface and bottom layers) flow regime (Figure 6). The depth of the interface between these layers varies spatially and temporally. Further data analyses indicated that the Dalyan Lagoon is well-mixed laterally, but poorly mixed longitudinally and vertically.

Bathymetry of the Sulungur Lake is illustrated in Figure 7. The area of each iso-depth line and the related volume are given in Table 1.

Sulungur Lake has saline water for most of the year. Data from the lake were collected 5 times over a 639-day period between July 1998 and March

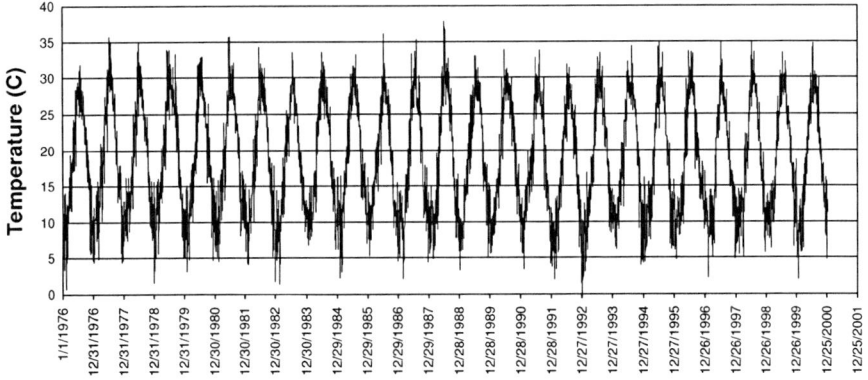

Figure 4. Daily average air temperature

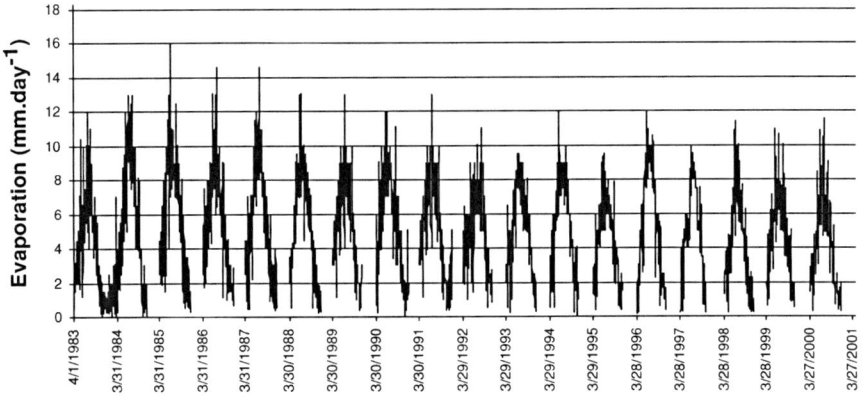

Figure 5. Daily average evaporation

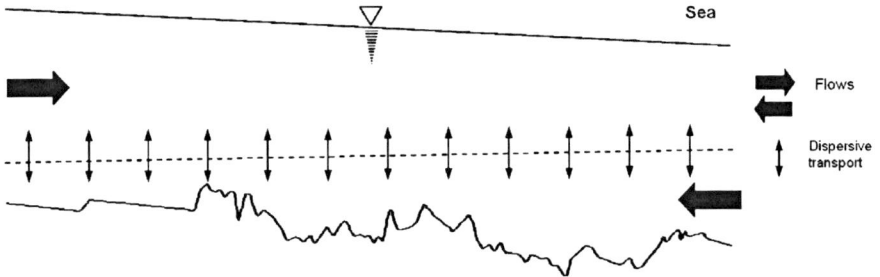

Figure 6. Sketch of the two layer flow along the main channel of Dalyan Lagoon

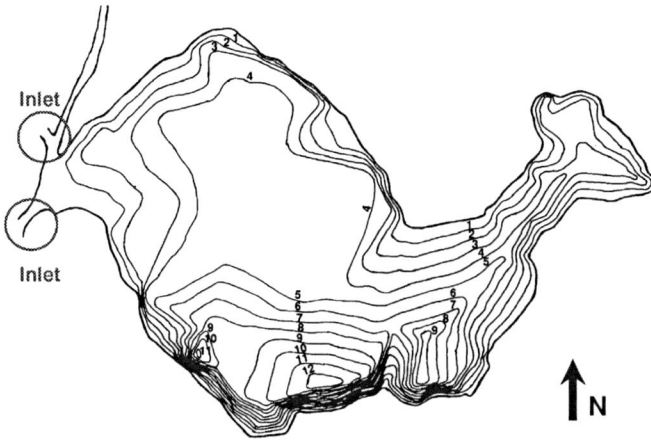

Figure 7. Bathymetry of Sulungur Lake

2000 (Gurel, 2005). The salinity of Sulungur Lake is close to that of the Mediterranean sea, and in the summer, as the less saline surface water flow from Koycegiz Lake decreases, more sea water intrudes into Sulungur Lake, causing high levels of salinity throughout the whole lake. Field measurements from August 1999 show lake salinity above 25 ppt at all depths.

Salinity data for Sulungur Lake are illustrated in Figures 8 and 9. In autumn and winter, high outflows from Koycegiz Lake and high precipitation rates decrease the salinity of the Sulungur Lake. Field measurements in March 2000 show salinity approaching 10 ppt at the surface of the lake. These data indicate a stratification depth of 3 m separating a surface and a bottom layer. The upper layer with a volume of 8,230,000 m^3 is dynamic, whereas the lower

TABLE 1. Sulungur Isodepths and volumes

Upper isodepth (m)	Lower isodepth (m)	Average area (m^2)	Volume (m^3)
0	1	2,925,000	2,925,000
1	2	2,755,000	2,755,000
2	3	2,550,000	2,550,000
3	4	2,170,000	2,170,000
4	5	1,580,000	1,580,000
5	6	1,070,000	1,070,000
6	7	830,000	830,000
7	8	725,000	725,000
8	9	650,000	650,000
9	10	570,000	570,000
10	11	460,000	460,000
11	12	340,000	340,000

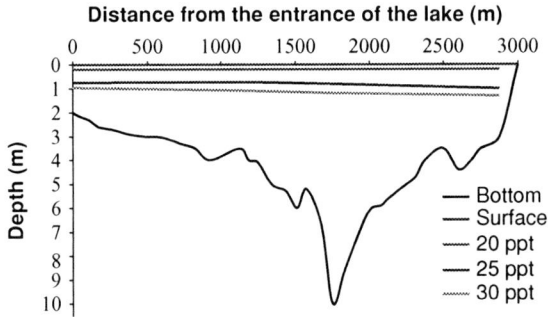

Figure 8. Salinity distribution in Sulungur Lake in December 1998 (Erturk, 2002)

layer with a volume of 8,395,000 m^3 has relatively calm conditions. Further data analyses have shown that water drained from the watershed will always be less dense than the upper layer of the lake.

Estimated monthly averaged drainage flows are given in Table 2. The average inflow from Dalyan Lagoon, estimated to be 2 m^3/s, carries salinity inflow concentrations that vary annually (Table 3). Watershed nutrient loads are provided in Table 4.

17.1.3. ORGANIZATION OF TEAMS

Three student teams were organized for this exercise—Hydrodynamics, Ecology, and Emergency Response. Background instruction and specific training

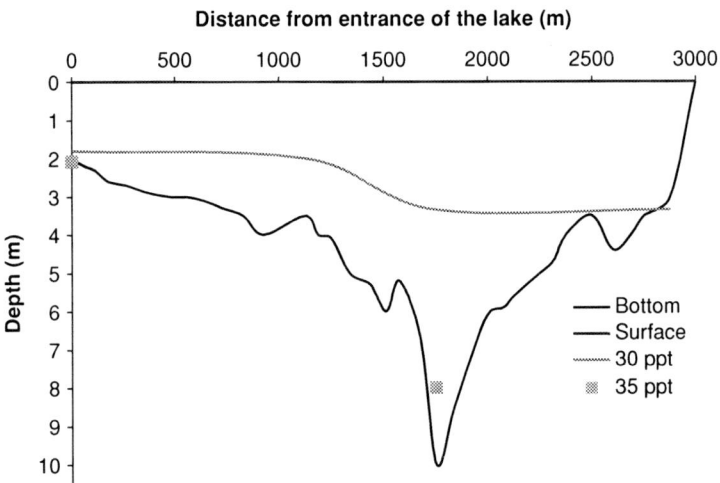

Figure 9. Distribution of salinity in Sulungur Lake in August 1999 (Erturk, 2002)

TABLE 2. Estimated monthly
average flow rates from the drainage
area of Sulungur Lake

Month	Flow rate ($m^3 s^{-1}$)
January	1.66
February	1.24
March	0.72
April	0.41
May	0.22
June	0.08
July	0.01
August	0.04
September	0.07
October	0.61
November	1.12
December	1.88

on tools and techniques were provided immediately prior to this exercise. The objectives of the Hydrodynamics Team were to develop an operational transport model representing dieldrin transport through the Dalyan Lagoon system, and to provide information about the spatial and temporal evolution of dieldrin to the Ecology Team and the Emergency Response Teams. Beginning with the Dalyan Lagoon transport model, the Ecology Team was required to develop

TABLE 3. Monthly average
salinity of inflows into Sulungur
Lake from two inlets shown in
Figure 7.

Month	Salinity (ppt)
January	4.3
February	7.3
March	10.6
April	13.8
May	9.8
June	20.9
July	32.4
August	32.1
September	27.9
October	17.5
November	7.4
December	0.97

Hydrodynamics Team	Ecology Team
Armir Aliev, Azerbeijan	A. Aliyeva, Azerbaijan
Marco Bajo, Italy	A. T. Attokurov, Kyrgyz Republic
Francesca De Pascalis, Italy	L. Bliudziute, Lithuania
Christian Ferrarin, Italy	C. Cazacu, Romania
Debora Bellafiore, Italy	S. Datcu, Romania
Gulshaan Abdykaimovna Ergeshova,	H. Dokmeci, Turkey
Kyrgyz Republic	A. Egorov, Russia
Igamberdiev Rakhmatullo Mamirovich,	A. Ekdal, Turkey
Kyrgyz Republic	M. Gurel, Turkey
Karim Hilmi, Morocco	T. Guyonet, Canada
Luis Daniel Fachada Fernandes, Portugal	Y. Karaaslan, Turkey
Maria Madalena dos Santos Malhadas,	I. Kokorite, Latvia
Portugal	F. Maps, Canada
Francisco Javier Campuzano Guillén,	A. Mamedova1, Azerbaijan
Spain	A. Miltenyte, Lithuania
Natalia Yurievna Demchenko, Russia	A. Ongen, Turkey
Olga Chubarenko, Russia	A. A. Orozumbekov, Kyrgyz
Otuzbay Geldieyew, Turkmenistan	Republic
Abderrahmen Yassin Hamouda, Tunisia	R. Pilkaityte, Lithuania
Alena Paliy, Russia	N. Reznichenko, Russia
Boukadi Khanes, Tunisia	I. Steinberga, Latvia
Loreta Kelpsaite, Lithuania	K. Świtek, Poland
	Z. A. Teshebaeva, Kyrgyz Republic

Emergency Response Team

M. T. Tomczak, Poland
M. Zalewski, Poland

Ilias Baimirzaevich Aitiev, Kyrgyz Republic
Kathryn Briggman , United States of America
Maria Magdalena Bucur, Romania
Andrea Critto, Italy
Dimitriy Dommin, Russia
Evgenia Gurova, Russia
William Matthew Henderson, United States of America
Alina Mockute, Lithuania
Teodora Alexandra Palarie, Romania Kimberly Smith, United States of America

and run operational toxicant fate models, predict human and ecological effects of the terrorist attack on the Dalyan Lagoon System, and make appropriate recommendations to the Emergency Response team. The Emergency Response Team was responsible for developing the short-term emergency response plan, and for providing medium and long-term recommendations for mitigation of damages and recovery of the resource.

17.2. Tools and Techniques

Two primary models were used in this exercise—the Water Quality Analysis Simulation Program, WASP7 (Di Toro et al., 1983; Ambrose et al.,

TABLE 4. Estimated monthly average nutrient loads, in kg/month

Month	NH$_4$–N	NO$_3$–N	ORG–N	PO$_4$–P	ORG–P	CBOD
January	296	72	18	64	32	924
February	657	49	5	107	1	924
March	347	53	10	82	1	924
April	176	66	9	34	17	924
May	441	77	1	58	1	924
June	264	77	1	49	3	924
July	377	108	4	47	1	924
August	208	71	4	48	1	924
September	123	39	9	26	13	924
October	426	59	6	77	1	924
November	709	134	5	100	1	924
December	315	76	19	69	35	924

1988; Wool et al., 2001), and the Aquatic Ecosystem Simulation Model, AQUATOX (Park et al., 2004). These models are available free of charge and are supported by the U.S. Environmental Protection Agency (U.S. EPA). The current versions of WASP and AQUATOX, along with documentation and training material can be accessed and downloaded from the Watershed and Water Quality Modeling Technical Support Center (WWQTSC), at http://www.epa.gov/athens/wwqtsc. They are described more completely in earlier sections of this book and in Appendix 1. A secondary model was used by the authors as an addendum to this exercise—the Environmental Fluid Dynamics Code, EFDC (Hamrick, 1996). The hydrodynamics-only version of EFDC is also available at the WWQTSC.

17.2.1. THE WASP7 SURFACE WATER BODY MODEL

WASP is a general dynamic mass balance framework for modeling contaminant fate and transport in surface waters. Based on the flexible compartment modeling approach, WASP can be applied in one, two, or three dimensions with advective and dispersive transport between discrete physical compartments, or "segments." A body of water is represented in WASP as a series of discrete computational elements or segments. Environmental properties and chemical concentrations may vary spatially among the segments. Each variable is advected and dispersed among water segments, and exchanged with surficial benthic segments by diffusive mixing. Sorbed or particulate fractions may settle through water column segments and deposit to or erode from surficial benthic segments. Within the bed, dissolved variables may migrate downward or upward through percolation and pore-water diffusion. Sorbed variables may migrate downward or upward through net sedimentation or erosion.

WASP is designed to permit substitution of different water quality kinetics code into the program structure to form different water quality modules. Two classes of modules are provided with WASP. The toxicant WASP modules combine a kinetic structure initially adapted from EXAMS (Burns et al., 1982) with the WASP transport structure and simple sediment balance algorithms to predict dissolved and sorbed chemical concentrations in the water and underlying sediment bed. The eutrophication WASP module combines a kinetic structure initially adapted from the Potomac Eutrophication Model (Thomann and Fitzpatrick, 1982) with the WASP transport structure to predict nutrients, phytoplankton, periphyton, organic matter, and dissolved oxygen dynamics.

WASP7 includes a Windows-based interface for constructing input datasets and managing simulations. Data can be copied and pasted from spreadsheets. A Windows-based post-processor allows the user to plot or animate model output. Output is also provided as comma-delimited files for import to spreadsheets.

More details on WASP7 are provided in Appendix 1.

17.2.2. THE AQUATOX SURFACE WATER ECOLOGY MODEL

AQUATOX is a process-based simulation model for aquatic systems. This model predicts the fate of various pollutants, such as nutrients and organic chemicals, and their effects on the ecosystem, including fish, invertebrates, and aquatic plants.

AQUATOX simulates the transfer of biomass, energy and chemicals from one compartment of the ecosystem to another by simultaneously computing each of the most important chemical or biological processes for each day of the simulation period. It can predict the environmental fate of chemicals in aquatic ecosystems, their direct and indirect effects on the chemical water quality and biological response and aquatic life uses.

As a general ecological risk model, AQUATOX represents the combined environmental fate and effects of conventional pollutants, such as nutrients, sediments, and toxic organic chemicals in aquatic ecosystems. It considers several trophic levels, including attached and planktonic algae and submerged aquatic vegetation, invertebrates, and forage, bottom-feeding, and game fish.

More details on AQUATOX are provided in Chapter 2.

17.2.3. THE ENVIRONMENTAL FLUID DYNAMICS CODE, EFDC

The Environmental Fluid Dynamics Code (EFDC Hydro) is an orthogonal, curvilinear grid hydrodynamic model that can be used to simulate aquatic systems in one, two, and three dimensions. EFDC can solve the circulation and transport of material in complex surface water environments including estuaries, coastal embayments, lakes, and offshore. EFDC uses stretched or sigma vertical coordinates and Cartesian or curvilinear, orthogonal horizontal coordinates to represent the physical characteristics of a water body. It solves three-dimensional, vertically hydrostatic, free surface, turbulent averaged equations of motion for a variable-density fluid. Dynamically-coupled transport equations for turbulent kinetic energy, turbulent length scale, salinity and temperature are also solved. EFDC allows for drying and wetting in shallow areas by a mass conservation scheme. The physics of EFDC and many aspects of the computational scheme are equivalent to the widely used Blumberg–Mellor model and U.S. Army Corps of Engineers' Chesapeake Bay model. EFDC produces a special hydrodynamic output file that can be selected for input to WASP. This file includes network segmentation and time-varying flows, velocities, depths, volumes, salinity, and temperature.

17.3. Results and Discussion

The approach to developing the maximum amount of information quickly was to proceed with three modeling efforts. First, adapt the existing WASP7 model of the Dalyan Lagoon System from its eutrophication module to its toxicant model, and investigate the transport of the spill through Sulungur Lake and the Dalyan Lagoon. Second, set up AQUATOX for Sulungur Lake, and investigate the medium and long-term toxicant fate and ecological effects within Sulungur Lake. Third, set up the EFDC model for Sulungur Lake to investigate the short-term spill transport dynamics in more detail. The full WASP7 Dalyan Lagoon model is used to provide boundary conditions for the AQUATOX and EFDC Sulungur Lake models.

On the second Monday morning of the workshop, the Hydrodynamics team was presented with the attack scenario, along with the previously-developed WASP7 eutrophication model of the Dalyan Lagoon system. This team was given a day to adapt the model to simulate the evolution of the toxicant spill, to investigate the importance of different transport factors controlling dieldrin attenuation, and to recommend mitigation options. The Hydrodynamics Team produced a 27-page technical report, "Hydrodynamic investigation on the spreading of a pollutant following a terrorist attack in the

Dalyan Lagoon—Technical Report," and a 5-page public report "Information on the spreading of the pollutant following the terrorist attack in the Dalyan Lagoon—Public Report."

On Tuesday morning, the Ecology Team was presented with the hydrodynamics technical report and the WASP7 spill model of the Dalyan Lagoon System. This team was given a day to add relevant chemical transformation processes, to investigate the influence of environmental factors, and to compare results with human and ecological standards. In addition, the Ecology Team was charged with setting up AQUATOX on Sulungur Lake, investigating the probable ecological response in the lake to the spill, and recommend mitigation options. The Ecology Team produced a 23-page technical report, "Modeling of the ecotoxicological effects of dieldrin discharged in the Dalyan Lagoon System—Technical Report," and a 10-page public report "Modeling of the ecotoxicological effects of dieldrin discharged in the Dalyan Lagoon System—Public Report."

The EFDC model could not be set up and run in the time frame of this workshop. The EFDC simulations described below were completed by the authors subsequent to the workshop.

On Wednesday morning, the Emergency Response Team was presented with executive summaries of the hydrodynamics and ecology technical reports, along with the reports and an oral briefing. The team was given a day to prepare an emergency action plan and provide recommendations for medium and longer term remediation of the resource. The ERT produced a 30-page report, "Emergency Response Plan for Dieldrin Contamination in the Sulungur Lake," along with public announcements and briefing materials.

Results from the student reports are summarized below, augmented by additional simulations, analysis, and interpretation from the authors.

17.3.1. WASP7 SIMULATION RESULTS

The WASP7 Dalyan Lagoon eutrophication model was composed of 49 surface water segments and 49 subsurface water segments based on the bathymetry, geomorphologic and salinity data (Figure 10). For this toxicant spill exercise, this model network was imported into the WASP7 simple toxicant module. For the initial setup and transport investigation, pulse loading and flows were applied with no chemical reactions. The simulation period started on 25 February 2006, four days before the terrorist attack, and finished at 1 April 2006.

Principal boundary conditions for the Dalyan Lagoon system are the Mediterranean inflow into the subsurface of the lagoon inlet, and the seasonal

Figure 10. Dalyan Lagoon Model Network—surface segments

fresh water inflows from Koycegiz, Dalyan and Sulungur Lakes (Figure 11). According to the scenario of this case study, a pulse load of dieldrin was released at midnight on 01 March 2006 at a constant rate of 15,000 kg/day for a period of 8 h (Figure 12) into surface segment 20, located in the eastern portion of Sulungur Lake. By 0800, a total of 5000 kg of active ingredient had entered the system.

Assuming average flow conditions, three loading patterns were investi-gated, representing discharge to the 0.5 m lake surface layer, discharge to the 4 m lake bottom layer, and discharge averaged between the surface and bottom layers. Alternative high and low flow scenarios were investigated as-suming discharge to the lake surface. The high flow scenario represents the spill response under high precipitation conditions. The fresh water inflows were doubled and the saline water inflows were halved. The low flow scenario represents the spill response under the spill response under dry, high evapo-ration conditions. The saline water inflows were doubled, and the fresh water inflows were halved.

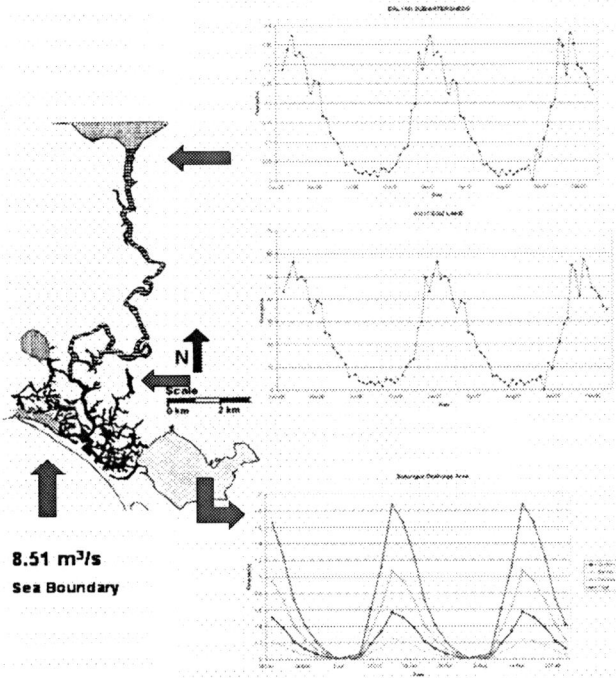

Figure 11. Flow time series at the boundaries

Time series plots and tables are provided for the following locations:

- *Segment 20:* Substance release area.
- *Segment 16:* Boundary between Sulungur Lake and Dalyan Lagoon.
- *Segment 25:* Dalyan channel system.

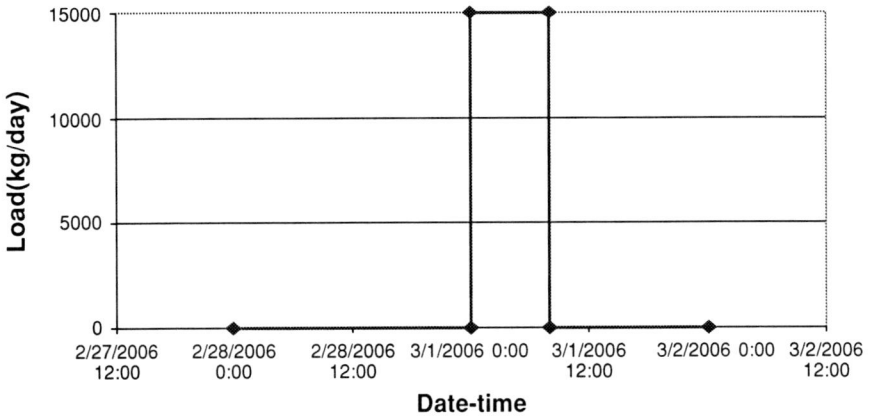

Figure 12. Load into Sulungur Lake

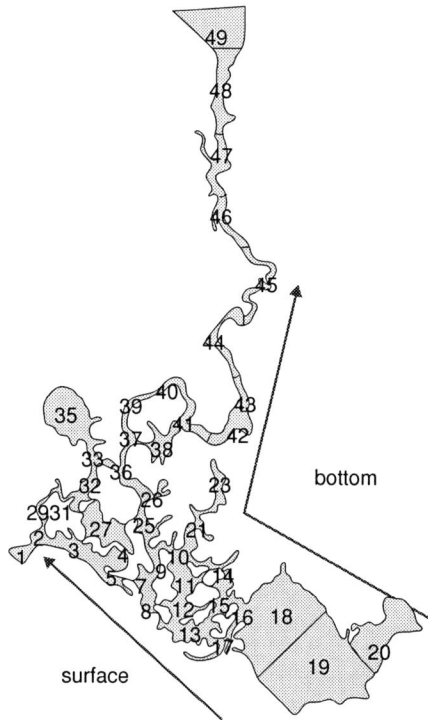

Figure 13. Schematic pathways for surface and low water layers.

- *Segment 3:* In the proximity of the Mediterranean Sea.
- *Segment 45:* In the channel that connects Dalyan Lagoon with Koycegiz Lake in the proximity of a town.

Previous studies of the Dalyan Lagoon system have described two different pathways for spreading of the toxicant in the lagoon (Figure 13). Flow pathways and vertical mixing coefficients were calculated in those studies using salinity as a tracer (Erturk, 2002). Under normal flow conditions, pollutant in surface water will reach the sea within days, whereas pollutant in the more saline bottom water will spread northward through the system.

For average flow conditions, the conservative toxicant concentration responses to the three loading scenarios are captured in Tables 5–7. Due to mixing and dilution, rapid attenuation of the concentration peak occurs over the first three days, with an effective half-life of about 1 day. After this initial period, dilution proceeds more slowly. Differences among the three loading scenarios are most pronounced in the early days of the simulations and in the segments closest to the release. The different loading simulations give similar results after about 10 days, as illustrated in Figure 14 for the surface and bottom layer releases of dieldrin.

TABLE 5. Concentration (μg/l) evolution in the surface (sup) and bottom (bot) layers. Default case with release in the surface layer

	Segments									
	20		16		25		3		45	
Time (days)	sup	bot	sup	bot	sup	bot	sup	bot	sup	bot
0.5	4411.899	95.827	8.7104	1.6018	0	0	0.7178	0.008	0	0
1	2705.594	123.2433	38.6459	19.7247	0	0.0018	7.5512	0.1755	0	0
3	520.9058	138.3111	96.2734	122.6234	0.0359	0.6969	28.9358	1.3119	0.0001	0.0006
5	315.444	147.6217	76.4123	136.2258	0.14	1.934	22.6534	1.0494	0.0016	0.0056
10	147.9558	127.464	44.033	117.9554	0.5233	4.4642	12.429	0.5679	0.0133	0.043
20	62.6791	68.5183	19.5241	63.2726	0.5496	3.8271	5.7175	0.2598	0.019	0.0611
40	15.7625	17.4353	5.536	16.1571	0.199	1.1551	1.481	0.0666	0.0079	0.0221
80	1.3079	1.3653	0.6611	1.2877	0.0267	0.0932	0.1237	0.0054	0.0018	0.0034

TABLE 6. Concentration (μg/l) evolution in the surface (sup) and bottom (bot) layers. Default case with release in the bottom layer

Time (days)	Segments									
	20		16		25		3		45	
	sup	bot	sup	bot	sup	bot	sup	bot	sup	bot
0.5	102.5162	1503.344	2.9937	50.7023	0	0	0.1726	0.0017	0	0
1	131.2813	634.8602	16.4934	206.3134	0	0.0015	2.6	0.0574	0	0
3	140.0455	279.8993	47.1741	256.5476	0.0677	1.7854	12.7312	0.5555	0.0002	0.0007
5	147.4516	233.2304	48.5622	214.5102	0.3268	4.7484	13.4478	0.599	0.0032	0.0115
10	126.5242	156.8956	41.5307	144.8023	0.9699	7.5961	11.7388	0.5287	0.0274	0.0882
20	68.6932	77.4633	21.6819	71.5355	0.6899	4.694	6.3942	0.2903	0.0247	0.0792
40	17.6885	19.569	6.2185	18.1349	0.2282	1.3226	1.667	0.0749	0.009	0.0254
80	1.468	1.5326	0.7421	1.4455	0.0301	0.1049	0.1389	0.0061	0.002	0.0038

TABLE 7. Concentration (μg/l) evolution in the surface (sup) and bottom (bot) layers. Default case with release in both (surface and bottom) layers

Time (days)	Segments									
	20		16		25		3		45	
	sup	bot	sup	bot	sup	bot	sup	bot	sup	bot
0.5	2257.208	799.5856	5.8521	26.1521	0	0	0.4452	0.0048	0	0
1	1418.437	379.0519	27.5697	111.519	0	0.0017	5.0756	0.1165	0	0
3	330.4754	209.1052	71.7236	189.5854	0.0518	1.2412	20.8335	0.9337	0.0002	0.0007
5	231.4476	190.4261	62.4872	175.3681	0.2334	3.3412	18.0506	0.8242	0.0024	0.0085
10	137.24	142.1797	42.7819	131.3788	0.7466	6.0301	12.0839	0.5483	0.0203	0.0656
20	65.6862	72.9908	20.603	67.4075	0.6197	4.2606	6.0558	0.275	0.0219	0.0702
40	16.7255	18.5022	5.8773	17.146	0.2136	1.2388	1.574	0.0707	0.0084	0.0238
80	1.388	1.4489	0.7016	1.3666	0.0284	0.099	0.1313	0.0058	0.0019	0.0036

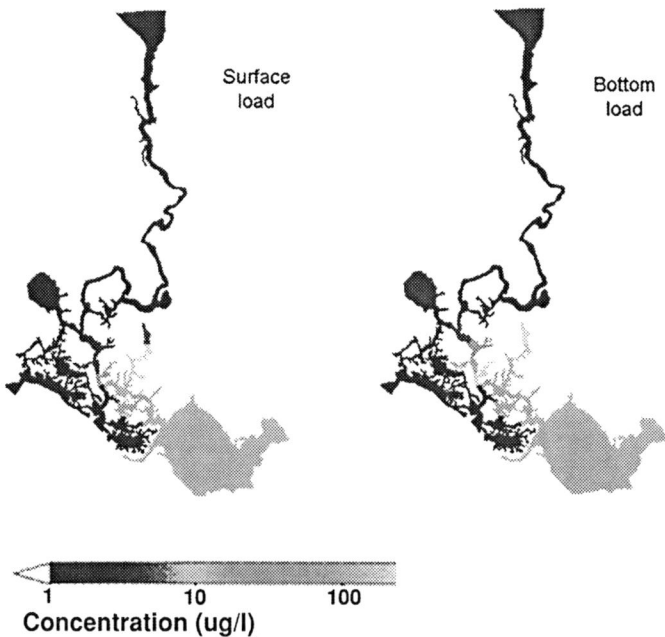

Figure 14. Spatial distribution of the concentration after 10 days is presented for the cases of release in the surface and the bottom layer

Following this initial investigation of conservative spill transport, the WASP7 Dalyan Lagoon spill model was refined to include key chemical properties of dieldrin. These properties are summarized in Table 8.

First, it should be noted that the solubility of dieldrin is 200 µg/l under the typical conditions of Sulungur Lake, which is much lower than predicted concentrations in upper Sulungur Lake during the first days after the spill. Total calculated concentration, then, includes suspended dioxin powder as well as aqueous and sorbed chemical phases.

The photolysis rate constant of 0.108 day^{-1} gives a nominal half-life of about 7 days. Since this loss rate is driven by UV light (Appendix 1), the light attenuation with depth should reduce this rate significantly. For chlorophyll concentrations of 5 µg/l, solids concentrations of 2 mg/l, and DOC of 1 mg/l, the light extinction coefficient for 300 nm ultraviolet light is calculated to be 7.4 m^{-1}. For depths of 0.3 to 2.5 m, calculated photolysis rate constants in surface segments varied between 0.003 and 0.025 day^{-1} (half lives between 1 and 8 months). The photolytic half-life in bottom waters should be much longer.

The volatilization loss rate for a wind speed of 2–5 m/s, a temperature of 20 °C, and depth of 0.5 m is about 0.035 day^{-1}, giving a half-life in surface

TABLE 8. Specified properties for dieldrin

Parameter	Value	Explanation, reference
Sorption		
Chemical 1 log 10 of the octanol–water partition coefficient	5.48	AQUATOX library
Volatilization		
Water body type	1	Wind data are present
Air temperature	1	Temperature data are present
Henry's law constant	5.41×10^{-6}	AQUATOX library
Molecular weight	380.9	AQUATOX library
Chem k_1 option 1 = input K_V as parameter	5	
Solubility (mg/l)	0.2	AQUATOX library
Photolysis		
Chemical 1 photolysis option	0	
Chemical k_P overall depth averaged photolysis rate constant, 1/day	0.108	AQUATOX library

segments of about 20 days. Volatilization rates apply to the dissolved chemical phase only. Since much of the dieldrin shortly after the spill in upper Sulungur Lake will not have dissolved, calculated volatilization rates in this area will be overestimated.

Biodegradation, oxidation, reduction and hydrolysis were considered to be insignificant, because of the physical-chemical properties of dieldrin (see Appendix 2). The reference safety standard for dieldrin set by OSHA and NIOSH for human skin contact is 0.25 µg/l (see Appendix 2).

The following time series (Figure 15) show the predicted evolution of dieldrin in the surface layer of the significant segments for different flow scenarios. For surface loading under higher fresh water inflow conditions, dieldrin is about 30% more diluted in Sulungur Lake, and is transported more rapidly to the open sea boundary. The peak concentration near the sea is about half of the peak concentration at the lagoon-lake boundary. Significantly less dieldrin is transported northward with bottom flow. The peak concentration in the middle Dalyan channel system is more than 200 times less than the peak concentration near the sea.

On the other hand, for surface loading under lower fresh water inflow conditions, dieldrin is about 30% less diluted in Sulungur Lake, and is transported more slowly to the open sea boundary. The peak concentration near the sea is about 6 times less than the peak concentration at the lagoon-lake boundary. More dieldrin is mixed with the bottom layer and transported northward with bottom flow. Still, the peak concentration in the middle Dalyan channel system is about 10 times less than the peak concentration near the sea.

The next time series (Figure 16) show the evolution of dieldrin in the surface layer of the significant segments for different loading scenarios under average flow conditions. For combined surface—subsurface loading, surface dieldrin concentrations are half of the surface loading values at the spill site, and about 30% more diluted at the Dalyan—Sulungur Lake boundary and near the Mediterranean Sea. The peak concentration near the sea is about a third of the peak concentration at the lagoon-lake boundary. About a third more dieldrin is transported northward with bottom flow. The peak concentration in

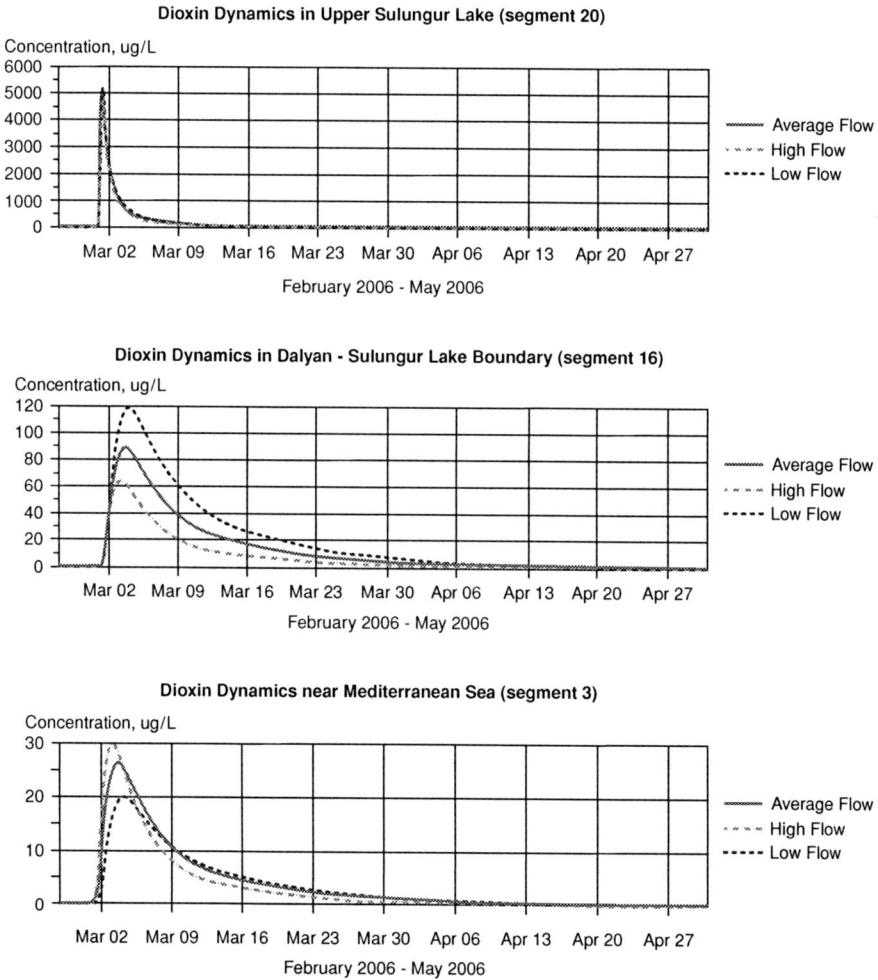

Figure 15. Concentration dynamics by location for different flow scenarios

Dioxin Dynamics in Middle Dalyan Lagoon (segment 25)

Concentration, ug/L

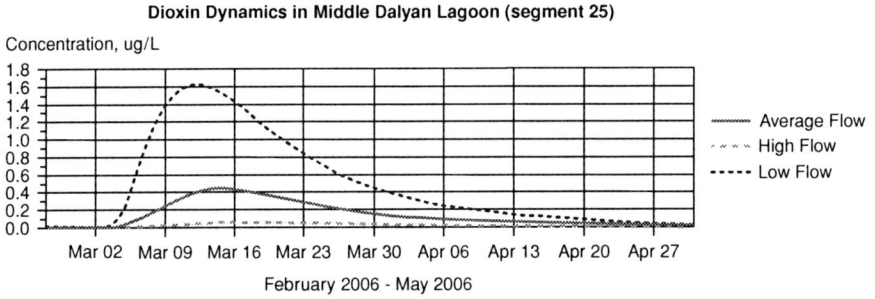

February 2006 - May 2006

Dioxin Dynamics in Upper Dalyan Lagoon (segment 45)

Concentration, ug/L

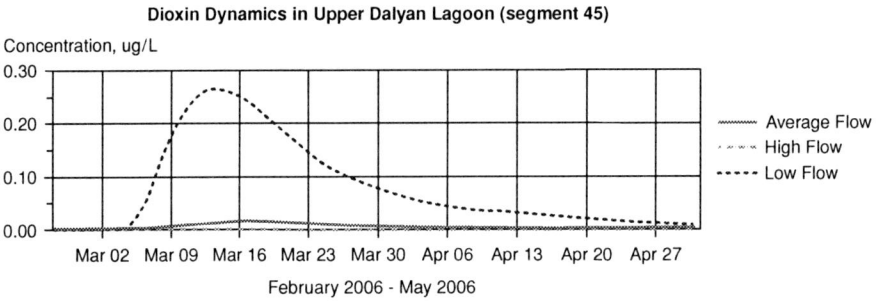

February 2006 - May 2006

Figure 15. (*Continued*)

the middle Dalyan channel system is more than 200 times less than the peak concentration near the sea.

For bottom loading under average fresh water inflow conditions, the peak surface dieldrin concentration in Sulungur Lake is less than half of that for surface loading, and is transported more slowly to the open sea boundary. The peak concentration near the sea is about 4 times less than the peak concentration at the lagoon-lake boundary. More toxicant is mixed with the bottom layer and transported northward with bottom flow. Still, the peak concentration in the middle Dalyan channel system about 12 times less than the peak concentration near the sea.

The final set of time series (Figure 17) show the effects of volatilization and photolysis on the evolution of dieldrin in the surface layer of the significant segments under average flow conditions. Very little effect is evident at the spill site; not enough time is available for kinetics to reduce the peak concentrations. Due to longer travel times, kinetic reactions reduce the peak concentration about 15% at the Dalyan–Sulungur Lake boundary, about 20% near the Mediterranean Sea, and about 35% in the middle and north Dalyan channel system.

An additional simulation was run to test the influence of mixing with a 2 cm surficial benthic layer. Settling and resuspension velocities for detrital solids were set at 0.3 and 0.00003 m/day, respectively. While calculated

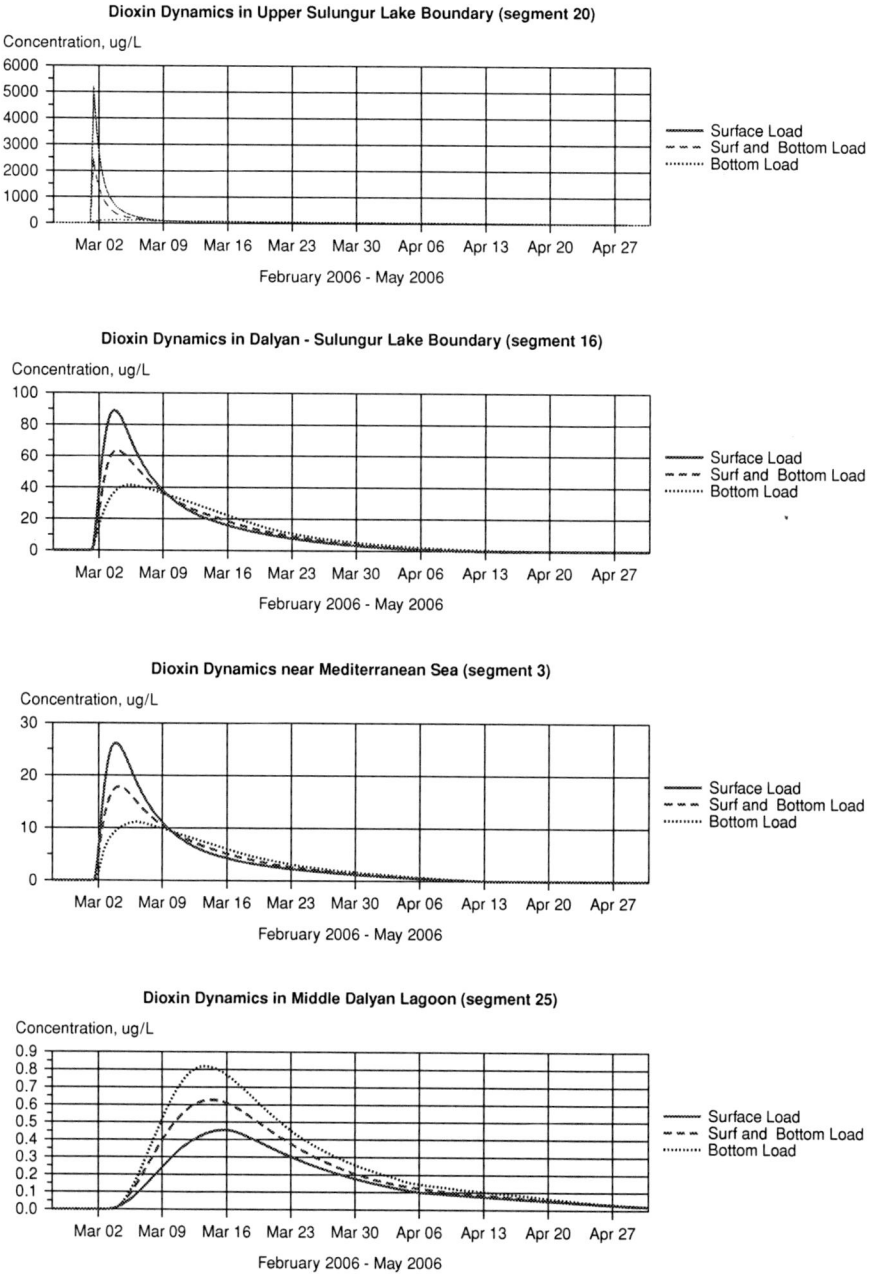

Figure 16. Concentration dynamics for different loading scenarios

Dioxin Dynamics in Upper Dalyan Lagoon (segment 45)

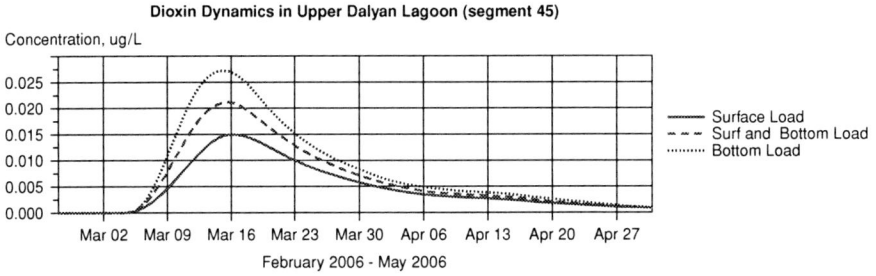

Concentration, ug/L

Surface Load
Surf and Bottom Load
Bottom Load

Mar 02 Mar 09 Mar 16 Mar 23 Mar 30 Apr 06 Apr 13 Apr 20 Apr 27

February 2006 - May 2006

Figure 16. (Continued)

dieldrin concentrations reach a maximum of 400 μg/g in the sediments of upper Sulungur Lake, the added sediment interaction caused very little change in calculated surface water concentrations throughout Dalyan Lagoon.

For surface brackish waters moving seaward, the maximum concentration of dieldrin will exceed the maximum allowed concentration for environmental safety considerations by 1.5–4 orders of magnitude (Figure 18). These maximum values are reached within 3 days at most. The concentration of toxicant will remain above the standard in these channels for at least 30–50 days (Figure 19). The inlet area, open to the sea, is the first to experience concentrations below the safety threshold (after about 22 days).

For bottom saline water, the maximum concentrations of the pollutant will also exceed (1–4 orders of magnitude) the safety threshold, except for the two last segments considered in the analysis (segments 93 and 94), which represent the Dalyan town area (Figure 20). In these segments, the maximum concentration value will be less than the safety threshold. The maximum values will be reached after 12–14 days. The concentrations will decrease steadily thereafter, and it will take about 10–30 days for concentrations to fall below the safety threshold again (Figure 21).

Dioxin Dynamics in Upper Sulungur Lake (segment 20)

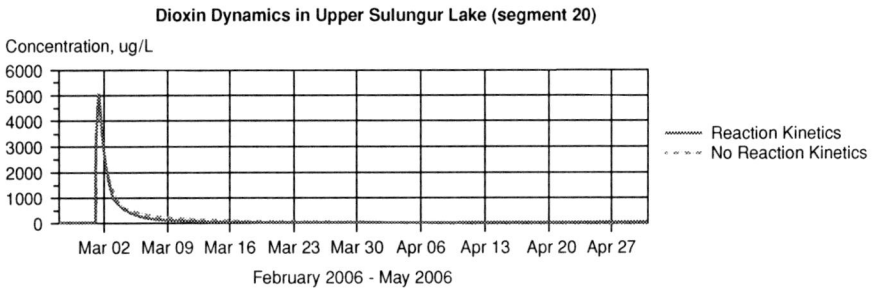

Concentration, ug/L

Reaction Kinetics
No Reaction Kinetics

Mar 02 Mar 09 Mar 16 Mar 23 Mar 30 Apr 06 Apr 13 Apr 20 Apr 27

February 2006 - May 2006

Figure 17. Concentration dynamics by location for different kinetic reaction scenarios

Dioxin Dynamics in Dalyan - Sulungur Lake Boundary (segment 16)

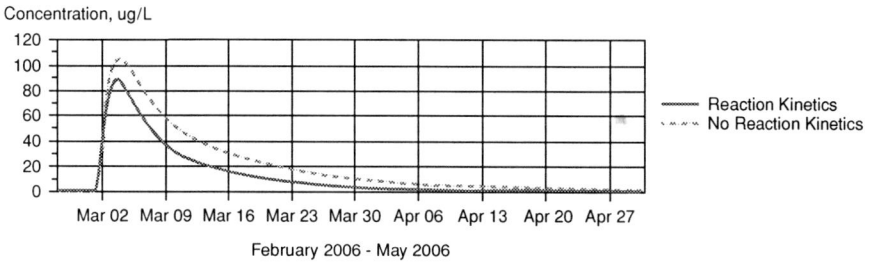

Concentration, ug/L

— Reaction Kinetics
- - - No Reaction Kinetics

February 2006 - May 2006

Dioxin Dynamics near Mediterranean Sea (segment 3)

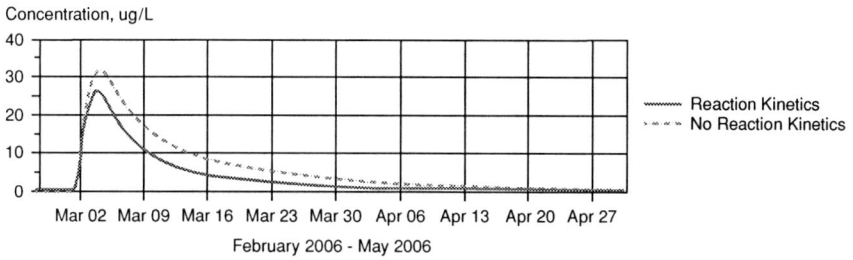

Concentration, ug/L

— Reaction Kinetics
- - - No Reaction Kinetics

February 2006 - May 2006

Dioxin Dynamics in Middle Dalyan Lagoon (segment 25)

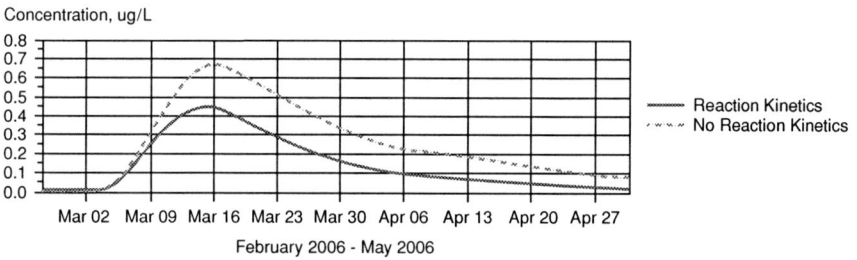

Concentration, ug/L

— Reaction Kinetics
- - - No Reaction Kinetics

February 2006 - May 2006

Dioxin Dynamics in Upper Dalyan Lagoon (segment 45)

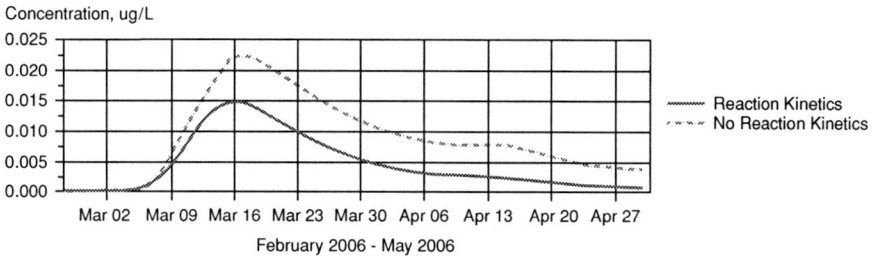

Concentration, ug/L

— Reaction Kinetics
- - - No Reaction Kinetics

February 2006 - May 2006

Figure 17. (Continued)

Maximum concentration of dieldrin in surface water segments towards the sea

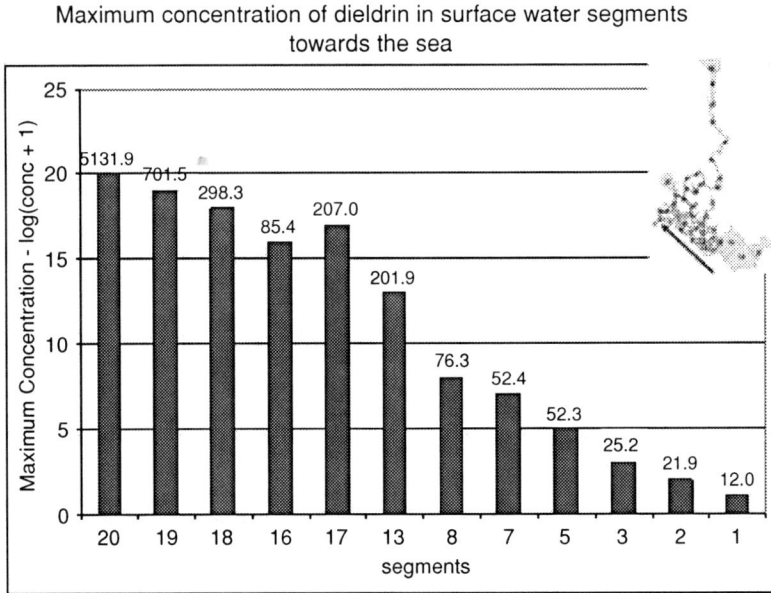

Figure 18. Maximum concentration of dieldrin in surface water (value in µg/l at top of bars)

Clearance time for surface water segments towards the sea

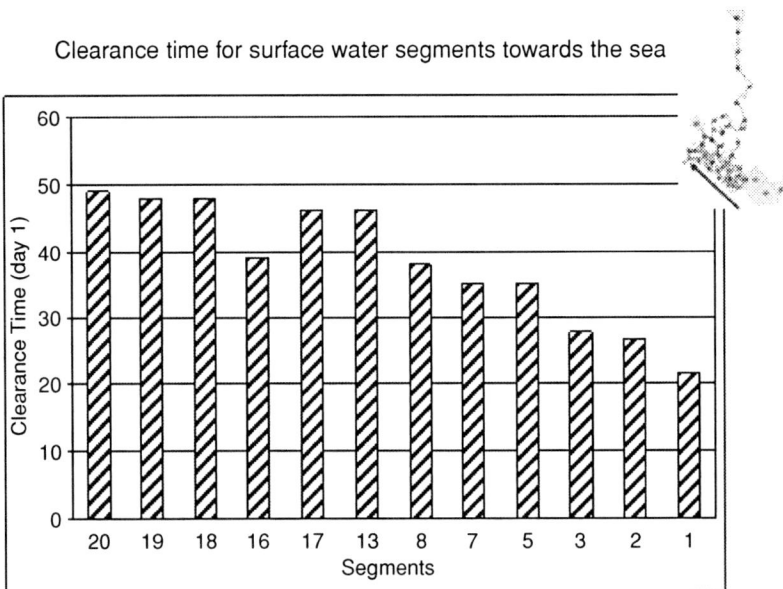

Figure 19. Clearance time for surface water

Maximum concentration of dieldrin in bottom water segments towards the town

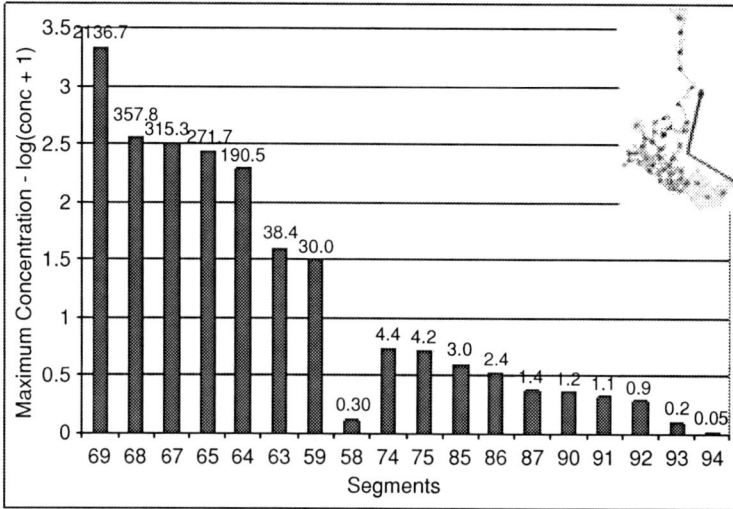

Figure 20. Maximum concentration of dieldrin in bottom water (value in μg/l on top of bars)

Clearance time for bottom water segments towards the town

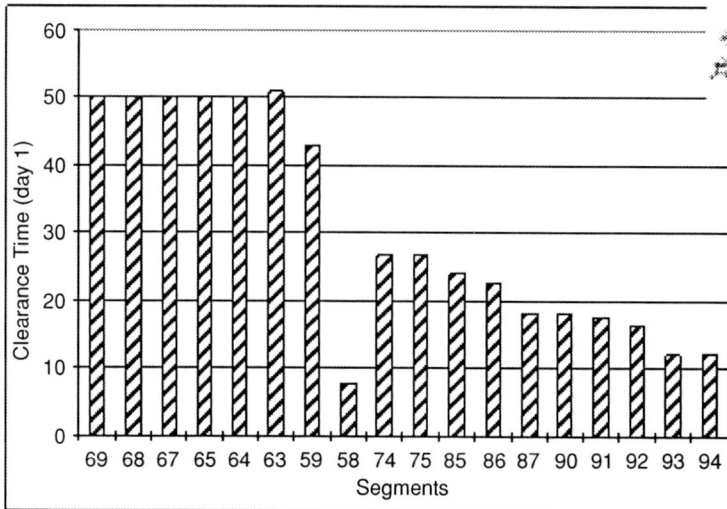

Figure 21. Clearance time for bottom water

17.3.2. AQUATOX SIMULATION RESULTS

To gain insight into the potential ecosystem response to the dieldrin spill, the AQUATOX model (Park et al., 2004) was set up to simulate Sulungur Lake as a whole. Lake geometry, light input, freshwater inflows, and inputs of CBOD, phytoplankton, phosphorus, ammonia, and nitrate were obtained from previous WASP eutrophication simulations. The AQUATOX simulation period was set to 10 years. Constant values for the whole simulation period were used for water pH (7.8), wind speed (1.2 m/s), dissolved oxygen (6 mg/l), carbon dioxide (1 mg/l), and evaporation rate (57.5 inches/year). Initial conditions were set for refractory (150 g/m^2), and labile (200 g/m^2) sedimented detritus. The system was allowed to stratify, with the epilimnion mean temperature of 20°C and a range of 18°C, and the hypolimnion mean temperature of 16°C and a range of 14°C. In the modeled food web, three types of algae (Diatoms, blue–green algae, Dinoflagellate), three species of invertebrates (Amphipod, Copepod, Rotifer), and one species of fish (Mullet) were included. The terrorist attack was represented as a one-day load of 5000 kg of dieldrin occurring on 3/1/2006.

Dieldrin Fate: Simulation results show that the mean diedrin concentration in the Sulungur Lake epilimnion water peaks at 250 ug/l then declines rapidly (Figure 22). Due to the high octanol–water partition coefficient of dieldrin (log $K_{ow} = 5$), accumulation of toxin occurs mainly in the detrital fraction of the sediment. According to model results, the concentrations of dieldrin in refractory and labile sedimented detritus peak in the months following the terrorist attack and then slowly decline over two years. Therefore, invertebrate

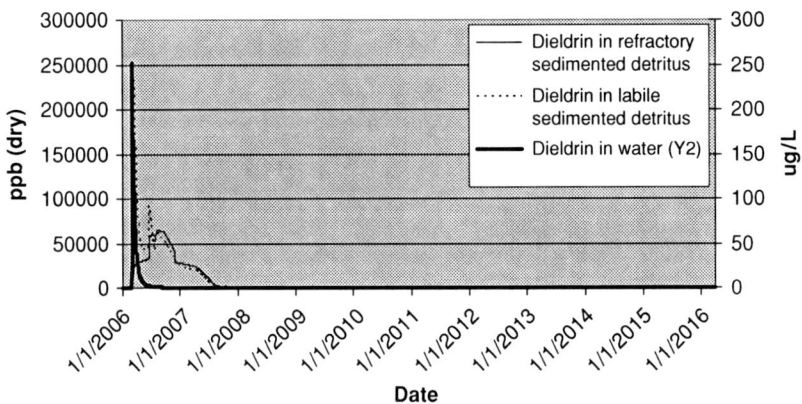

Figure 22. Dieldrin in sedimented detritus and epilimnetic water (Y2)

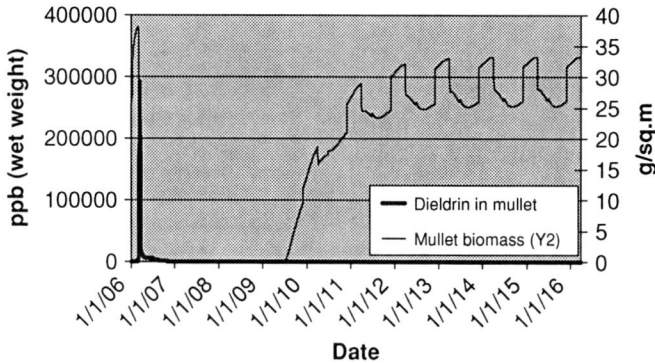

Figure 23. Toxin concentration in mullet tissue and mullet biomass after the attack

aquaculture and human contact with the upper layer of sediment should be safe after 2 years.

Fishery: Fishery activities are carried out in Koycegiz, Alagol, and Sulungur Lakes, and along the Iztuzu coast by means of a cooperative (DALKO), which has 507 shareholders and approximately 60 employees. Fish production fluctuates. Mullet, which has economic value, is found in the sea near the coast, but during the breeding period adult mullet travel to brackish streams and lagoons and their offspring grow here. One third of adults are caught as they travel by brackish waters, and they are used in caviar production (Gurel et al., 2005).

AQUATOX simulations were run to investigate how the terrorist attack scenario would affect the mullet's biomass and how dieldrin would accumulate in the fish. Results show that the total mullet biomass drops to zero approximately nine days after the terrorist attack (figure 23). Fish stock was predicted to recover after 3–4 years to the same level as before the attack, and the recovered mullet biomass shows similar population dynamics as before the attack (Figures 23 and 24). The recovery is due to the small (10^{-3}g/m^2 d) input load of biomass that is assumed to occur. Because of the rapid die-off of mullet, there is no long-term bioaccumulation of dieldrin predicted by the model (Figure 23). Therefore, the exploitable stock will be safe for human consumption after stock recovery.

The medium-term time scale management scenario of artificial introduction of mullet to Sulungur Lake was also tested. The introduction of 1000 kg of mullet was assumed to occur on 3/1/2008, two years after the terrorist attack. The reintroduction of fish results in the recovery of the stock to within 10% of background biomass by May 2010. Hence, the reintroduction would be successful, and would shorten the stock recovery time, as compared to the

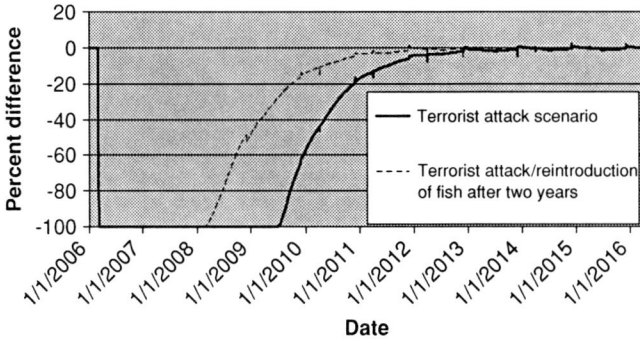

Figure 24. Percent difference in biomass of mullet in the terrorist attack scenario and terrorist attack with reintroduction of fish after two years, as compared to background conditions

natural recovery of biomass, which occurs within 10% of background by July 2011 (figure 24).

The response of sediment feeders (amphipods) to the terrorist attack was also examined. Amphipods in the background simulation show a cyclic pattern in response to temperature (Figure 25). The AQUATOX model predicted that amphipod biomass will decline in response to the attack, which is expects because of their exposure to dieldrin through the sediment (Figure 25). Recovery time for this population was about 5–6 years.

17.3.3. RESULTS FROM EMERGENCY RESPONSE TEAM

The Emergency Decision Response Team (ERT) was responsible for developing the short-term emergency response plan, and for providing medium and long-term recommendations for mitigation of damages and recovery of

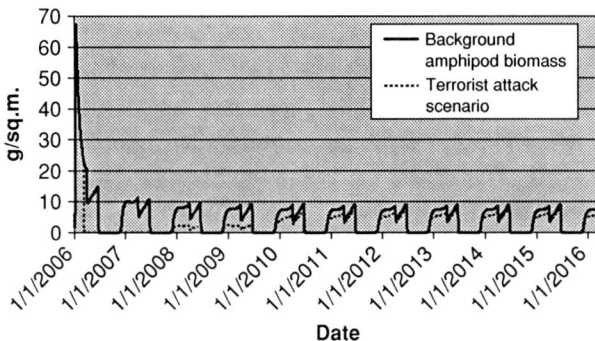

Figure 25. Percent difference in amphipod biomass in the terrorist attack scenario as compared to background conditions

the resource. The ERT first listed explicitly the information provided by the hydrodynamic and ecology teams:

- Dieldrin is an organochlorine insecticide; it is a dangerous chemical that exhibits neurotoxic and carcinogenic properties. Dieldrin has low solubility in water and tends to bioaccumulate.
- The maximum concentration of dieldrin in surface water segments towards the sea will exceed any allowed water quality criteria within about 3 days.
- The maximum concentration of dieldrin in the vicinity of the Dalyan town will not exceed the allowed water quality criteria.
- Dieldrin concentrations in the lagoon will slowly decline, reaching acceptable levels after almost twenty days in upstream segments and thirty days in downstream segments. It will take about 50 days to reach acceptable levels in Sulungur Lake.
- Human contact with the sediment will be safe after 2 years.
- Fish stock can recover to the same level as before the attack in 3–4 years.

The ERT highlighted the need for additional information, especially regarding socio-economic aspects and more detailed data on potential endangered and/or sensitive species. The ERT defined the following assumptions that supported the development of the proposed action plan:

- Dieldrin will not be transported to the land.
- 50% of the contaminant powder is not toxic.
- There would not be drought conditions during the 80 days period after the incident.
- The terrorists were not caught immediately after the attack.
- Well-equipped laboratory in Izmir (1 h away by flight) is prepared for analyzing the samples containing dieldrin.

The objectives of the proposed action plan were to minimize and mitigate the short, medium and long-term effects of the toxic chemical on the Dalyan lagoon system.

The decision-making model used in this case study includes five steps, as discussed in Chapter 4:

1. Identify the problem
2. Explore alternatives
   (a) Generate alternatives
   (b) Evaluate alternatives
3. Select an alternative
4. Implement the solution
   (a) Develop an action plan
   (b) Determine objectives

(c) Identify needed resources
(d) Build a plan
(e) Implement the plan
5. Evaluate the situation
(a) Monitor the process
(b) Evaluate the results

Short-Term Action Plan: Following this model, a short-term action plan was developed with 7 elements:

• Implement a communication plan.
• Evacuate and isolate the target area.
• Set up and initiate monitoring.
• Identify sensitive habitats.
• Construct artificial waterfall barriers.
• Remove dead fish and wildlife.
• Notify regional and national governments of terrorist activity.

1. Implement communication plan: As soon as possible, a brief meeting will be held of the ER Lead, Alternate ER Lead, ER Team, ER Communications Coordinator, and representatives from local police, fire, government, hospital, hydrodynamic modelers, and ecosystem modelers. The ER Lead will summarize the available information and provide directions on actions to be immediately taken.

Within 24 h, the ERT will go to site and directly notify fishermen and those living in the vicinity of the lake. The ERT will post warning signs on all access points, such as docks and beaches. The designated ER Spokesperson will give a television and radio broadcast, and address the incident as clearly, honestly and accurately as possible. Police or volunteers will post warning notices in hotels, restaurants, religious establishments, schools, and retailers in the area.

2. Evacuate and isolate target area: The ERT deployed to evaluate the situation will immediately contact the emergency operations manager of area(s) to evacuate and isolate. The emergency operations manager will alert all necessary officials including police chief, fire chief, search and rescue team and first aid team to mobilize their units to the specific site as designated by initial emergency response team report. A local evacuation plan will be activated. Search and rescue team (wearing personal protective equipment with first aid supplies) will locate persons in or near the target area including any boats in the lake. Fire mobilization team will evacuate residential and commercial buildings located in or near the target area. A first aid team will assist any distressed persons immediately on site and alert local hospital to send ambulances if needed.

A police mobilization team will close to the public any access roads or waterways leading to the target area, and allow access only to persons directly involved with the emergency response. The immediate target area will be roped off, and a stationary post next to the target area will be established to monitor and assist in controlling emergency actions and prevent the public from entering the area.

3. Monitoring and model validation: The emergency operations manager (ERM) will contact the analytical lab in Izmir (1 h distant by flight) to prepare for analyzing water samples for dieldrin. The ERM will contact and direct appropriate local environmental scientists to collect water samples safely from the target site and other locations suggested by the Hydrodynamics Team. These samples will be placed in sealed hazardous waste containers and transported immediately to lab for analysis. The lab response team will begin analyzing samples and the lab manager will contact the ERM with estimated time for results. Lab data will be relayed to the Hydrodynamics Team and the Ecological Team for model testing and refinement.

4. Identification of the sensitive habitats: Even though the first priority is to protect human lives, as the incident occurred in a Protected Area one of the ERT's main concerns is related to the sensitive ecosystems (especially sensitive species) that are going to be affected by the dieldrin spreading into the system.

The liaison of the ERT in charge with communicating with the Ecology Team has established a list of actions for that team's contact point (see Table 9).

TABLE 9. List of actions for Ecology Team

Actions	Information needed	Information resources
Provide a list of sensitive ecosystems	Ecosystem location Ecosystem description	Existing databases Scientific reports
Provide a list of sensitive species	Species name Species stock	Expertise
Provide an environmental sensitivity index map	Map where the identified sensitive species are located, in order to develop specific actions for protecting them	
Provide hot spots recommendation	Provide alternatives to actions for all identified hot spots	

Contact of the Ecology Team liaison has been done according to the Communication Strategy within 8 h. The decision makers requested that the ecological experts supply additional information about the ecology of sea turtles. This information, which was needed to identify priority areas for protection efforts, is given below:

Most scientists recognize 8 species of sea turtles living on earth today and five of these species (*Caretta caretta, Chelonia mydas, Dermochelys coriacea, Eretmochelys imbricata* and *Lepidochelys kempii*) are found in the Mediterranean. However, the only species nesting regularly in the Mediterranean are *Caretta caretta* and *Chelonia mydas*. According to Caldwell (1962) and Uchida (1967), estimated age of maturity for *Caretta caretta* raised in captivity is 6–7 years. Various studies of wild turtles gave estimates for the age of maturity as 10–15 years (Mendonca, 1981), 14–19 years (Zug et al., 1983), 22 years (Frazer, 1983) and 12–30 years (Frazer and Ehrhart, 1985). Mating of *C. caretta* occurs along the way to the nesting beach for several weeks prior to the onset of nesting, or in specific aggregation areas. Mating pairs are frequently seen at the surface, although there are reports of submerged copulations (Dodd, 1988).

Nesting season in the northern hemisphere is generally from May through August, and it is from October through March in the southern hemisphere. Laying eggs usually takes place at night. As the female approaches the beach, she raises her head to view the beach. At this stage, she is most sensitive to disturbance and will rapidly swim away in case of danger. While laying eggs, the female is less sensitive to disturbance. When the female finishes depositing eggs, the eggs are gently covered with the moist sand and the sand is pressed with the rear flippers. Finally, the front flippers begin to throw sand backwards as the turtle slowly moves forward, forming a camouflaged area to hide the egg chamber. After this is completed, the female returns rapidly to the sea.

5. Artificial waterfalls: ERT applied the Precautionary Principle and gave special attention to protection actions for the Dalyan town and even though the Ecology Team model predicted that concentration of the chemical in the vicinity of the town will not exceed safely limits, still special care was paid to ensure that dieldrin would not spread into the north area. Considering results of the model and expertise provided by Hydrodynamics Team, ERT decided to construct artificial waterfalls in order to stop the spread of dieldrin through the bottom layer to the north. The decision took into account the lack of information regarding short-term (flooding of the surrounding area) and long-term (perturbation of the entire Dalyan Lagoon system) impacts of building the barriers and also time restriction; therefore military bridges were designated to serve this purpose. Another reason for this decision was that

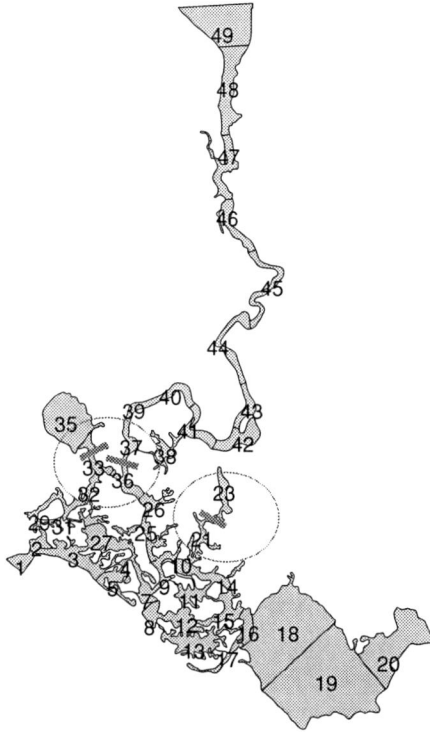

Figure 26. Recommended Segments for building artificial waterfalls

the removal of the military bridges will be easily done when agreed after validating data from monitoring and modeling.

The first action done was to contact Public Works to supply hydrodynamic engineering expertise to identify the location to emplace artificial waterfalls. Based on recommendations from the Hydrodynamics Team three segments (Figure 26) were considered critical. The reasoning is that 2 segments are on the channels connecting the south and north areas and the other segment should prevent infiltration from the contaminated water into the north area.

The second action was *identification of the military facility in Dalyan.* This is going to help emplace the military bridges used as artificial waterfalls. The representative for the military facility was *given information about the terrorist attack* and of the need for technical assistance for building the barriers in the designated locations. He was *instructed to organize three teams* for building the waterfalls for efficiency and timeliness. He was given the *task to communicate with the communication officer* about the process of installing and securing the military bridges. The process of installing the military bridges took place within 8 h after the military representative was informed and in 23 h after the terrorist attack.

6. Removal of dead fish and wildlife: The ERM will arrange for a hazardous waste team to enter the target area and remove all dead animals according to standard collection protocols. The ERM will arrange for boats with the Cooperative Boat Conveyors of Dalyan. Once all dead animals are safely collected, they will be transported by hazardous waste trucks to the designated hazardous waste site. At this site, the hazardous waste team will determine the best course of action for safely eliminating dead species with minimal effect to the environment, and inform the ERM of specific actions to be taken.

7. Notification of Regional and National Governments of Terrorist Activity: The ERM will contact government liaisons to inform them of the situation and coordinate efforts to track and capture terrorist perpetrators, including any support from regional and national governments.

Medium-Term Recommendations

Monitoring and modeling updating. In order to ensure human safety in the affected area, a medium-term monitoring plan has to be developed and implemented continuously. This plan should be developed accordingly with datasets requested by the modelers and expert team recommendations and to be updated in the long-term.

Notification of the population: Due to the fact that the information is changing continuously, there is a need for updating it, in order to keep the public informed about the conditions to which they are exposed and keep it updated in the long term.

Security: In order to ensure the safety of the recreation and tourism areas the emergency team will be established.

Inspection plan development: For the inspection of the area the incoming trucks in access roads will be checked for the prevention of possible attacks.

Emergency response plan: The lesson learned from this experience is the importance of developing an emergency response plan so that the decision makers, together with the expert teams and the population, will be prepared to react efficiently in similar situations. Consequently, the emergency response plan will be developed and implemented.

Social rehabilitation: Taking into account the fact that the terrorist attack diminished the resources and services provided by the system, the local government must develop alternative solutions for those who were directly and indirectly affected.

Barriers for the fish: Due to the fact that in medium term the water will be still contaminated, there is a need to build protection barriers so that the fish would not enter the contaminated area. The science experts will decide which segments are the most suitable for the barriers.

Sea turtle protection plan: Taking into account that the sea turtle is a protected species in this area and exposure to a contaminated environment will have a negative effect on them, there is a need to develop and implement a plan for their protection.

Wildlife rehabilitation: After the terrorist attack the local wildlife community will be affected, so in addition to the sea turtle protection plan there is a need for developing and implementing a medium and long-term wildlife restoration plan.

Fish and wildlife disposal: Considering that there will be a large number of carcasses in the area, the hazardous waste team will continually clean up the area.

Target protection for sensitive ecosystems: The results received from the expert teams showed which ecosystems are the most sensitive and in need of protection.

Fishery management service: In order to ensure the welfare of the fishermen, the fisheries management plan is to be developed as well as regulations to restore the fish population. Due to the fact that most of the fishermen lost their jobs after the contamination, there is a need to determine the feasibility of fish culture as socioeconomic compensation.

Long-Term Recommendations:

Monitoring and modeling updating: In order to ensure the safety of the population from the affected area, a long-term monitoring plan has to be developed and implemented. This plan should consider the datasets requested by the modelers and expert team recommendations.

Education and research: A long-term education plan based on lessons learned from this experience will be useful to train the population to be prepared in similar emergency situations. A long-term research plan can be developed having as its basis this situation in order to observe the ecosystem's reaction to extreme conditions (huge quantity of chemical release). These long-term plans will be developed and implemented by the Ministry for Education and Research.

Further communication: Due to the fact that the information is changing continuously, there is a need for updating it, in order to keep the public informed about the conditions they are exposed to.

Social rehabilitation: Taking into account the fact that the resources and services provided by the system were diminished by the terrorist attack, the local government must develop alternative solutions for those who were directly and indirectly affected.

Emergency response plan—evaluation and updating: The main lesson learned from this experience is the importance of developing an emergency response plan so that the decision makers, together with the expert teams and the population, will be prepared to react efficiently in similar situations.

Ecosystem restoration: In order to obtain a state closer to that existing before the terrorist attack, the development and implementation of a long-term restoration plan is needed. Based on the already implemented monitoring plan, the experts could choose the most efficient way to restore the affected system. The barriers prevented the migration of fish into the polluted system will be removed.

Policy/legislation recommendations: In order to try to prevent such extreme terrorist attacks, stricter legislation regarding the manipulation and commercialization of hazardous substances is needed.

17.4. Conclusions and Recommendations

17.4.1. CONCLUSIONS FROM STUDENT TEAMS

Hydrodynamics Team: A model was prepared to simulate the transport of a conservative substance released in the eastern reach of Sulungur Lake. The simulation results show the presence in the Dalyan Lagoon system of two flow patterns: a predominant surface flow that tends to exit the system by the sea boundary, and a secondary bottom flow that is directed northward. As a

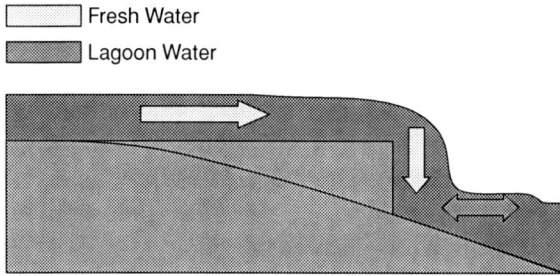

Figure 27. Waterfall-like barrier

consequence, residence time of the contaminant is higher in the bottom layer of the lake. So, if the substance is released in the surface layer, it tends to be transported to the sea; if it is released in the bottom layer, it tends to be transported to the inner part of the Dalyan Lagoon channel system.

Compared with the average flow situation, high fresh water input reduces the residence times in the lagoon, and transports the toxicant more rapidly to the sea. On the other hand, decreasing the fresh water inflow and increasing the bottom salt water flow increases the residence time in the system and results in relatively higher concentrations in upstream locations.

Toxic concentrations are reduced to less than 1 µg/l after 80 days throughout the lagoon system. The contaminant reaches the sea boundary after 24 h in relatively small concentrations. Due to its rapid transport and mixing, no practical solution could be applied to avoid the toxicant's reaching the sea. Since the contaminant could, even in small concentration, be transported northward to Koycegiz Lake, scientists suggest blocking the path of the bottom flux to prevent this from occurring. The adequate places for building barriers have been identified in the area in segments 33, 23 and 37 (see figure 26 in the previous section).

On segment 37 the barrier should stop the bottom flow directed to Koycegiz Lake but at the same time let the fresh water pass to avoid flooding of the surroundings (Figure 27).

Ecology Team: The results obtained from the modeling studies were analyzed and the following conclusions were drawn:

- Transport in the surface layer was predominantly towards the sea boundary.
- Transport of the toxicant in the bottom layer was towards the central areas of the Lagoon system.
- Increasing the fresh water input reduces the concentrations, mainly in the surface layers. The opposite situation can be observed when reducing fresh water inputs.

- With higher fresh water inflow, the toxicant reaches surrounding areas faster.
- The contaminant reaches the sea boundary after 24 h but in relatively small concentrations.
- Residence time of the contaminant is higher in the bottom layer of the lake.
- The toxicant containing dieldrin which is 50% of purity will be reduced, in the lagoon system, to the maximum acceptable limit in maximum 50 days.
- The highest concentrations obtained from the simulations were 5000 μg/l in the surface layer, in the vicinity of the spill.
- The aquatic ecosystem will recover in 3–4 years after the attack.
- No bioaccumulation will occur as the entire fish stock collapses (ca. 9 days after the spill).

Recommendations for Action Plans:

Short-term recommendations

1. Detection of dieldrin for monitoring and model calibration;
2. Removal of the dead fish—incineration should be controlled, best available technology should be used for ash disposal strategies, ash should be transported to hazardous disposal sites.

Medium-term recommendations:

3. Monitoring (develop a plan) and modeling future developing perspectives;
4. Development of a sea turtle protection plan;
5. Development of a rehabilitation plan for the wildlife (including all sensitive species);
6. Ecosystem recovery by artificial fish introduction.

Long-term recommendations:

7. Continuing of monitoring and modeling; education and research;
8. Ecosystem restoration; potential fish repopulation.

Emergency Response Team: The fast response and results from the Hydro-dynamics Team and Ecology Team enabled the ERT to promptly develop and implement a short-term action plan and recommendations for medium and long term management. These important expert teams provided us with accurate and fast information that allowed ERT to make quick decisions. Our recommendation is to implement their recommendations to continue monitoring and modeling this water system. We also recognize the need for additional assessments and recommend socio-economic assessments.

The preparedness of the ERT proved to be an important factor in the quick response provided. This enabled us to inventory available tools, information and resources and to identify and prioritize knowledge and information gaps. However, we also learned that the action plan depended on all sorts of data and information we received and that it constantly needed to be updated on all levels. A critical aspect of the process of decision making was the effective transfer of information and knowledge between interdisciplinary teams and also between scientists and decision makers. Therefore, we recognize the need for further efforts in creating a common language and understanding between parties involved (scientists with different backgrounds and decision makers) and even consider designing an interface in terms of a decision support system.

The approach used for our action plan was based on considering the complexity of the socio-ecological systems and therefore it is one of the most important inputs for the decision making process that we recommend to be used in future emergency response plans. Another long-term recommendation is related to the necessity of designing legislation to track hazardous materials to avoid future chemical terrorist attacks.

Constant preparedness for emergency situations is a key element for fast and good responses; therefore, for the medium and long term we consider necessary further steps of gathering data and information to complete and implement the ERT recommendations.

17.4.2. POST-MORTEM ANALYSIS OF EXERCISE

The WASP7 Dalyan Lagoon computational network and transport calibration was developed to study the longer-term nutrient and eutrophic status of Dalyan Lagoon system. In this exercise, the existing model was adapted to quickly investigate the short to medium-term toxicant dynamics in the system. During the ASI, there was no time to set up and run a more physically-based hydrodynamic model with a finer-scale computational network. Following this exercise, the authors applied the EFDC and WASP7 models to a finer scale three-dimensional computational network in Sulungur Lake to investigate the short-term spill dynamics in greater detail than was possible with the larger scale network. The objective was to evaluate how well the coarse scale model reproduced the essential features of the spill.

Sulungur Lake bathymetry is illustrated in Figure 28. The EFDC model can run on either curvilinear or Cartesian computational networks. At the time of this exercise, an EFDC grid generator was not available, so we set up a medium scale Cartesian network with 300 meter square cells, as illustrated in Figure 29. Cells are labeled with east-west i-coordinates, north-south j-coordinates, and vertical k-coordinates. The EFDC grid extends past the lake–lagoon boundary. Although there are two tidal inlets from the main lagoon network to the lake,

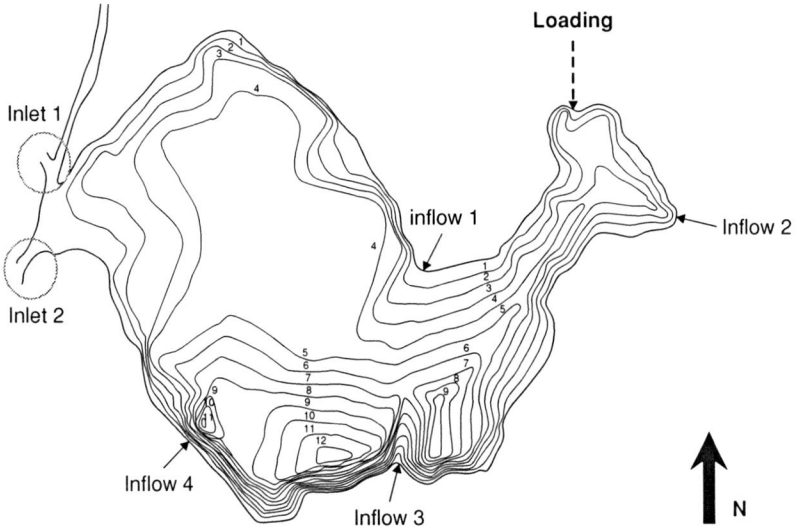

Figure 28. Sulungur Lake Bathymetry

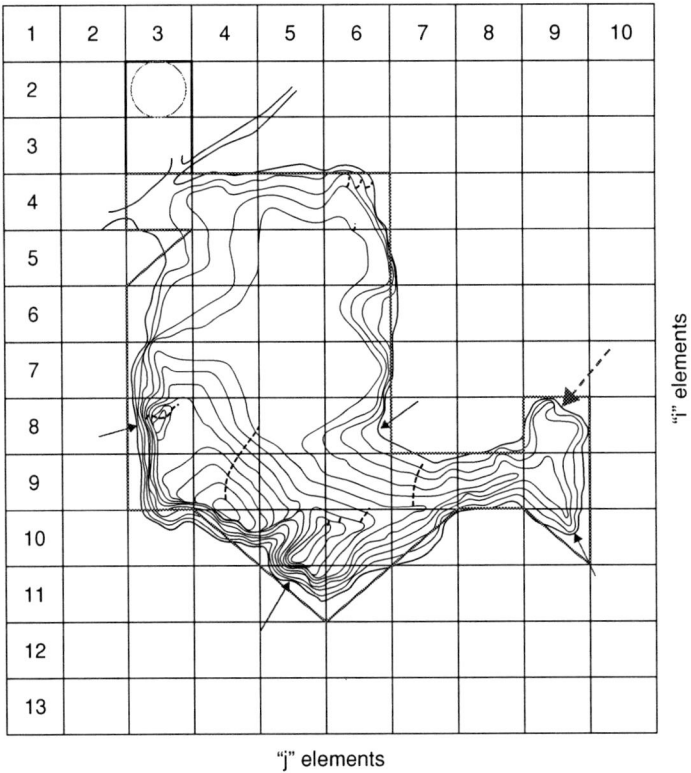

Figure 29. Sulungur Lake Computational Grid

0	9	6	9	0	0	0	0	0	0
0	9	5	9	0	0	0	0	0	0
0	9	5	9	6	6	9	0	0	0
0	9	5	5	5	5	9	0	0	0
0	9	3	5	5	5	9	0	0	0
0	9	5	5	5	5	9	0	0	0
0	9	5	5	5	5	9	9	9	9
0	9	5	5	5	5	9	9	5	9
0	9	5	5	5	5	5	5	5	9
0	9	9	2	5	5	1	9	2	9
0	0	9	9	2	1	9	9	9	9
0	0	0	9	9	9	9	0	0	0
0	0	0	0	0	0	0	0	0	0

"i" elements

"j" elements

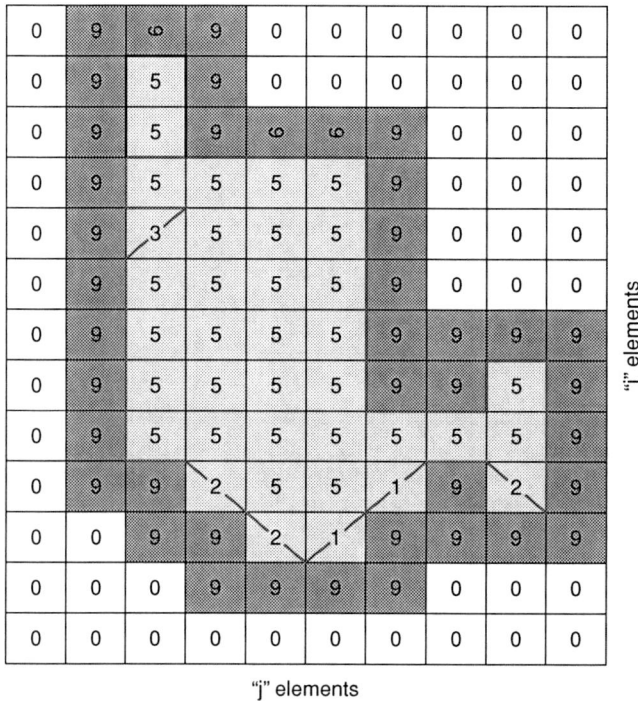

Figure 30. Sulungur Lake Cell Types

they are close together and with this relatively coarse grid we chose to lump them into a single tidal boundary at cell (3, 3). The WASP grid begins 2 columns into the EFDC grid, with a boundary at cell (4, 4). EFDC cells are categorized in Figure 30. Type 0 designates cells are dry and do not contact the water body. Type 9 cells are also dry, but border the water body. Type 5 designates square water body cells. Types 1, 2, 3, and 4 are triangular water body cells with dry land to the upper right, lower right, lower left, and upper left, respectively. Superimposing this grid on the lake bathymetry in Figure 28, we estimated average cell depths to the nearest half meter (Figure 31).

The lagoon circulation is forced by tide, wind, precipitation, evaporation, and freshwater inflow. We specified a regular, M2 tide with an amplitude of 10 cm at the lake–lagoon boundary. The monthly-average drainage flow (Table 2) was divided approximately by subwatershed surface area and allocated to four input locations—35% to cell (8, 6), 45% to cell (9, 9), 10% to cell (10, 5), and 10% to cell (8, 3), as illustrated in 28. Wind speed and direction, precipitation, and evaporation measured at the closest meteorological station were applied to Sulungur Lake.

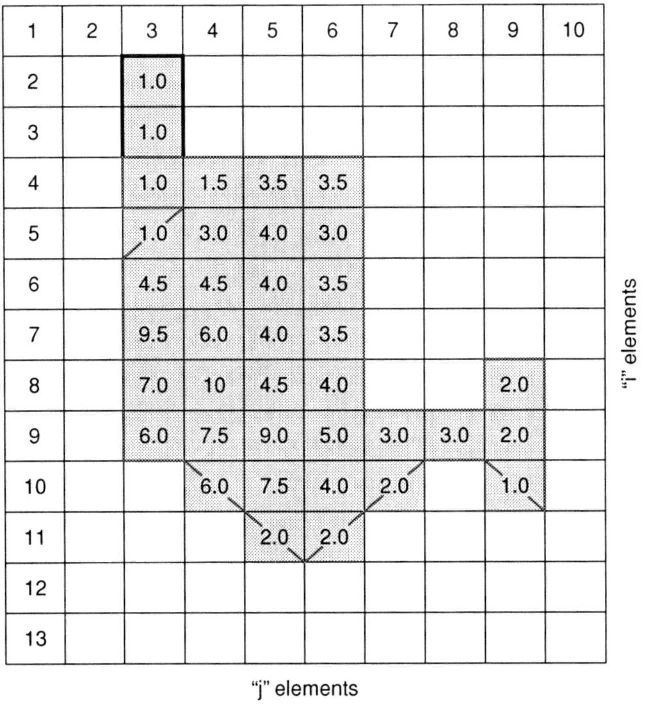

	1	2	3	4	5	6	7	8	9	10
2			1.0							
3			1.0							
4			1.0	1.5	3.5	3.5				
5			1.0	3.0	4.0	3.0				
6			4.5	4.5	4.0	3.5				
7			9.5	6.0	4.0	3.5				
8			7.0	10	4.5	4.0			2.0	
9			6.0	7.5	9.0	5.0	3.0	3.0	2.0	
10				6.0	7.5	4.0	2.0		1.0	
11					2.0	2.0				
12										
13										

"i" elements

"j" elements

Figure 31. Sulungur Lake Cell Depths

Zero current velocities and a flat water surface were assumed for initial conditions. Initial salinity concentrations were specified based on data from 3 monitoring stations. The hydrodynamic simulation was run with a 30 s time step from January 1–April 1, 1998, allowing for a 2 month "spin-up" before the dieldrin spill on March 1. The hydrodynamic file for WASP began on February 25 with output every 300 s.

Before simulating the spill, we tested the numerical stability of the network and time step with a conservative tracer simulation. Precipitation and evaporation were set to 0, and initial concentrations and boundary concentrations were set to 1 mg/l. WASP tracer output oscillates approximately 0.1% around 1000 μg/l near the lagoon boundary, and approximately 0.2% around 999.5 μg/l near the upstream boundary (Figure 32). While this oscillation could be reduced with smaller time steps, we considered it accurate enough for the spill simulation.

Next, we simulated the dieldrin spill. As in the student WASP simulations, a pulse load of dieldrin was released at midnight, March 1 at a constant rate of 15,000 kg/day for a period of 8 h (Figure 12) into surface segment 30 (cell

Dye Surface Profile along Axis

S#6: I=9 J=3 K
S#8: I=5 J=4 K
S#16: I=6 J=5 K
S#18: I=8 J=5 K
S#26: I=8 J=6 K
S#30: I=9 J=7 K
S#31: I=10 J=7 K

Mon 02   Tue 03   Wed 04   Thu 05   Fri 06   Sat 07   Sun 08

January 2006

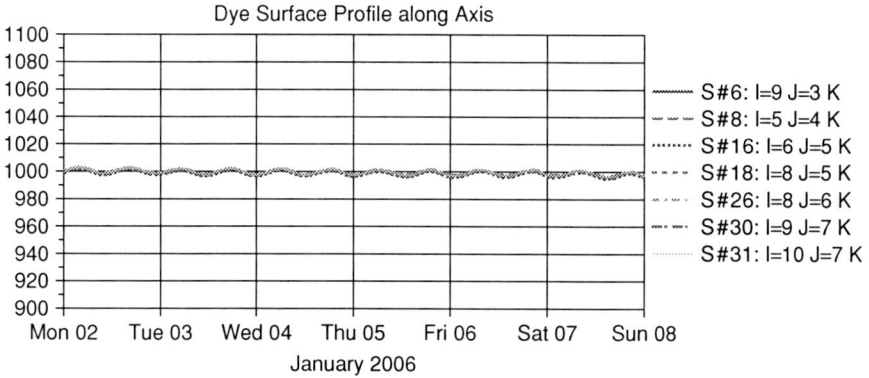

Figure 32. Boundary Tracer Accuracy Test

$i8$–$j9$), located in the eastern portion of Sulungur Lake. The concentration dynamics in the surface and bottom segments of the upper lake are illustrated in Figures 33–37. Calculated peak concentrations at the surface of the spill site reach 80,000 µg/l near the end of the release. Lateral mixing, and, to a lesser extent, vertical mixing dilute the high concentrations in this cell, with dilution half-lives of 1–2 days following the spill. Concentrations in adjacent surface cells increase over the first 1–2 days following the spill, then begin to decline as dieldrin is transported farther through the lake. Surface concentrations in the upper lake are laterally well mixed in about three weeks. Concentrations in adjacent bottom cells are significantly lower, and increase over a 1–2 week

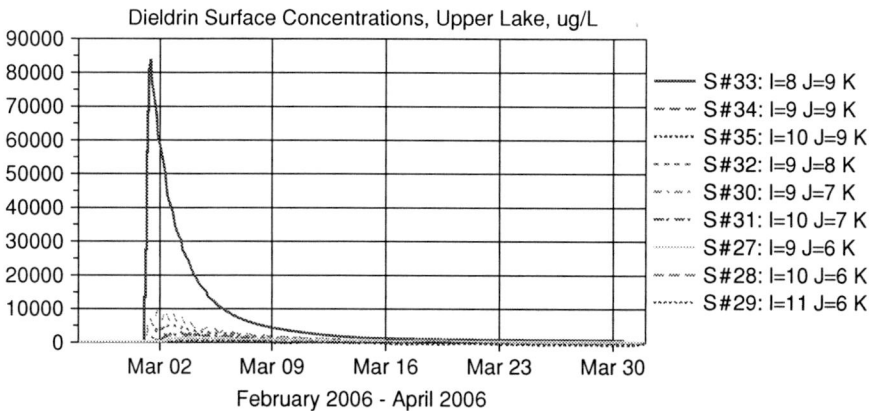

Dieldrin Surface Concentrations, Upper Lake, ug/L

S#33: I=8 J=9 K
S#34: I=9 J=9 K
S#35: I=10 J=9 K
S#32: I=9 J=8 K
S#30: I=9 J=7 K
S#31: I=10 J=7 K
S#27: I=9 J=6 K
S#28: I=10 J=6 K
S#29: I=11 J=6 K

Mar 02   Mar 09   Mar 16   Mar 23   Mar 30

February 2006 - April 2006

Figure 33. Dieldrin Surface Concentrations in Upper Lake

Dieldrin Surface Concentrations, Upper Lake, ug/L

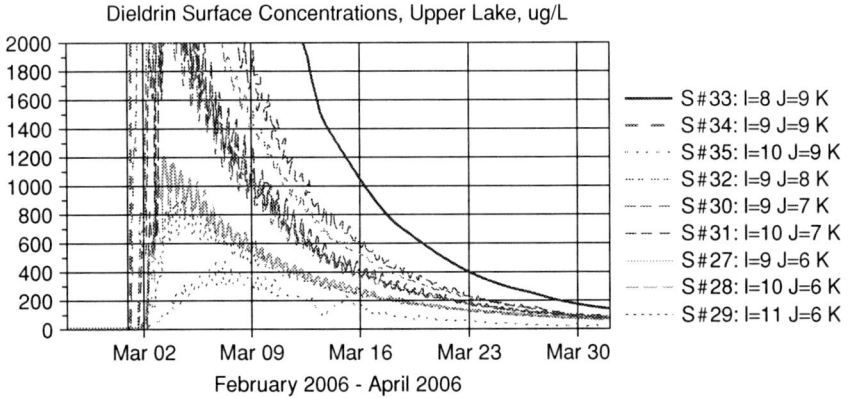

Legend:
- S#33: I=8 J=9 K
- S#34: I=9 J=9 K
- S#35: I=10 J=9 K
- S#32: I=9 J=8 K
- S#30: I=9 J=7 K
- S#31: I=10 J=7 K
- S#27: I=9 J=6 K
- S#28: I=10 J=6 K
- S#29: I=11 J=6 K

February 2006 - April 2006

Figure 34. Dieldrin Surface Concentrations in Upper Lake—detail

period following the spill. Dieldrin is well-mixed vertically within two weeks following the spill.

Simulated transport of the spill pulse through the middle and lower reaches of the lake are shown in Figures 38 and 39. Surface concentrations in mid-lake cells peak between 450 and 325 µg/l from 3.5 to 6 days after the spill. At the lagoon-lake boundary, surface concentrations at the end of ebb tide peak at 225–250 µg/l from 6 to 9 days after the spill, while bottom concentrations reach 180 µg/l 9 to12 days after the spill.

These short-term dynamics are controlled mostly by advection, mixing, and dilution of the dieldrin mass. Two loss processes were treated in this simulation—volatilization and photolysis. Volatilization is controlled by chemical properties, such as partition coefficient and Henry's Law constant,

Dieldrin Bottom Concentrations, Upper Lake, ug/L

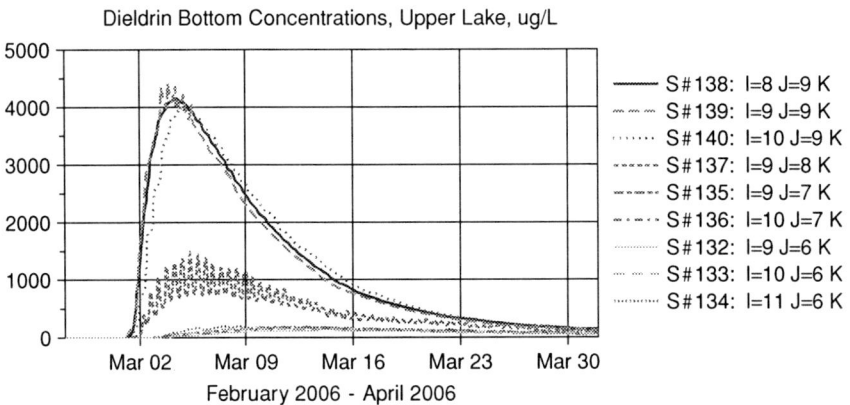

Legend:
- S#138: I=8 J=9 K
- S#139: I=9 J=9 K
- S#140: I=10 J=9 K
- S#137: I=9 J=8 K
- S#135: I=9 J=7 K
- S#136: I=10 J=7 K
- S#132: I=9 J=6 K
- S#133: I=10 J=6 K
- S#134: I=11 J=6 K

February 2006 - April 2006

Figure 35. Dieldrin Bottom Concentrations in Upper Lake

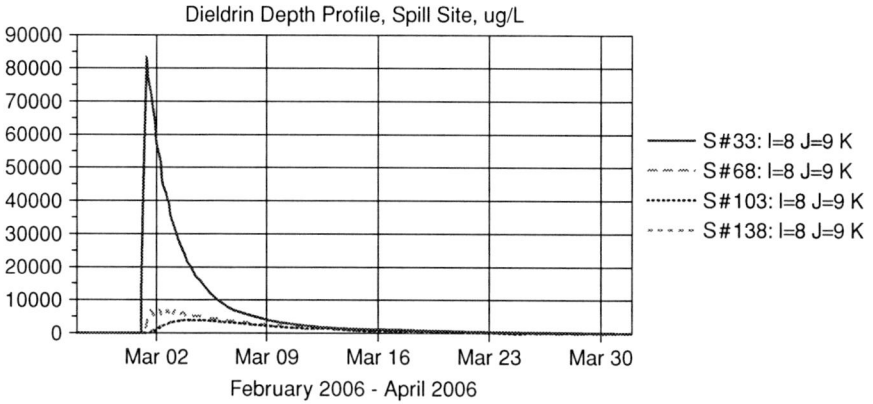

Figure 36. Dieldrin Depth Profile at Spill Site

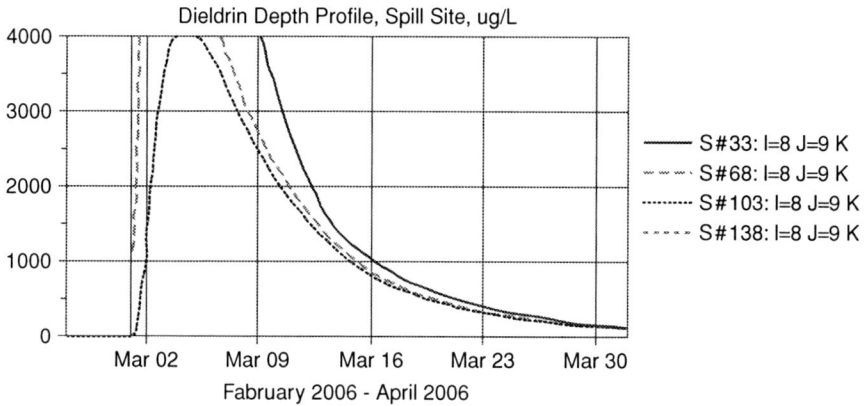

Figure 37. Detail of Dieldrin Depth Profile at Spill Site

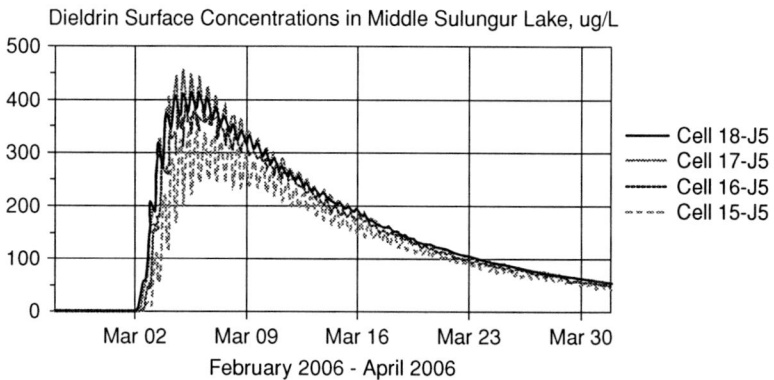

Figure 38. Dieldrin Spill Dynamics in Middle-Lower Lake

Dieldrin Concentration at Lagoon-Lake Boundary

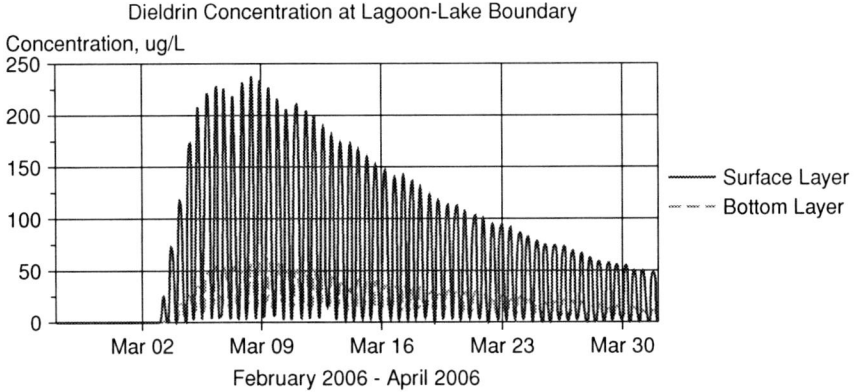

Figure 39. Dieldrin Dynamics at Lagoon-Lake Boundary

and by environmental properties, including wind speed and segment depth. During the simulation period, wind speed varied between 0 and 5 m/s. Surface segment depths vary between 0.3 and 2.5 m. Calculated volatilization loss constants varied between 0.01 and 0.035 day^{-1}, giving half lives between two weeks and two months (Figure 40). Photolysis is controlled by the surface rate constant, and by sunlight, segment depth, and light extinction, which is a function of solids, DOC, and chlorophyll a. The reported photolysis rate constant of 0.108 day^{-1} was assumed to reflect mid-summer surface conditions (e.g., 600 ly/day), giving a second-order surface rate constant of 0.00018 ly^{-1}. During the simulation period, sunlight varied between 350 and 420 ly/day. Solids, DOC, and chlorophyll a concentrations were assumed to be 2 mg/l, 1 mg/l, and 5 µg/l, respectively, giving a light extinction coefficient

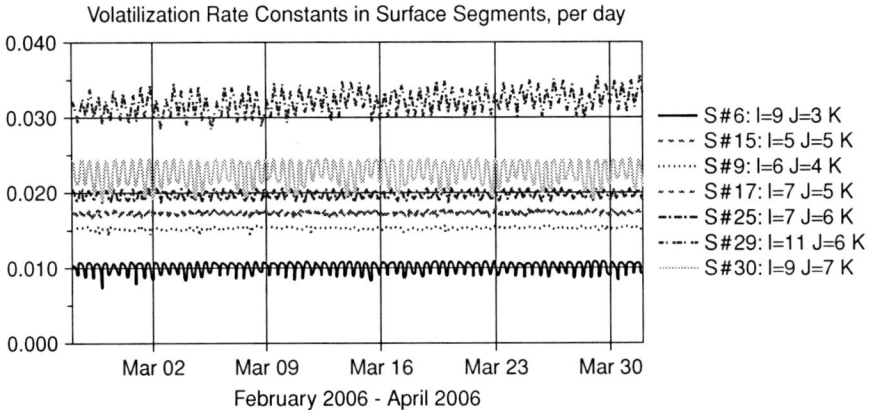

Figure 40. Volatilization Rate Constants in Surface Segments, per day

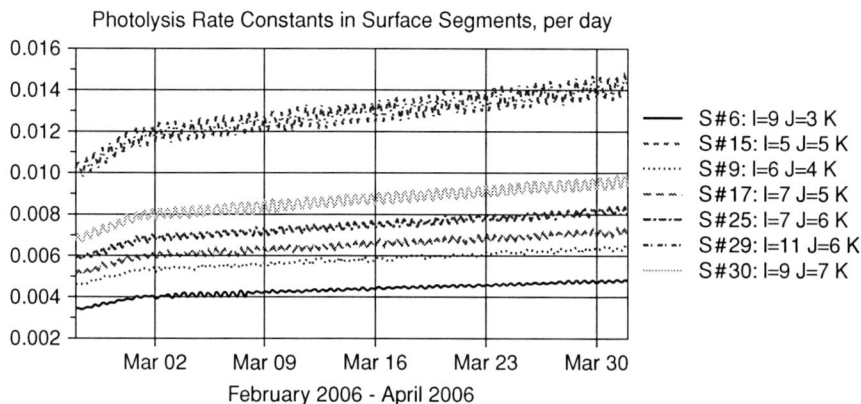

Figure 41. Photolysis Rate Constants in Surface Segments, per day

for 300 nm ultraviolet light of 7.4 m^{-1}. Calculated photolysis rate constants in surface segments varied between 0.004 and 0.014 day^{-1}, giving half lives between 7 weeks and 6 months (Figure 41).

To compare these results with the coarse-scale WASP simulation, concentrations were volumetrically-averaged over the surface and bottom layers in the upper lake reach and compared with results for segment 20 and 69 in the coarse scale network (Figure 42). The peak concentration in the coarse-scale network was 5000 μg/l, about a third less than the averaged peak for the fine scale network, which reached a value just over 7400 μg/l. Attenuation half-lives for the averaged concentrations in the fine-scale network are 2.5 days, a bit more than twice the half lives in the coarse scale network. Bottom concentrations in the fine scale network reach 1000 μg/l, about 10 times the bottom concentrations in the coarse scale network.

Results for representative cells in the middle lake reach are compared with coarse network segments in Figure 43. In mid-lake, the predicted surface peak concentration for the fine scale network is about half of that for the coarse scale network, and it arrives two days later. Attenuation half lives are much longer—9–10 days versus 2–3 days in the coarse network. The predicted bottom peak concentrations for the fine and coarse networks are similar, but the fine network peak arrives about 8 days later. Attenuation half lives are slightly longer in the medium scale network—13–14 days versus 10–11 days.

Results for representative cells in the lower lake reach are compared with coarse network segments in Figure 44. In the lower lake, the predicted surface and bottom peak concentrations for the fine scale network are similar to those in the coarse scale network. The fine scale peak arrives two days later in the surface, and 3–4 days later in the bottom. Attenuation half lives are longer. Surface attenuation half lives in the fine scale network are

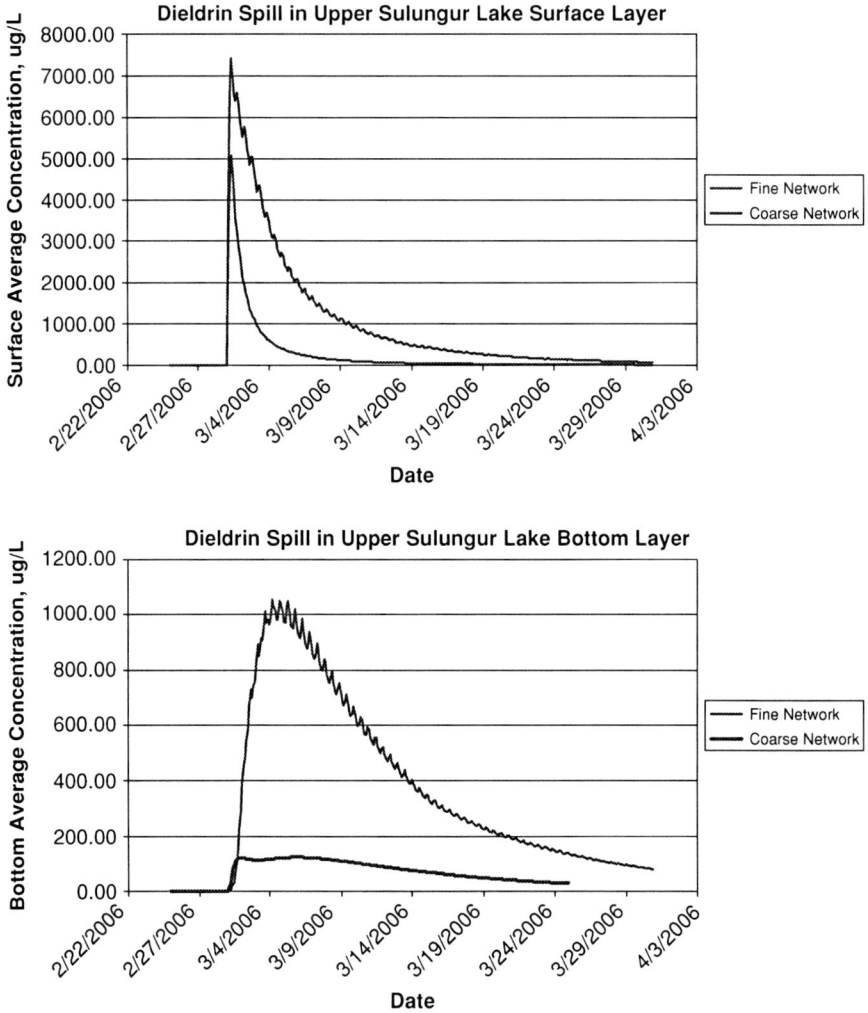

Figure 42. Averaged Dieldrin Dynamics in Upper Lake

10–11 days versus 4 days in the coarse network. Bottom attenuation half lives in the fine scale network are 13–14 days versus 10 days in the coarse network.

17.4.3. FINAL REMARKS

This case study exercise proved to be a useful focal point in training more than 50 students on advanced modeling techniques for rapidly assessing the effects of CBRN agents on water resources. This exercise enabled students to place

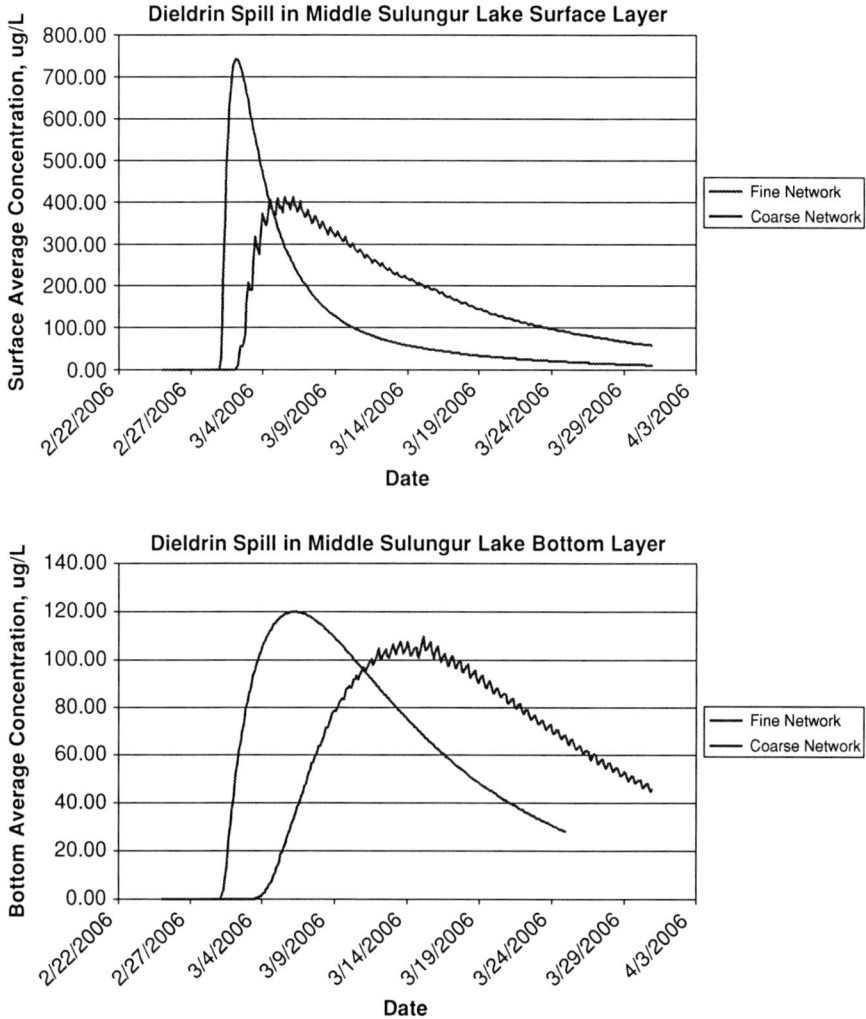

Figure 43. Averaged Dieldrin Dynamics in Middle Lake

their theoretical knowledge to work in a simulated real-world environment where time is an issue and effective teamwork combined with practical implementation become as important as theoretical rigor. In addition, this exercise demonstrated to the authors some practical shortcomings of present modeling technology when applied to highly visible, highly urgent problems requiring rapid, transparent, and defensible results.

Remarks on the modeling frameworks. This case study required the modeling of chemical transport, transformation, and ecological response to a pulse-loading event. Each of the modeling frameworks employed in this case

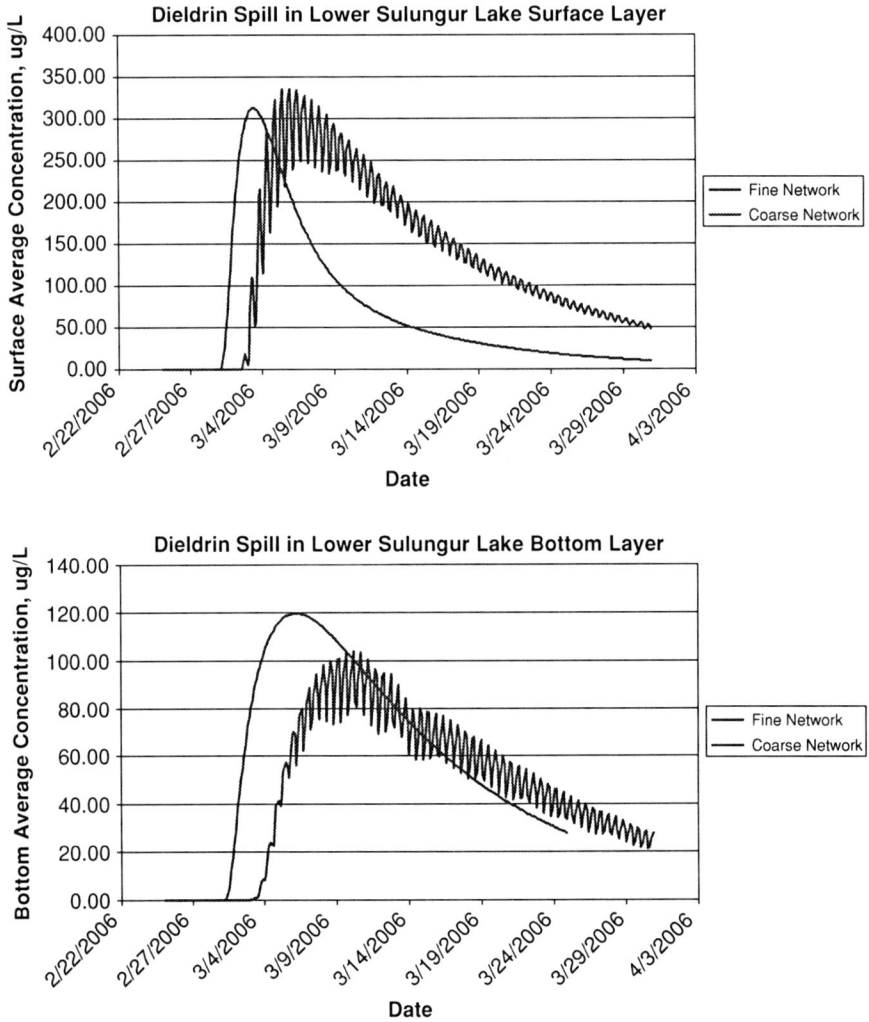

Figure 44. Averaged Dieldrin Dynamics in Lower Lake

study—WASP, AQUATOX, and EFDC—offered a different set of advantages and limitations. We found that the WASP7 modeling framework provided useful transport modeling capabilities along with toxic chemical modules that can effectively simulate the important physical and chemical processes controlling the fate of a spilled chemical. A separate eutrophication module handled the simulation of nutrients, algae, and detritus. These separate modules, however, do not simulate the ecological effects of spilled chemicals. AQUATOX, on the other hand, offered a very detailed theoretical construct for simulating ecological effects of toxic chemicals. The transport scheme, however, implemented a very simple well-mixed or two-layer water column with underlying benthic layers that could not be applied to the entire lagoon system. A multi-segment version of AQUATOX, under development at the time of this NATO-ASI course, should offer improved transport capabilities for more realistic applications. EFDC provided a more realistic three-dimensional transport simulation. Lack of support software, however, prevented the application of a curvilinear grid, which would be required to simulate the entire lagoon system. Very long run times were encountered even for the simulation of Sulungur Lake alone.

To address the entire simulation problem, we ran these three models separately, providing both manual and automatic input–output linkage. During the workshop, we concentrated on WASP and AQUATOX. A WASP computational network was already set up for the entire lagoon system, set up to investigate eutrophication dynamics. Though not optimized for chemical spill dynamics, this network proved useful in investigating the transport of dieldrin throughout the lagoon system. The calculated flows and concentrations at the Dalyan Lagoon—Sulungur Lake interface were extracted and used as boundary forcing for the Sulungur Lake AQUATOX application. AQUATOX was then used to simulate the long-term chemical and ecological response of Sulungur Lake as a whole.

To accomplish this WASP—AQUATOX linkage manually was difficult and time consuming. Even with the necessary simplifications, the linkage required more than 8 h of combined effort from three modeling experts. We conclude that a modeling framework that combines the chemical transport and fate capabilities of WASP with the chemical fate and ecological response capabilities of AQUATOX would be useful in real life emergency cases, when time is utmost importance. Such a modeling framework could be built with either external or internal linkages.

The external linkage option would require WASP to automatically generate an AQUATOX boundary input file with agreed-upon content and format, allowing the efficient transfer of information to drive a more detailed eco-dynamic simulation at a location within the WASP network. While this arrangement offers some advantages, there will be potential compatibility problems if the AQUATOX chemical fate equations and parameterization do not match those in the WASP application. This potential incompatibility could be minimized if the linkage file includes the WASP chemical fate parameters, which AQUATOX would use to drive its fate calculations.

The internal linkage option could be built within the WASP framework. This would require a new WASP water quality module that combines its existing toxicant module with its eutrophication module, with some extensions to higher level organisms but not as advanced at AQUATOX. Such a module would have been useful for this exercise and similar cases requiring rapid short and medium-range evaluations for emergency response. Evaluating the long-term ecodynamics throughout an entire lagoon system of complex shape would require the full capabilities of both WASP and AQUATOX. It is not clear whether this could best be accomplished through external or internal linkages.

Following the workshop, we investigated the linkage of WASP and EFDC to simulate the short-term chemical transport and transformation dynamics in Sulungur Lake. Using a higher resolution spatial network and a physically-based hydrodynamic model should produce more realistic short-term simulations of spill dynamics. While EFDC and WASP are separate models, they have been equipped for automated linkage. When instructed, EFDC generates a binary hydrodynamics result file, and WASP automatically uses the stored computational network, time step, and detailed advective and dispersive flows. To date, this linkage has been used in long-term modeling projects that employ experts in both models (Wool et al., 2003). Here we tested our ability to generate the EFDC—WASP linkage under simulated emergency conditions without an EFDC specialist available.

The development of an EFDC model network with relevant model inputs for the post mortem analysis of the terrorist attack took us many hours. This effort involved generating the network and basic model input files, finding and correcting errors in the input data set, running and analyzing preliminary simulations, and finally running the formal scenarios for linkage with WASP. While we required a significant amount of time to become familiar with EFDC, the setup time alone would prevent timely emergency response even for an experienced user.

Clearly the EFDC model needs additional support software, including a grid generator and a graphical user interface-based input data processor. A Windows-based preprocessor was available for earlier versions of EFDC, but model evolution has made that software out of date. Some of this software is under development and should be available. Analysis of EFDC simulation results was aided tremendously by the use of MOVEM, the general WASP postprocessor. Even with a complete set of pre- and post-processing software, however, EFDC would have to be set up for a water body ahead of time before it could be used in an emergency.

Computational time is also of great importance when dealing with emergency cases. EFDC is an extensive model with a large computational burden. Even if all the model input data is ready, a model run with one year simulation time can take from hours to days. In this study, we ran EFDC for 3 months (January 1–April 1) on a computational network of 37 reaches and 4 layers (148 cells) with a time step of 30 s. EFDC solved for flows, volumes, salinity, and dye tracer, requiring 4–5 h on a standard desktop computer, and 12–15 h on a laptop.

There are several methods to cope with this computational burden. If EFDC and WASP are used together, the simulations can be decoupled. First, flow fields are generated by EFDC and saved to a binary hydrodynamics file. This file is read by WASP during its chemical transport and fate simulations, saving the hydrodynamic simulation time for all scenarios with the same flow conditions. In this study, we ran WASP linked to an EFDC hydrodynamics file for just over 1 month (February 25–April 1). WASP used the EFDC network (minus two seaward reaches, totaling 140 segments) and time step (30 s). Solving for only dieldrin, each WASP simulation required just $2\frac{1}{2}$ min on a laptop. When scenarios involve differing physical forcing factors (e.g., meteorology, tide, inflow), additional hydrodynamic simulations must be generated for all combinations of physical forcing for use by WASP before any emergency case occurs.

Depending on the extent and complexity of the system, WASP simulations alone may take a long computational time. In this case, it would be useful to generate a WASP model network that uses a coarser spatial resolution than the underlying EFDC hydrodynamic network. The spill area could be modeled using a high spatial resolution, with decreasing spatial resolution farther from the spill site. In this case, utilities that convert an EFDC model network to such a WASP model network and remap all the flows accordingly would be useful. For such an option to be successful, potential numerical solution issues must be identified and resolved.

Remarks on preparation. For this training exercise, many details were organized ahead of time for the students. Model input files had been prepared for eutrophication in the Dalyan Lagoon system. These simulations were optimized with respect to the spatial network and time step for rapid simulations, which required only a few minutes for the period of interest. During a previous study of Dalyan Lagoon, it took one of the authors several hundred hours over several months to set up, optimize, and verify the model network and validate the transport parameters for WASP.

Given this case study preparation, the Hydrodynamics Team produced a report on conservative toxicant spill results for different scenarios after the first day of work. The Ecology Team then produced a report on chemical fate and ecological response results for different scenarios at the end of the second day of work. With these reports in hand, the Emergency Response Team generated recommendations for action during the third day of work. In total, the student teams required three days (and nights) to complete the recommendations for emergency response and remedial actions.

One obvious lesson is that for timely results in an emergency situation, the models should be set up and ready to run ahead of time. In a real emergency case, there will not be enough time to generate all model inputs. This is particularly true regarding the computational grid and associated transport parameters.

Even determining an appropriate set of chemical fate constants could be problematic during an emergency. Many thousands of chemicals with different properties could be introduced to aquatic ecosystems, whether deliberately or by accident. AQUATOX has an internal database, but this database is quite limited when compared to the number of toxic substances that could potentially be spilled to aquatic ecosystems. An extensive model support database integrating the chemical and ecotoxicological properties, as well as the methods for monitoring and analysis for these chemicals, should be developed and kept ready for use in the emergency situations. Support software should be developed to link this model database with external chemical databases, replicating information regularly. The internal structure of this database might usefully be compatible with Hazardous Materials Data Sheets.

Modelers should also be organized for efficient emergency response before an emergency case occurs. Maps of aquatic ecosystems under high risk of accidental spills and/or terrorist attack should be prepared for each country and organization expected to respond to emergency cases for these systems. Ideally, under emergency circumstances, the number of variables and uncertainties will be reduced for modeling and decision making, and only two basic variables will be left: identification of the chemicals that are spilled and characterization of their mass loading.

As a final note, modelers should be trained not only in basic scientific and engineering principles, but also in operational modeling techniques. Training courses like this NATO Advanced Study Institute could be organized in different regions around the world to share and enhance the existing knowledge on modeling the fate and ecosystem response of toxic spills under emergency conditions.

Disclaimer: This paper has been reviewed in accordance with the U.S. Environmental Protection Agency's peer and administrative review policies and approved for publication. Mention of trade names or commercial products does not constitute endorsement or recommendation for use.

Acknowledgements

This chapter reflects the combined efforts of many who participated in the NATO-CCMS workshop. The students worked long hours in their teams gaining a working knowledge of the models, then applying them and reporting results in their case study roles. Their energy and good humor will serve our profession well. We acknowledge the important contributions of our colleagues, Alpaslan Ekdal and Melike Gurel, who supported this case study by assisting in our preparation, then assisting the student teams with timely advice. Rosemarie Russo worked with the student team reports, portions of which are included in this chapter. Earl Hayter provided useful help in the post-workshop application of EFDC, and Craig Barber assisted in refining and interpreting the application of AQUATOX.

References

Ambrose RB, Wool TA, Connolly JP, Schanz RW (1988) WASP4, A Hydrodynamic and Water Quality Model–Model Theory, User's Manual, and Programmer's Guide. EPA/600/3-87-039, U.S. Environmental Protection Agency, Athens, GA

Burns LA, Cline DM, Lassiter RR (1982) Exposure Analysis Modeling System (EXAMS): User Manual and System Documentation. EPA/600/3-82/023, U.S. EPA, Athens, GA

Caldwell DK, Caldwell MC (1962) The black steer of the Gulf of California. Los Angeles County Museum Contributions to Science 61:1–31

Connolly JP, Winfield R (1984) A User's Guide for WASTOX, a Framework for Modeling the Fate of Toxic Chemicals in Aquatic Environments. Part 1: Exposure Concentration. EPA-600/3-84-077, U.S. Environmental Protection Agency, Gulf Breeze, FL

Di Toro DM, Fitzpatrick JJ, Thomann RV (1983) Water Quality Analysis Simulation Program (WASP) and Model Verification Program (MVP)—Documentation. Contract No. 68-01-3872, Hydroscience Inc., Westwood, NY, for U.S. EPA, Duluth, MN

Dodd Jr CK (1988) Disease and population declines in the flattened musk turtle *Sternotherus depressus*. American Midland Naturalist 119:394–401

Erturk A (2002) Hydraulic Modelling of the Koycegiz Dalyan Lagoon System. MSc Thesis, Istanbul Technical University, Institute of Science and Technology, Istanbul

Frazer NB (1983) Survivorship of adult female loggerhead sea turtles, *Caretta caretta*, nesting on Little Cumberland Island, Georgia, USA. Herpetologica 39:436–447

Frazer NB, Ehrhart LM 1985. Preliminary growth models for green, *Chelonia mydas*, and loggerhead, *Caretta caretta*, turtles in the wild. Copeia 1:73–79

Gurel M, Tanik A, Erturk A, Dogan E, Okus E, Seker DZ, Ekdal A, Yuceil K, Baykal BB, Gonenc IE (2005) Koycegiz-Dalyan lagoon: a case study for sustainable use and development. In: Gonenc IE, Wolflin JP (eds) Coastal lagoons: ecosystem processes and modeling for sustainable use and development, CRC Press, Boca Raton, pp. 440–474

Hamrick JM (1996) Users manual for the environmental fluid dynamics computer code. Special Rep. 331 in Applied Marine Science and Ocean Engineering, Virginia Institute of Marine Science, College of William and Mary, Virginia

Mendonca M (1981) Comparative growth rates of wild immature *Chelonia mydas* and *Caretta caretta* in Florida. Journal of Herpetology 15:447–451

Park RA, Clough JS, Wellman MC (2004) AQUATOX (Release 2) Volume 1: User's Manual. EPA-823-R-04-001, Office of Water, Washington, DC

Thomann RV, Fitzpatrick JJ (1982) Calibration and Verification of a Mathematical Model of the Eutrophication of the Potomac Estuary. Prepared for Dept. of Environmental Services, Government of the District of Columbia, Washington, DC

Uchida I (1967) On the growth of the loggerhead turtle, *Caretta caretta*, under rearing conditions. Bull. Jap. Soc. Scient. Fish. 33:497–506

U.S. Environmental Protection Agency. Ambient Water Quality Criteria for Aldrin/Dieldrin, Office of Water, Washington DC

Wool TA, Ambrose Jr RB, Martin JL, Comer EA (2001) The Water Quality Analysis Simulation Program, WASP6; Part A: Model Documentation. U.S. Environmental Protection Agency, Center for Exposure Assessment Modeling, Athens, GA

Wool TA, Davie SR, Rodriguez HN (2003) Development of Three Dimensional Hydrodynamic and Water Quality Model to Support Total Maximum Daily Load Decision Process for the Neuse River Estuary, North Carolina. American Society of Civil Engineers, Journal of Water Resources Planning and Management 129(4)

Zug GR, Wynn A, Ruckdeschel C (1983) Age estimates of Cumberland Island loggerhead sea turtles.

APPENDIX 1.   THE WASP7 SURFACE WATER BODY MODEL

1. Introduction: The Water Quality Analysis Simulation Program, WASP7 (Di Toro et al., 1983; Ambrose et al., 1988; Wool et al., 2001) is a general dynamic mass balance framework for modeling contaminant fate and transport in surface waters. Based on the flexible compartment modeling approach, WASP can be applied in one, two, or three dimensions with advective and dispersive transport between discrete physical compartments, or "segments." WASP provides a selection of modules to allow the simulation of conventional water quality variables as well as toxicants.

WASP has a long history of application to a variety of water bodies for a variety of water quality problems. Earlier versions of WASP were used to examine eutrophication and PCB pollution of the Great Lakes (Thomann, 1975; Thomann et al., 1976; Thomann et al., 1979; Di Toro and Connolly, 1980), eutrophication of the Potomac Estuary (Thomann and Fitzpatrick, 1982), kepone pollution of the James River Estuary (O'Connor et al., 1983), heavy metal pollution of the Deep River, North Carolina (JRB, 1984), and volatile organic pollution of the Delaware River Estuary (Ambrose, 1987). In addition to these, numerous applications are listed in Di Toro et al. (1983).

Published applications of more recent versions of WASP include eutrophication and mixing in Prince William Sound embayments (Lung et al., 1993), eutrophication in the inner shelf of the Gulf of Mexico (Bierman et al., 1994), eutrophication in the Mississippi River and Lake Pepin (Lung and Larson, 1995), water quality of the Speed River (Gualtieri and Rotono, 1996a, 1996b), metam spill in the Sacramento River (Wang et al., 1997), pollutant loading for the Black and Chehalis Rivers in Washington (Pickett, 1997), hydrodynamics and water quality in a large South Carolina reservoir (Tufford and McKellar, 1999; Tufford et al., 1999), eutrophication in the Neuse River Estuary (Wool et al., 2003), eutrophication of Tampa Bay (Wang et al., 1999), metals in Upper Tenmile Creek, Montana (Caruso, 2003, 2004, and 2005), and mercury in a south Georgia (USA) river basin (Ambrose et al., 2005).

WASP is designed to permit substitution of different water quality kinetics code into the program structure to form different water quality modules. Two classes of modules are provided with WASP. The toxicant WASP modules combine a kinetic structure initially adapted from EXAMS (Burns et al., 1982) with the WASP transport structure and simple sediment balance algorithms to predict dissolved and sorbed chemical concentrations in the water and underlying sediment bed. The eutrophication WASP module combines a kinetic structure initially adapted from the Potomac Eutrophication Model (Thomann and Fitzpatrick, 1982) with the WASP transport structure to predict

387

I. E. Gonenc et al. (eds.), Assessment of the Fate and Effects of Toxic Agents on Water Resources, 387–394.
© 2007 *Springer.*

nutrients, phytoplankton, periphyton, organic matter, and dissolved oxygen dynamics.

WASP7 includes a Windows-based interface for constructing input datasets and managing simulations. Data can be copied and pasted from spreadsheets. A Windows-based post-processor allows the user to plot or animate model output. Output is also provided as comma-delimited files for import to spreadsheets.

WASP is available free of charge and is supported by the U.S. Environmental Protection Agency. The current version of WASP, along with documentation and training material can be accessed and downloaded from the Watershed and Water Quality Modeling Technical Support Center, at http://www.epa.gov/athens/wwqtsc.

2. *WASP Transport Options:* A body of water is represented in WASP as a series of discrete computational elements or segments. Environmental properties and chemical concentrations may vary spatially among the segments. Each variable is advected and dispersed among water segments, and exchanged with surficial benthic segments by diffusive mixing. Sorbed or particulate fractions may settle through water column segments and deposit to or erode from surficial benthic segments. Within the bed, dissolved variables may migrate downward or upward through percolation and pore-water diffusion. Sorbed variables may migrate downward or upward through net sedimentation or erosion.

Advective water column flows directly control the transport of dissolved and particulate pollutants in many water bodies. In WASP7, four water column flow options are available for selection in the Data Input screen. Circulation patterns may be specified in the input dataset (Net Flow or Gross Flow options), simulated internally using the kinematic wave equation (1D Network Kinematic Wave option, for streams and rivers only), or simulated by an external hydrodynamic model, such as DYNHYD or EFDC (Hydrodynamic Linkage option).

Specified Flows: For specified flows, the user must supply a *continuity function* and a *time function* for each inflow to the network. WASP7 tracks each separate inflow from its point of origin through the model network to its point of exit. The actual flow between segments that results from the inflow is the product of the time function and the continuity function.

The *continuity function* describes the unit flow response through the network from origin to exit. The *time function* describes the inflow as it varies in time. During the simulation, WASP7 interpolates between points in the time function to obtain flows for each time step. If several inflow functions are specified, then the total flow between segments is the sum of the individual flow functions. Segment volumes are adjusted to maintain continuity. In this manner, the effect of several tributaries, density currents, and wind-induced gyres can be described.

Two specified flow options are available—Net Flow and Gross Flow. In the Net Flow Option, WASP7 sums all the flows crossing each segment interface to determine their net direction and strength, and then moves mass in the *ONE* direction with the net flow. In Gross Flow Option, WASP7 moves mass with each separate flow, independently of net flow. For example, if opposite flows are specified at a segment interface, WASP7 will move mass in *BOTH* directions. This option allows the user to describe large dispersive circulation patterns.

For unsteady flow in long networks, lag times may become significant, and hydrodynamic simulations may be necessary to obtain sufficient accuracy. Internal kinematic wave flow routing and external hydrodynamic model linkage are available as options.

Internal Kinematic Wave Flow Routing: For one-dimensional, branching streams or rivers, kinematic wave flow routing is a simple but realistic option to drive advective transport. The kinematic wave equation calculates flow wave propagation and resulting variations in flows, volumes, depths, and velocities throughout a stream network. This well-known equation is a solution of the one-dimensional continuity equation and a simplified form of the momentum equation that considers the effects of gravity and friction. WASP7 solves this kinematic wave equation for each segment in a stream network using a 4-step Runga–Kutta numerical technique.

External Hydrodynamic Linkage: Realistic simulations of unsteady transport in rivers, reservoirs, and estuaries can be accomplished by linking WASP7 to a compatible hydrodynamic simulation. This linkage is accomplished through an external "hyd" file chosen by the user at simulation time. The hydrodynamic file contains segment volumes at the beginning of each time step, and average segment interfacial flows during each time step. WASP7 uses the interfacial flows to calculate mass transport, and the volumes to calculate constituent concentrations. Segment depths, velocities, and salinity concentrations are also contained in the hydrodynamic file.

Before using hydrodynamic linkage files with WASP, a compatible hydrodynamic model must be set up for the water body and run successfully, creating a hydrodynamic linkage file with the extension of *.hyd. This is an important step in the development of the WASP input file because the hydrodynamic linkage file contains all necessary network and flow information for the WASP simulation.

WASP has the ability to get hydrodynamic information from different hydrodynamic models. The hydrodynamic models that currently support the WASP7 file format are: EFDC (three dimensions), DYNHYD (one dimension branching), RIVMOD (one dimension no branching), CE-QUAL-RIV1 (one dimension branching), SWMM/Transport (one dimension branching, SWMM/Extran (one dimension branching).

The first step in the hydrodynamic linkage is to develop a hydrodynamic calculational network that is compatible with the WASP7 network. Each WASP segment corresponds exactly to a hydrodynamic volume element, or node. Each WASP segment interface corresponds exactly to a hydrodynamic link, or volume element interface. The hydrodynamic model network must have additional nodes outside of the WASP network corresponding to WASP boundaries. These extra hydrodynamic nodes are necessary to provide WASP boundary flows from outside its network.

3. WASP Toxicant Modules: The primary WASP toxicant code, TOXI, is combined with separate databases to support four modules—simple toxicant, non-ionizing toxicants, organic toxicants, and mercury. TOXI modules simulate the transport and transformation of one to three chemicals and one to three types of particulate material. The three chemicals may be independent, such as isomers of PCB, or they may be linked with reaction yields, such as a parent compound-daughter product sequence. Each chemical exists as a neutral compound and up to four ionic species. The neutral and ionic species can exist in five phases: dissolved, sorbed to dissolved organic carbon (DOC), and sorbed to each of the up to three types of solids. Local equilibrium is assumed so that the distribution of the chemical between each of the species and phases is defined by distribution or partition coefficients. Several kinetic transformation processes affect the fate of organic chemicals in the aquatic environment. WASP includes algorithms for volatilization, biodegradation, hydrolysis, photolysis, oxidation and reduction. A choice of simple first-order kinetics or more mechanistic second-order kinetics is available to the user.

Properties of the three TOXI solids variables—inorganic silt, sand, and biotic solids—are defined by the user. Some applications use a single variable for total solids, whereas others use all three solids variables. The solids types are operationally defined by the specified properties, including particle density, and spatially and temporally-variable settling and re-suspension velocities. Biotic solids are further defined by spatially and temporally-variable, temperature-corrected production rates, spatially variable, temperature-corrected dissolution rate constants, and by fraction ash-free dry weight. Net burial rates are calculated internally from solids mass balances, assuming constant bulk density for each sediment layer.

The WASP mercury module simulates elemental mercury, Hg^0, inorganic divalent mercury, Hg(II), and monomethyl mercury, MeHg. Mercury species are subject to several transformation reactions, including oxidation of Hg^0 in the water column, reduction and methylation of Hg(II) in the water column and sediment layers, and demethylation of MeHg in the water column and sediment layers. These transformation processes are represented as first-order reactions operating on the total pool of the reactants (i.e., no difference in reactivity of recently deposited Hg(II) and that deposited weeks to years earlier) with rate constants that can vary spatially and temperature correction coefficients that adjust the rates with variations in water temperature. Water column reduction and demethylation reactions are driven by sunlight, and so their input surface rate constants are attenuated through the water column using specified light extinction coefficients. Hg^0 is subject to volatile exchange between the water column the atmosphere governed by a transfer rate calculated from velocity and depth, and by its Henry's Law constant. Rate constants can be applied to the dissolved, DOC-complexed, and solids-sorbed phases at varying strengths (0–1), as specified by the user.

The environmental and chemical database required for TOXI simulations is standard for chemical-specific models. Those data required for a particular simulation depend upon which transformation processes are important. Up to 20 spatially-variable environmental parameters, such as pH and light extinction, may be specified as needed. The chemical properties of the toxicant control what transformation processes are important in a particular environment. Molecular weight and solubility should always be specified. Other influential chemical properties should be specified as needed. Up to 19 time-variable functions can be used to study diurnal or seasonal effects on pollutant behavior. Some of these time functions are multiplied by spatially variable parameters within TOXI to produce time- and spatially-variable environmental conditions. If no time variability is required, the time functions may be omitted.

Although the amount and variety of data potentially used by the TOXI modules is large, data requirements for any particular simulation can be quite small. Usually only sorption and one or two transformation processes will significantly affect a particular chemical. To simulate the transport of many soluble compounds in the water column, even sorption can often be disregarded. Indeed, for empirical studies, all chemical constants, time functions, and environmental parameters can be ignored except a user-specified first order transformation rate constant and the partition coefficient.

TOXI provides calculated sediment and chemical concentrations for every segment at each print interval. Chemical concentrations are reported for the dissolved, sediment sorbed, and DOC sorbed phases, and as neutral and ionic concentrations. In addition, calculated transformation and transfer rates are reported.

4. WASP Eutrophication Module: The WASP eutrophication code, EUTRO, simulates the transport and transformation of up to 16 state variables, including dissolved oxygen, 3 reactive classes of carbonaceous biochemical oxygen demand, phytoplankton, periphyton, detrital carbon, nitrogen, and phosphorus, ammonia, nitrate, dissolved organic nitrogen, dissolved organic phosphorus, orthophosphate, inorganic solids, and salinity. EUTRO can be used to simulate any or all of these variables and the interactions between them. Each variable may exist in both dissolved and particulate phases, as specified by the user for each segment. Solids transport fields describing settling and resuspension of phytoplankton, detritus, and inorganic solids may be specified. The particulate concentration of each variable will be transported by the appropriate solids field.

By bypassing variables and computations, EUTRO can be operated at various levels of complexity. The simplest level, equivalent to an enhanced "Streeter-Phelps" equation, includes ultimate BOD, DO, and sediment oxygen demand (SOD). The reaeration rate constant is specified. The next level divides BOD into carbonaceous (CBOD) and nitrogenous (NBOD) fractions. Level 3 is a linear DO balance influenced by photosynthesis and respiration of user-specified phytoplankton and nitrification of ammonia to nitrate, as well as CBOD and SOD. The reaeration rate is computed from velocity, depth, and wind speed. Level 4 adds the phosphorus cycle and simulates phytoplankton dynamics subject to Michaelis–Menten nutrient limitation and light limitation. Level 5 adds periphyton and/or additional nonlinear kinetics.

The environmental and chemical database for EUTRO is standard for DO and eutrophication simulations. Those data required for a particular simulation depend upon which level of complexity is chosen. Up to 23 spatially variable environmental parameters and 100 rate constants in EUTRO may be specified as needed. Up to 22 time-variable functions in EUTRO can be used to study diurnal or seasonal effects on pollutant behavior. Some of these time functions are multiplied by spatially variable parameters within EUTRO to produce time- and spatially-variable environmental conditions. If no time variability is required, the time functions may be omitted.

EUTRO output reports state variable concentrations, along with key forcing functions and process rates for every segment at user-specified print intervals. Forcing functions, parameters, and rates include depth, temperature, wind speed, water velocity, reaeration rate, oxygen saturation, sediment oxygen demand, deoxygenation rate, phytoplankton growth and death rates, phytoplankton DO production and consumption, light and nutrient growth limitation factors, ambient light conditions, saturating light intensity, and carbon-chlorophyll ratio.

References for Appendix 1

Ambrose, R.B., 1987. Modeling volatile organics in the Delaware Estuary, Am. Soc. Civil Eng. J. Environ. Eng., 113, 703–721.

Ambrose, R.B., T.A. Wool, J.P. Connolly, and R.W. Schanz, 1988. WASP4, A Hydrodynamic and Water Quality Model–Model Theory, User's Manual, and Programmer's Guide. EPA/600/3-87-039, U.S. Environmental Protection Agency, Athens, GA.

Bierman, Jr., V.J., S.C. Hinz, D.W. Zhu, W.J. Wiseman, Jr., N.N. Rabalais, and R.E. Turner, 1994. A preliminary mass balance model of primary productivity and dissolved oxygen in the Mississippi River Plume/Inner Gulf Shelf Region, Estuaries, 17 (4), 886–899.

Burns, L.A., D.M. Cline, and R.R. Lassiter, 1982. Exposure Analysis Modeling System (EX-AMS): User Manual and System Documentation. EPA/600/3-82/023, U.S. EPA, Athens, GA.

Caruso, B.S., 2003. Water quality simulation for planning restoration of a mined watershed. Water, Air and Soil Pollution, 150(1-4), 221–234.

Caruso, B.S., 2004. Modeling metals transport and sediment/water interactions in a mining impacted mountain stream. Journal of the American Water Resources Association, 40 (6), 1603–1615.

Caruso, B.S., 2005. Simulation of metals total maximum daily loads and remediation in a mining-impacted stream. ASCE Journal of Environmental Engineering, 131 (5), 777–789.

Connolly, J.P., and R. Winfield, 1984. A User's Guide for WASTOX, a Framework for Modeling the Fate of Toxic Chemicals in Aquatic Environments. Part 1: Exposure Concentration. EPA-600/3-84-077, U.S. Environmental Protection Agency, Gulf Breeze, FL.

Di Toro, D.M., and J.P. Connolly, 1980. Mathematical Models of Water Quality in Large Lakes, Part 2: Lake Erie. EPA-600/3-80-065, pp. 90–101.

Di Toro, D.M., and W.F. Matystik, 1980. Mathematical Models of Water Quality in Large Lakes, Part 1: Lake Huron and Saginaw Bay. EPA-600/3-80-056, pp. 28–30.

Di Toro, D.M., J.J. Fitzpatrick, and R.V. Thomann, 1983. Water Quality Analysis Simulation Program (WASP) and Model Verification Program (MVP) – Documentation. Contract No. 68-01-3872, Hydroscience Inc., Westwood, NY, for U.S. EPA, Duluth, MN.

Gualtieri, C., and G. Rotondo, 1996a. Water Quality Modeling of Speed River. Part One: Model Calibration (In Italian). Ingegneria Sanitaria 2:March/April.

Gualtieri, C., and G. Rotondo, 1996b. Water Quality Modeling of Speed River. Part Two: Model Validation and Application (In Italian). Ingegneria Sanitaria 5/6:November/December.

Gurel, M., 2000. Nutrient dynamics in coastal lagoons: Dalyan Lagoon case study, Ph.D. Thesis, Istanbul Technical University, Institute of Science and Technology, Department of Environmental Engineering, Istanbul, Turkey.

Gurel, M., A. Tanik, A. Erturk, E. Dogan, E. Okus, D.Z. Seker, A. Ekdal, K. Yuceil, B.B. Baykal, and I.E. Gonenc, 2005. Koycegiz-Dalyan lagoon: A case study for sustainable use and development, edited by I.E. Gonenc, and J.P. Wolflin, Coastal Lagoons: Ecosystem Processes and Modeling for Sustainable Use and Development, CRC Press, Boca Raton, pp. 440–474.

Hamrick, J.M., 1996. Users manual for the environmental fluid dynamics computer code, Special Rep. 331 in Applied Marine Science and Ocean Engineering, Virginia Institute of Marine Science, College of William and Mary, Virginia.

JRB Inc., 1984. Development of Heavy Metal Waste Load Allocations for the Deep River, North Carolina. JRB Associates, McLean, VA, for U.S. EPA Office of Water Enforcement and Permits, Washington, DC.

Lung, W.S., J.L. Martin, and S.C. McCutcheon, 1993. Eutrophication and mixing analysis of embayments in Prince William sound, Alaska, ASCE J. Environ. Eng., 119 (5), 811–824.

Lung, W.S., and C.E. Larson, 1995. Water Quality Modeling of the Upper Mississippi River and Lake Pepin. ASCE J. Environ. Eng., 121 (10), 691–699.

O'Connor, D.J., J.A. Mueller, and K.J. Farley, 1983. Distribution of Kepone in the James River Estuary, J. Environ. Eng. Div., ASCE, 109 (EE2), 396–413.

Pickett, P.J., 1997. Pollutant Loading Capacity for the Black River, Chehalis River System, Washington, JAWRA, 33 (2), 465–480.

Thomann, R.V., 1975. Mathematical Modeling of Phytoplankton in Lake Ontario, 1. Model Development and Verification. EPA-600/3-75-005, U.S. Environmental Protection Agency, Corvallis, OR.

Thomann, R.V., R.P. Winfield, D.M. Di Toro, and D.J. O'Connor, 1976. Mathematical Modeling of Phytoplankton in Lake Ontario, 2. Simulations Using LAKE 1 Model. EPA-600/3-76-065, U.S. Environmental Protection Agency, Grosse Ile, MI.

Thomann, R.V., R.P. Winfield, and J.J. Segna, 1979. Verification Analysis of Lake Ontario and Rochester Embayment Three Dimensional Eutrophication Models. EPA-600/3-79-094, U.S. Environmental Protection Agency, Grosse Ile, MI.

Thomann, R.V., and J.J. Fitzpatrick, 1982. Calibration and Verification of a Mathematical Model of the Eutrophication of the Potomac Estuary, Prepared for Department of Environmental Services, Government of the District of Columbia, Washington, DC.

Tufford, D.L., and H.N. McKellar, 1999. Spatial and temporal hydrodynamic and water quality models of a large reservoir on the South Carolina (USA) coastal plain, Ecol. Modelling, 114 (2/3), 137–173.

Tufford, D.L., H.N. McKellar, J.R.V. Flora, and M.E. Meadows, 1999. A reservoir model for use in regional water resources management, Lake Reservoir Manage., 15 (3), 220–230.

Wang, P.F., J. Martin, and G. Morrison, 1999. Water quality and eutrophication in Tampa Bay, Florida, Estuarine Coastal Shelf Sci., 49, 1–20.

Wool, T.A., R.B. Ambrose, Jr., J.L. Martin, and E.A. Comer, 2001. The Water Quality Analysis Simulation Program, WASP6; Part A: Model Documentation. U.S. Environmental Protection Agency, Center for Exposure Assessment Modeling, Athens, GA.

Wool, T.A., R.B. Ambrose, Jr., and J.L. Martin, 2001. Water quality analysis simulation program (WASP) Version 6.0., United States Environmental Protection Agency, Region 4, Atlanta, GA.

Wool, T.A., S.R. Davie, and H.N. Rodriguez, 2003. Development of three dimensional hydrodynamic and water quality model to support total maximum daily load decision process for the Neuse River Estuary, North Carolina. American Society of Civil Engineers, J. Water Resources Planning Manage., 129 (4), 295–306.

APPENDIX 2

1. *Dieldrin properties:*

Dieldrin [CAS 60-57-1] $C_{12}H_8Cl_6O$

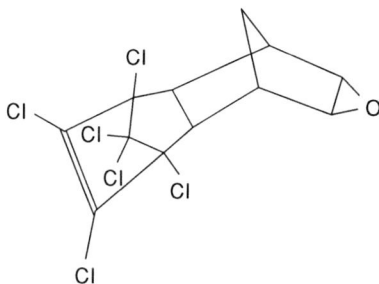

Molecular weight: 380.9126 g.

Melting point: 176 °C.

Boiling point: 385 °C (decomposition).

Water solubility: Insoluble. 0.195 mg/l (25 °C).

K_{ow}: 5.40.

Henry's law constant: 1E-005 atm m^3/mole.

Temp: 25 Colorless to light tan solid with a mild, chemical odor.

Organochlorine compound. Used as an insecticide used in particular to protect cotton plants.

Exposure Limits

OSHA GENERAL INDUSTRY PEL: 0.25 mg/m^3 (Skin).

OSHA CONSTRUCTION INDUSTRY PEL: 0.25 mg/m^{3}TWA (Skin).

ACGIH TLV: 0.25 mg/m^3 TWA (Skin); Appendix A4 (Not Classifiable as a Human Carcinogen).

NIOSH REL: 0.25 mg/m^3 TWA (Skin), Potential Carcinogen.

Health Factors—IARC: Group 3, not classifiable as to its carcinogenicity to humans SYMPTOM(s): Headaches; dizziness; nausea, vomiting, malaise; sweating; myoclonic limbjerks; clonic, tonic convulsions; coma; (carcinogenic) In animals: liver, kidney damage.

HEALTH EFFECTS: Cumulative liver damage (HE3) Suspect teratogen (HE5); LD50 (oral, rat) 46 mg/kg.

ORGAN: CNS, liver, kidneys, skin.

Monitoring—PRIMARY SAMPLING/ANALYTICAL METHOD (SLC1)— MEDIA: Glass Fiber Filter (37 mm).

ANL SOLVENT: Isooctane.

MAX V: 180 l MAX F: 1.5 l/min.

I. E. Gonenc et al. (eds.), Assessment of the Fate and Effects of Toxic Agents on Water Resources, 395–398.
© 2007 *Springer.*

ANL 1: Gas Chromatography; GC/ECD.
REF: 1 (NIOSH S-283).
SAE: 0.14.
CLASS: Fully Validated.
NOTE: Within 1 h after the sample has been collected, transfer the filter to
    a clean screw cap vial.

2. Toxicity

Humans	LD50 64 mg/kg (estimated)	acc. MERCIER, 1981
Mammals		
Rat	LD50 46–63 mg/kg, oral	acc. VERSCHUEREN, 1983
	LD50 52–117 mg/kg, dermal	acc. VERSCHUEREN, 1983
Mouse	LD50 38–77 mg/kg, oral	acc. MERCIER, 1981
Dog	LD50 56–120 mg/kg, oral	acc. MERCIER, 1981
Rabbit	LD50 45–50 mg/kg, oral	acc. MERCIER, 1981
Cow	LD50 25 mg/kg, oral	acc. MERCIER, 1981
Aquatic organisms		
Cyprinodont	LC50 5 ppb (96 h)	acc. VERSCHUEREN, 1983
Mugilidae	LC50 23 ppb (96 h)	acc. VERSCHUEREN, 1983
American minnow	LC50 16 mg/l (96 h)	acc. VERSCHUEREN, 1983
Blue perch	LC50 8 mg/l (96 h)	acc. VERSCHUEREN, 1983
Rainbow trout	LC50 10 mg/l (96 h)	acc. VERSCHUEREN, 1983
Water flea	LC50 250 mg/l (48 h)	acc. VERSCHUEREN, 1983
Crawfish	LC50 460 mg/l (96 h)	acc. VERSCHUEREN, 1983
Insects		
Pteronarcys California	LC50 0.5–39 µg/l (96 h)	acc. VERSCHUEREN, 1983

Characteristic effects
Humans/mammals: Dieldrin can cause poisoning following resorption via the skin, oral intake or inhalation. It acts as a stimulant to the central nervous system and accumulates in fatty tissue causing severe damage to the liver and kidneys. Animal experiments have revealed a carcinogenic effect, but as yet no teratogenic action.
Plants: Dieldrin does not have a toxic effect on plants (MERCIER, 1981).

Environmental Behavior
Water: Accumulation takes place in aqueous systems on account of the good solubility of the substance. Dieldrin is assigned in Germany to water hazard class 3 (highly water-polluting) as a result of its high toxicity to aquatic organisms.

Soil: Dieldrin accumulates in soils depending on their texture and water content.

Half-life: Roughly 95% of an applied quantity of between 3.1 and 5.6 kg/ha disappears from the soil after 12.8 years on average. Only some 9% evaporates out of loamy or sandy soils in 60 days. Between 75 and 100% of dieldrin is degraded or decomposed in 3–25 years (Verschueren, 1983).

Environmental behavior: Degradation UV light causes decomposition to form CO_2.

Food chain: Dieldrin accumulates in fatty tissue and in the mammary glands of human (WIRTH, 1981).

Environmental Impact: Dieldrin has been used extensively in the past as an insecticide for corn and for termite control, although it is no longer registered for general use. Dieldrin is extremely persistent, but it is known to slowly photorearrange to photodieldrin (water half-life—4 months). Dieldrin released to soil will persist for long periods (>7 year), will reach the air either through slow evaporation or adsorption on dust particles, will not leach, and will reach surface water with surface runoff. Once dieldrin reaches surface waters it will adsorb strongly to sediments, bioconcentrate in fish and slowly photodegrade. Biodegradation and hydrolysis are unimportant fate processes. The fate of dieldrin in the atmosphere is unknown but monitoring data have demonstrated that it can be carried long distances. Monitoring data demonstrate that dieldrin continues to be a contaminant in air, water, sediment, soil, fish, and other aquatic organisms, wildlife, foods, and humans. Human exposure appears to come mostly from food.

Assessment/comments: Dieldrin is a highly toxic substance towards aquatic organisms and is highly persistent in the environment. In addition, it accumulates in fatty tissue and can also cause severe poisoning in humans. Therefore, its use should be restricted as far as possible.

Environmental Standards

Medium/ acceptor	Sector	Country or organization	Status	Value	Category	Remarks	Source
Water	Drinkw	D	L	0.1 µg/l		Single substance	acc. RIPPEN, 1992
	Drinkw	D	L	0.5 µg/l		Sum of pesticides	acc. RIPPEN, 1992

Medium/ acceptor	Sector	Country or organization	Status	Value	Category	Remarks	Source
	Drinkw	EC	L	0.1 µg/l		single substance	acc. RIPPEN, 1992
	Drinkw	EC	L	0.5 µg/l		Sum of pesticides	acc. RIPPEN, 1992
	Drinkw	USA	G	1 µg/l		In state of Illinois	acc. WAITE, 1984
Soil		NL	G	0.5 µg/kg		Single substance, reference	acc. TERRA TECH 6/94
		NL	L	2.5 µg/kg		Aldrin+ dieldrin+ endrin intervention	acc. TERRA TECH 6/94
Air	Workp	D	L	0.25 mg/m^3	MAK	Skin	DFG, 1989
	Workp	SU	(L)	0.01 mg/m^3		Skin	acc. KETTNER, 1979
	Workp	USA	(L)	0.25 mg/m^3	TWA	Skin	ACGIH, 1986

References

http://chemfinder.cambridgesoft.com/result.asp.

U.S. Department of Labor Occupational Safety & Health Administration http://www.osha. gov/dts/chemicalsampling/data/CH_234600.html, http://www.pesticideinfo.org/Detail_ Chemical.jsp?Rec_Id=PC33416#Water http://www.speclab.com/compound/c60571.htm.

EPA, 1992. National Recommended Water Quality Criteria. Federal Register 57-60848 (December 22, 1992).

Jones, D.S., G.W. Suter, and R.N. Hull, 1997. Toxicological Benchmarks for Screening Contaminants of Potential Concern for Effects on Sediment-Associated Biota: 1997 Revision. Oak Ridge National Laboratory, ES/ER/TM-95/R4, Oak Ridge National Laboratory, Oak Ridge, TN.

Verschueren, K., 1983. Handbook of Environmental Data on Organic Chemicals, 2nd edn, New York.

LIST OF PARTICIPANTS

İ. Ethem Gonenc
IGEM Research and Consulting
Orman Kent Sitesi 1 Villa No. 713
Sırapınar Koyu, Istanbul

John Wolflin
US Fish and Wildlife Service
Chesapeake Bay Office
Annapolis
MD 21617, USA

Karen Terwilliger
Terwilliger Consulting, Inc.
28295 Burton Shore Rd.
Locustville
VA 23404, USA

Rosemarie C. Russo
Ecosystems Research Division
U.S.EPA
960 College Station Road
Athens
GA 30605-2700, USA

Irina P. Chuabrenko
Atlantic Branch of P.P. Shirshov
   Institute of Oceanology
Russian Academy of Sciences
Prospect Mira, 1,
Kaliningrad, 236000
Russian Federation

Georg Umgiesser
ISMAR-CNR, Istituto di Scienze
   Marine
S. Polo 1364
30125 Venezia, Italy

Boris V. Chubarenko
Atlantic Branch of P.P. Shirshov
   Institute of Oceanology
Russian Academy of Sciences
Prospect Mira, 1,
Kaliningrad, 236000
Russian Federation

Vladimir G. Koutitonsky
Institut des Sciences de la Mer de
   Rimouski (ISMER)
310 Allee des Ursulines
Rimouski, Quebec
G5L-3A1, Canada

Brenda Rashleigh
USEPA National Exposure Research
   Laboratory Ecosystems Research
Division 960 College Station Road
Athens
GA 30605–2700, USA

Angheluta Vadineanu
Department of System Ecology and
   Sustainability,
University of Bucharest SPL
Independentei 91-95 76201
   Bucharest
Romania

Eugeniusz Andrulewicz
Department of Fisheries
   Oceanography and Marine
   Ecology
Sea Fisheries Institute
81-332 Gdynia
Kollataja 1, Poland

Javier Gilabert
Department of Chemical and
  Environmental Engineering
Technical University of Cartagena
Alfonso XIII
44. 30202-Cartagena, Spain

Sofia Gamito
IMAR
Faculty of Marine and
  Environmental Sciences
University of Algarve
Campus de Gambelas
8005-139 Faro, Portugal

Arturas Razinkovas
Coastal Research and Planning
  Institute
Klaipeda University
Klaipeda University
Manto 84, Klaipeda,
LT-5808, Lithuania

Natalia N. Kazantseva
Federal State Unitary Enterprise
Keldysh Research Center of the
  Russian Aero-Space Agency
Onezthskaya Str., 8,
Moscow, Russian Federation

Robert B. Ambrose
USEPA National Exposure Research
  Laboratory Ecosystems Research
Division 960 College Station Road
Athens
GA 30605–2700, USA

Biymyrza Toktoraliev
Osh Technological University
Isanova Street 81
714018 Osh., Kyrgyzstan

Ramiro Neves
Instituto Superior Técnico Av.
  Rovisco Pais
1 1049-001 Lisboa
Portugal

Hanafi Menouar
The University of Sciences and
  Technology of Oran129/3 Cite
  500 Log Seddikia
ORAN 31025, Algeria

Amir Aliev
Sea Meteorological Center of the
  Ministry of Ecology and Natural
  Resources
Baku
370154, Azerbaijan

Ayten Mamedova
Baku State Universit
Zaxid Khalilov Street 23
Baku
Azerbaijan

Afat Aliyeva
Baku State Universit
Zaxid Khalilov Street 23
Baku
Azerbaijan

Frédéric Maps
Institut des sciences de la mer de
  Rimouski (ISMER) 310 des
  Ursulines
Rimouski, Québec
G5L 3A1, Canada

Thomas Guyondet
Insitut des Sciences de la Mer de
  Rimouski-310

Allée des Ursulines-Rimouski
Québec
G5L 3A1, Canada

Marco Bajo
ISMAR-CNR (National Research
   Council)
S. Polo 1364
30125 Venice, Italy

Francesca De Pascalis
ISMAR-CNR
1364 S. Polo
30125 Venezia, Italy

Christian Ferrarin
San Polo
1364-Palazzo Papadopoli
30125 Venice, Italy

Debora Bellafiore
San Polo
1364-Palazzo Papadopoli
30125 Venice, Italy

Andrea Critto
Interdepartmental Centre IDEAS
University of Venice
Calle Larga S. Marta
2137 I30123
Venice, Italy

**Gulshaan Abdykaimovna
Ergeshova**
18–28 Salieva Street
Osh, Kyrgystan

**Zulumkan Abdymanapovna
Teshebaeva**
93-61 Isanova Street
Osh, Kyrgystan

**Igamberdiev Rakhmatullo
Mamirovich**
86-43 Isanova Street
Osh, Kyrgystan

**Almazbek Anarbekovich
Orozumbekov**
Department of Ecology
Osh Technological University
81, Isanova Street
Osh Technological University
Osh, Kyrgystan

Aybek Toktosunovich Attokurov
Department of Ecology
Osh Technological University
81, Isanova street
Osh Technological University
Osh, Kyrgystan

Ilias Baimirzaevich Aitiev
86-9 Isanov Street
Osh, Kyrgystan

Iveta Steinberga
University of Latvia
Faculty of Geographical and Earth
   Sciences
10 Alberta Str., Riga
LV1010, Latvia

Ilga Kokorite
University of Latvia
Faculty of Geography and Earth
   Sciences
Alberta Street 10, Riga
LV-1010, Latvia

Alina Mockute
Coastal Research and Planning
   Institute

Klaipeda University H. Manto 84
LT-92294 Klaipeda, Lithuania

Lina Bliūdžiutė
Coastal Research and Planning
    Institute
Klaipėda University H.Manto 84
LT-5808, Klaipėda
Lithuania

Renata Pilkaitytė
Coastal Research and Planning
    Institute
Klaipeda University H.Manto 84
LT-92294, Klaipeda, Lithuania

Aistė Miltenytė
Coastal Research and Planning
    Institute
Klaipeda University H.Manto 84
LT-92294, Klaipeda
Lithuania

Loreta Kelpšaitė
Coastal Research and Planning
    Institute
Klaipeda University H.Manto 84
LT-92294 Klaipeda
Lithuania

Karim Hilmmii
Institut National de Recherche
    Halieutique (INRH) 02
Rue Tiznit
Casablanca 20 000
Morocco

Krzysztof Świtek
Sea Fisheries Institute
Department of Fisheries
    Oceanography and Marine
    Ecology

Kollataja Str. 1
81-332 Gdynia
Poland

Mariusz Zalewski
Sea Fisheries Institute Department
    of Fisheries Oceanography and
    Marine Ecology
Kollataja Str. 1
81-332 Gdynia
Poland

Maciej Tomasz Tomczak
Sea Fisheries Institute
Department of Fisheries
    Oceanography and Marine
    Ecology
Kollataja Str. 1
81-332 Gdynia, Poland

Luis Daniel Fachada Fernandes
IST–MARETEC Secção de
    Ambiente e Energia
Departamento de Engenharia
    Mecânica Av. Rovisco Pais
    1049-001 Lisboa
Portugal

**M. Madalena dos Santos
Malhadas**
IST–MARETEC Secção de
    Ambiente e Energia
Departamento de Engenharia
    Mecânica Av. Rovisco Pais
    1049-001 Lisboa
Portugal

Palarie Teodora-Alexandra
University of Bucharest
Department of Systems Ecology and
    Sustainability

Spl. Inde., 91-95
Sector 5, Bucharest
Romania

Maria—Magdalena Bucur
University of Bucharest
Department of Systems Ecology and
  Sustainability
Spl. Inde., 91-95
Sector 5, Bucharest
Romania

Sabina Datcu
University of Bucharest
Department of Systems Ecology
  and Sustainability
Spl. Inde., 91-95
Sector 5, Bucharest
Romania

Constantın Cazacu
University of Bucharest
Department of Systems Ecology and
  Sustainable Development
Spl Inde., 91-95 050095 Bucharest
Romania

Natalia Reznichenko
Laboratory for Coastal Systems
  Study
Atlantic Branch of P.P. Shirshov
  Institute of Oceanology of
  Russian Academy of Sciences
  Prospect Mira
1, Kaliningrad, 236000
Russia

Natalia Demchenko
Laboratory for Coastal Systems
  Study
P.P. Shirshov Institute of Oceanology
  of Russian Academy of Sciences
Atlantic Branch

Kaliningrad, 236000
Russia

Alexey Egorov
Federal State Unitary Enterprise
  "Keldysh Research Center" of the
  Russian Aero-Space Agency
Onezthskaya Str. 8 Moscow
125438, Russian Federation

Dmitriy Domnin
Laboratory for Coastal System
  Study
Atlantic Branch of P.P. Shirshov
  Institute of Oceanology
Russian Academy of Sciences
  Prospect Mira, 1,
Kaliningrad, 236000
Russia

Evgenia Gurova
Laboratory for Coastal System
  Study
Atlantic Branch of P.P. Shirshov
  Institute of Oceanology
Russian Academy of Sciences
  Prospect Mira, 1,
Kaliningrad, 236000
Russia

Alena Paliy
Laboratory for Coastal System
  Study
Atlantic Branch of P.P. Shirshov
  Institute of Oceanology
Russian Academy of Sciences
  Prospect Mira, 1,
Kaliningrad, 236000
Russia

Olga Chubarenko
Moscow Institute of Physics and
  Technology (State University)

Institutskiy Pereulok, 9,
Dolgoprudniy
Moscow Region, 141700
Russia

Francisco Javier Campuzano Guillén
Maretec, Instituto Superior Técnico
  (IST)
Universidad Técnica de Lisboa
Lisbon
Portugal (Spain)

Abderrahmen Yassin Hamouda
INSTM, Institut National des
  Sciences et Technologies de la
  Mer 28
rue du 2 mars 1934
2025 Salammbô
Tunisia

Boukadi Khanes
INSTM, Institut National des
  Sciences et Technologies de la
  Mer 28
rue du 2 mars 1934
2025 Salammbô
Tunisia

Atakan Öngen
Trakya University Corlu
  Engineering Faculty
59850, Çorlu/Tekirdağ
Turkey

Melike Gürel
Istanbul Technical University
Faculty of Civil Engineering
  Environmental Engineering
  Department
34469 Istanbul
Turkey

Alpaslan Ekdal
Istanbul Technical University
Faculty of Civil Engineering
  Environmental Engineering
  Department
34469 Istanbul
Turkey

Yakup Karaaslan
Environmental Engineering
Ministry of Environment and Forest
  General of Environment
  Management
Department of Soil and Water
Ankara, Turkey

Handan Dokmeci
Trakya University Corlu
  Engineering Faculty
59850, Çorlu/Tekirdağ
Turkey

Nusret Karakaya
Trakya University Corlu
  Engineering Faculty
59850, Çorlu/Tekirdağ
Turkey

Ali Ertürk
Istanbul Technical University
Faculty of Civil Engineering
  Environmental Engineering
  Department
34469 Istanbul
Turkey

Otuzbay Geldiyew
Scientific and Production
  Cooperation Ashgabat
Turkmenistan

William Matthew Henderson
USEPA/ORD/NERL/ERD
960 College Station Road Athens
GA 30605, USA

Kathryn B. Briggman
Johns Hopkins University
110 West University

Parkway E2 Baltimore
MD 21210, USA

Kimberly Smith
Emory University
Department of Chemistry
536 Lantern Wood Drive Scottdale
GA 30079, USA

Printed in the United States
85452LV00001B/144/A